Multiparticulate
Oral Drug Delivery

DRUGS AND THE PHARMACEUTICAL SCIENCES

A Series of Textbooks and Monographs

edited by

James Swarbrick
Applied Analytical Industries, Inc.
Wilmington, North Carolina

ADDITIONAL VOLUMES IN PREPARATION

Multiparticulate Oral Drug Delivery

edited by

Isaac Ghebre-Sellassie
Parke-Davis Pharmaceutical Research Division
Warner-Lambert Company
Morris Plains, New Jersey

CRC Press
Taylor & Francis Group
Boca Raton London New York

CRC Press is an imprint of the
Taylor & Francis Group, an **informa** business

CRC Press
Taylor & Francis Group
6000 Broken Sound Parkway NW, Suite 300
Boca Raton, FL 33487-2742

First issued in paperback 2019

© 1994 by Taylor & Francis Group, LLC
CRC Press is an imprint of Taylor & Francis Group, an Informa business

No claim to original U.S. Government works

ISBN-13: 978-0-8247-9191-9 (hbk)
ISBN-13: 978-0-367-40204-4 (pbk)

A CIP record for this book is available from the British Library.

Library of Congress Cataloging-in-Publication Data available on application

Visit the Taylor & Francis Web site at
http://www.taylorandfrancis.com

and the CRC Press Web site at
http://www.crcpress.com

Preface

Although multiparticulates, which comprise minitablets, pellets, and granules, have been taken by patients in one dosage form or another for quite some time, it is only in the last two decades that the full potential of multiparticulates as drug delivery systems has been realized in terms of both flexibility during formulation development and therapeutic benefits to patients. Not only can multiparticulates be divided into desired doses without formulation and process changes, but they can also be blended to deliver simultaneously incompatible bioactive agents, or particles with different release profiles at the same site or at different sites within the gastrointestinal tract. In addition, technological advances in dosage form design, the advent of highly specialized pieces of equipment, and the popularity of controlled-release dosage forms as a means of drug delivery have made multiparticulates a viable and attractive alternative to single-unit dosage forms.

Multiparticulates also have numerous therapeutic advantages over single units. When taken orally, multiparticulates generally disperse freely in the gastrointestinal tract, maximize absorption, minimize side effects, and reduce inter- and intrapatient variability. As a result, interest in multiparticulates as oral drug delivery systems has been growing steadily over the last few years, and there is a multimillion dollar market for Wurster coaters, rotary granulators, and extruders/spheronizers.

Multiparticulates are diverse in size and shape, and are manufactured by employing a variety of technologies, some of which have already been discussed extensively elsewhere (e.g., *Pharmaceutical Pelletization Technology* and *Microencapsulation*, Marcel Dekker, Inc.). The objective of the book is fourfold: to review additional manufacturing technologies for multiparticulates, to describe the various aspects of multiparticulate dosage form development, to assess the in vivo behavior and performance of multiparticulates, and to compare the market position of multiparticulates relative to other dosage forms.

Balling (spherical agglomeration), spray congealing, and cryopelletization are discussed in Chapters 1 through 3, respectively. While all three processes produce core pellets that may or may not require subsequent coating to generate a target release profile, the formulation and processing requirements are drastically different, and are discussed at length in the respective chapters. Coating of multiparticulates using polymeric solutions, polymeric dispersions, and molten materials is covered in Chapters 4 through 6. Both formulation factors and processing variables that affect film formation, integrity, and performance are critically discussed. In addition, conditions that minimize agglomeration of the particles during the coating process are described. After multiparticulates are manufactured, they may be filled into hard-shell gelatin capsules, compressed into tablets, suspended in liquids or packaged in sachets, as described in Chapters 7 through 9. Chapter 10 addresses key formulation and process factors that are critical to the development of controlled-release products. It highlights the various conditions that can adversely affect the functional performance of the rate-controlling membranes used in in vitro dissolution testing. Chapter 11 delineates the main chemical and physical forces that control the release of drugs from coated multiparticulate dosage forms. Chapters 12 and 13 discuss at length the biopharmaceutical aspects and in vivo performance of multiparticulates and the therapeutic advantages they offer over single units such as tablets. Chapter 14 describes in detail the intricate and well-controlled manufacturing process of hard-shell gelatin capsules. The composition of various packaging materials, a description of packaging machinery, and the importance of packaging during and beyond the development phase are articulated in Chapter 15. A thorough discussion on the assessment of the marketing potential of multiparticulate oral drug delivery system vis-à-vis other oral dosage forms is given in Chapter 16.

This book is intended to complement other volumes that deal with multiparticulate oral dosage forms. As such, it will provide scientists engaged in pharmaceutical research and development with the necessary information required to develop multiparticulate drug delivery systems systematically.

I would like to express my sincere thanks to the outstanding contributors, who devoted their time and effort to bring about the successful completion of the project. Thanks also to Dr. Mahdi B. Fawzi, Vice President, and Dr. Russell U. Nesbitt, Senior Director, Product Development, Parke-Davis Pharmaceutical Research Division, Warner-Lambert Company, for their continuous support and encouragement.

Isaac Ghebre-Sellassie

Contents

8. Multiparticulate Encapsulation Equipment and Process   159
   Donald K. Lightfoot

9. Compaction of Multiparticulate Oral Dosage Forms   181
   Metin Çelik

10. Key Factors in the Development of Modified-Release Pellets   217
    Stuart C. Porter and Isaac Ghebre-Sellassie

11. Mechanisms of Release from Coated Pellets   285
    *Jennifer B. Dressman, Bernhard Ø. Palsson, Asuman Ozturk,
    and Sadettin Ozturk*

12. Biopharmaceutical Aspects of Multiparticulates   307
    Johannes Krämer and Henning Blume

13. In Vivo Behavior of Multiparticulate Versus Single-Unit
    Dose Formulations   333
    George A. Digenis

14. Capsule Shell Composition and Manufacturing   357
    Ronnie Millender

15. Packaging of Multiparticulate Dosage Forms:
    Materials and Equipment   405
    K. S. Murthy, Franz Reiterer, and Jürgen Wendt

16. Marketing Considerations for Multiparticulate Drug
    Delivery Systems   457
    Roland Daumesnil

Index   475

Contributors

Henning Blume, Ph.D., D. Sc. Professor and Head, Zentrallaboratorium, Deutsches Arzneiprüfungsinstitut, Eschborn, Germany

Roland Bodmeier, Ph.D. Associate Professor of Pharmacy, Pharmaceutics Division, College of Pharmacy, The University of Texas at Austin, Austin, Texas

Metin Çelik Department of Pharmaceutics, College of Pharmacy, Rutgers —The State University of New Jersey, Piscataway, New Jersey

Roland Daumesnil Director, International Market Development, Capsugel AG, Arlesheim, Switzerland

George A. Digenis, Ph.D. Professor, Department of Medicinal Chemistry and Pharmaceutics, College of Pharmacy, University of Kentucky, Lexington, Kentucky

Jennifer B. Dressman* Associate Professor of Pharmaceutics, College of Pharmacy, The University of Michigan, Ann Arbor, Michigan

Yoshinobu Fukumori, Ph.D. Associate Professor, Faculty of Pharmaceutical Sciencies, Kobe Gakuin University, Arise, Ikawadani-cho, Nishi-ku, Kobe, Japan

Isaac Ghebre-Sellassie, B. Pharm., Ph.D. Associate Research Fellow, Pharmaceutical Product Development, Parke-Davis Pharmaceutical Research Division, Warner-Lambert Company, Morris Plains, New Jersey

A. Atillâ Hincal, Ph.D. Professor, Pharmaceutical Technology Department, Hacettepe University, Ankara, Turkey

---

Current affiliation: Institute for Pharmaceutical Technology, J.W. Goethe University, Frankfurt, Germany.

David M. Jones   Scientific Director, International Process Technology Development, Glatt Air Techniques, Inc., Ramsey, New Jersey

H. Süheyla Kaş, Ph.D.   Professor, Pharmaceutical Technology Department, Hacettepe University, Ankara, Turkey

Axel Knoch, Ph.D.   Pharmaceutical Development, Parke-Davis Pharmaceutical Research, Gödecke AG, Freiburg, Germany

Johannes Krämer   Apotheker, Head of Department of Biopharmacy, Deutsches Arzneiprüfungsinstitut, Eschborn, Germany

Klaus Lehmann, Ph.D.   Research and Development Pharma Polymers, Röhm GmbH, Darmstadt, Germany

Donald K. Lightfoot, B.A.   Manager of Clinical Supply Manufacturing, Pharmaceutical Technologies, SmithKline Beecham Pharmaceuticals, King of Prussia, Pennsylvania

Ronnie Millender   Director, Global Quality Systems, Capsugel, Division of Warner-Lambert Company, Greenwood, South Carolina

K. S. Murthy, Ph.D.   Senior Research Associate, Product Development Laboratories, Parke-Davis Pharmaceutical Research, Warner-Lambert Company, Morris Plains, New Jersey

Asuman Ozturk   Chiron Corporation, Emeryville, California

Sadettin Ozturk   Miles, Inc., Berkeley, California

Ornlaksana Paeratakul, Ph.D.   College of Pharmacy, The University of Texas at Austin, Austin, Texas

Bernhard Ø. Palsson   College of Engineering, The University of Michigan, Ann Arbor, Michigan

Phillip J. Percel, M.S.   Staff Research Associate, Research and Development Department, Eurand America, Inc., Vandalia, Ohio

Stuart C. Porter, B. Pharm., Ph.D.   Vice President, Scientific Services, Colorcon, West Point, Pennsylvania

Franz Reiterer   Head of Research and Development, Teich AG, Obergrafendorf-Mühlhofen, Austria

Lucy S. C. Wan, B. Pharm., Ph.D., D.Sc., D. Pharm.(hon)   Professor and Head of Department of Pharmacy, National University of Singapore, Singapore

Jürgen Wendt   Technical Manager, Research and Development, Teich AG, Obergrafendorf-Mühlhofen, Austria

Multiparticulate Oral Drug Delivery

1

Manufacture of Core Pellets by Balling

Lucy S. C. Wan

National University of Singapore
Singapore

I. INTRODUCTION

In a broad sense, pellets are very much like granules; the technique for producing pellets can also produce granules. Pellets are formed with the aid of a pelletizer. This machine is able to form approximately spherical bodies from a mass of finely divided particles continuously, by a rolling or a tumbling action on a flat or curved surface with the addition of a liquid [1]. Pelletizers can be classified based on the angle of their axis as (1) horizontal drum or (2) inclined dish [1]. Rotary fluidized-bed granulators are also used for pelletization.

II. PELLETIZERS

A. Horizontal Drum Pelletizer

The pelletizer is essentially a horizontal cylinder with a length/diameter ratio of 2:1 to 4:1. Generally, it is fitted with longitudinal ribs or lifters to promote tumbling action and with internal baffles to control pellet growth. This is achieved by dividing the pelletizer into zones. The pelletizer is usually powered by a fixed- or a variable-speed motor. The pitch of the cylinder (up to 10° from the horizontal) helps to transport the material down the length of the cylinder. A retaining ring is often fitted at the fixed end to prevent the spillback of feed material. A liquid phase may be introduced either before or immediately after the solids have been placed in the cylinder.

    The pelletizer offers flexibility and is well suited to processes involving chemical reactions. It has been found to be useful in many industrial processes, such as balling of iron ore, kaolin, bentonite, and fertilizers. A criticism of this type of cylinder is that the charge of material tends to slide over the smooth surface of the interior instead of rolling over it. A buildup of cake on the wall of the cylinder decreases the sliding action and thus helps to protect the shell from erosion.

In an ideal sense, the most important factor in determining the size of the granulating drum is the residence time required for granulation. Studies have shown that average granule size is a linear function of drum revolutions [2–4], and granule growth kinetics are affected by drum loading. The use of rotary granulators can avoid the residence-time approach since in such granulators the focus is on specific production area. However, there are different schools of thought regarding whether the cylindrical surface area should or should not be multiplied by the rotation speed. One school [5] advocates the former and another the latter [6]. Specific production area is a rather simple technique, but it may be a doubtful one for drum sizing. It does not consider the effect of total drum throughput. In addition, there are other uncertainties and hence the design of granulating drums is more dependent on experience of the use of such granulators rather than on theoretical principles.

Tumbling blenders can be adapted for granulation of fine powders. By means of a whirling agitator device, the granulating liquid can be introduced into the tumbling powder [7,8]. Blades are attached to the device, and the tip speeds of the blades vary from 900 to 1200 per minute. This results in voids being created in the tumbling powder. From the orifices in the blade tips, the atomized liquid is introduced into these voids.

B. Inclined Dish Pelletizer

The inclined dish pelletizer is known by many names, such as tilted pan, pan granulator, and granulation bowl plate. It consists mainly of a shallow cylindrical dish rotating about an inclined axis. The angle of the dish is adjustable from 40 to 70° to the horizontal. The speed of rotation is variable. A scraper is located at the top part of the dish to prevent a buildup of material and to direct the flow pattern of the tumbling powder. Dust covers are optional. Binding liquids can be added at various locations of the dish by means of spray nozzles.

An important feature of the inclined dish pelletizer is its ability to separate the pellets into various size ranges. The advantages of the dish pelletizer over the drum pelletizer include greater capacity, longer residence time for difficult materials, less sensitivity to the disruption of the process as a result of the dumping effects of a longer recirculating load, lower space requirements, lower operating cost, greater sensitivity to operating variables, and easy observation of the process.

C. Rotary Fluidized Bed Granulator

In simple terms, a rotary fluidized-bed drier is one modified to impart rotational movement to the powder bed. This is achieved by replacing the standard fluidized-bed drier bowl with one that has a rotating plate as an air distributor. The rotating plate and the air blown in through the gap between the wall of the fluidized-bed drier and this palte provide the rotational movement. For granulation, the binder liquid is sprayed via one or two binary nozzles located axially to the rotational movement of the powder bed. This operation results in the rounding of granules to approximate spherical pellets. This balling process can be influenced by operating conditions.

As a matter of interest, the rotary fluidized-bed granulator can be used to deposit a drug via a solution or suspension onto nonpareils, making it an alternative approach to the production of pellets. One advantage of the rotary fluidized-bed granulator is that it allows high rates of drug application with low drug losses. Furthermore, both aqueous and organic solvents can be used.

III. PELLET FORMATION AND KINETICS

Fine powders can readily be formed into agglomerates by the introduction of a liquid phase followed by suitable agitation or tumbling. The liquid and solid phases are brought into close contract; this allows binding forces to develop and bring about agglomeration. Growth of the particles occurs either by collisions and successful adherence of primary feed particles into discrete pellets or by the formation of a nucleus onto which particles collide and attach themselves. This results in growth formation [9,10].

When two particles come into close contact, the cohesive forces that hold the particles together [10,11] are:

1. *Intermolecular attractive forces.* These are very short range attractions, active up to a maximum of 10^3 Å. On the whole, van der Waals dispersive forces make the most significant contributions [11–13].
2. *Electrostatic attractive forces.* These are almost always present in particulate systems. They are produced primarily by interparticle friction. Although these forces are generally less than those experienced in other binding mechanisms, the net effect is to hold or orient particles in a contact region for sufficiently long for other, more dominant mechanisms to operate [14].
3. *Liquid bridge modes.* There are three physical situations in which the amount of liquid present produces cohesive forces between particles. The contributing mechanisms [9,10] are adsorbed liquid layers, mobile liquid bridges, and viscous or adhesive binders.

If only immobile adsorbed surface films are present, the magnitude of the cohesive bonds between the two particles [2,11,15–17] is affected in two ways; (1) surface imperfections are smoothed out, which increases the available particle–particle area, and (2) the effective interparticle distance is decreased. This allows van der Waals attractive or electrostatic forces to play a role in which the magnitude of the cohesive forces between particles is increased.

Once sufficient liquid is present to produce liquid bridges in an assembly of particles, the cohesive strength of the material increased. With a greater liquid content in the available pore space, the state of liquid saturation can be described [2] as (1) a pendular state, (2) a funicular state, or (3) a capillary state. The surface tension of the binding liquid is responsible for this phenomenon and the decrease in the total surface free energy of the system accompanying the reduction in air–liquid interfacial area is the driving force for nucleation of particles.

Beyond this nucleation phenomenon, the change in size can occur by a number of mechanisms. The prevailing mechanisms depend on factors such as the solid properties, liquid properties, and mode of operation. After nucleation has occurred, the predominating growth mechanisms are (1) coalescence, and (2) layering of either feed particles or fines from the breakdown of established agglomerates. Coalescence and layering can occur simultaneously, but generally one mechanism predominates.

IV. FACTORS AFFECTING PELLET GROWTH

A. Bridging/Binding Liquid Requirements

In any pelletization process, the moisture content of the feed material is very important in deciding the outcome of the process. In fact, in most continuous production processes, it is the principal variable used to control pellet growth. Moisture content is usually taken as a simple percentage by weight or volume of the mass. In the granulation process, Newitt and Conway-Jones [2] first observed that the critical moisture content required for granulation to occur correlated with 90% of the moisture required to saturate the voidage in the powder being granulated, as measured by its packed density. Capes and Dankwerts [4] found that sand was closed packed only at a low liquid content; at a higher liquid content they found that the moisture occupied 85 to 93% of the granule voids. Sherrington [18] used coarser sand to produce smaller granules and found that these granules were formed at a much lower liquid content than that required to saturate the voidage in the sand. A simple relationship between the liquid content in the granulator and the average size of the granules was proposed.

Many workers have observed that the rate of granule growth is very dependent on the liquid of the granulating mass [2–4,19–21]. This was attributed to increased granule plasticity or surface moisture at a higher liquid content, which then leads to a greater probability of the granules sticking together on cohesion. Kapur and Furstenau [19,20] reported an exponential dependence of rate parameters on moisture content. The effect of the moisture–temperature relationship is important in fertilizer granulations. When one or more of the components is water soluble, the total volume of the liquid or solution phase, rather than the moisture content per se, controls the granulation behavior [22–24]. The amount of solution phase, in turn, is a function of the temperature of the granulation.

Capes et al. [25] have derived equations for bridging liquid requirements of the feed material, a general equation being Eq. (1). The equations are approximations only, and two constants are required for particles above and below 30 μm.

$$W = \frac{EP_a}{EP_a + (1 - E)P_s} \tag{1}$$

For average feed particle diameters below 30 μm,

$$W = \frac{1}{1 + 1.85(P_s/P_a)} \tag{2}$$

For average feed particle diameters above 30 μm,

$$W = \frac{1}{1 + 2.17(P_s/P_a)} \tag{3}$$

where W is the weight fraction of liquid (wet basis), E the void fraction, and P_s and P_a are particle and liquid densities, respectively.

Pietsch [26] pointed out that moisture content also has an effect on the shape of the granule. Underwetting gives a "golf ball"-like structure, and overwetting gives a "blackberry"-like structure.

B.   Residence Time of Material in the Pelletizer

The most important factor in evaluating the performance of the drum pelletizer is the residence time required for the formation of pellets. Kapur and Furstenau [27] suggested that initially the coalesced pellets produced from limestone powder in a batch rotating drum were strong and would not break or abrade easily. This is due to the nature of the packing structure and the large number of point contacts attained. Granules in this form were observed to have excess liquid on the surface due the rearrangement of particles on continuous rolling as they compact with time. It was also shown that with increasing residence time, the average granule size increased. The growth rate passed through a maximum before falling off in the balling region due to a decrease in the successful collision rate.

The observations of Capes and Dankwerts [4] made from their study using narrow sieve cuts of sand were not in agreement with the explanation described above. In the case of narrow size distributions, owing to a sharp drop in the number of contact points as compared with those in a wide size spread, the mechanical strength of the granule is lower. The effects of dilation [28], which can occur with a narrow size distribution due to shearing and/or tearing forces in the drum or pan granulators, cause liquid to move into the granule as it expands in volume. Hence the external surface of the granule appears dry. Thus optimum conditions for coalescence are not achieved. As a result, there is breakage of granules to fragments and fines, due to crushing and surface abrasion. With increasing time, the fragments and fines attach themselves to the larger of the remaining granules. These have greater collision cross-sectional area, which continues to grow by means of a layering mechanism, although at a diminishing growth rate. In contrast, in studies in which very much coarser starting materials were used [18,29] the findings indicated that granule growth was very rapid up to a limiting size. This was a function of the liquid content. The granule size was effectively independent of residence time.

Bhrany et al. [30] also investigated the rate of granule growth in a dish pelletizer. It was found that the percentage by weight of the product larger than a given size increased linearly with residence time to a maximum value and then remained constant. In their investigation Fogel [3] and Kayatz [31] found that there was a linear dependence of granule size on residence time. Kapur [32] derived an exponential relation for growth kinetics from the random collision mechanism, which was assumed to apply during the nuclei growth, and a power law relation for a nonrandom collision process assumed during ball growth. Log-log plots appeared to fit the results for the ball growth region [32,33]. On the other hand, Kanetkar et al. [34] employed log-linear plots to linearize the granule growth plots.

C. Speed and Angle of Inclination of the Pelletizer

Agitation speed is the number of revolutions of the drum or dish pelletizer per unit time. It is usually expressed as the number of revolutions per minute. The rate of growth of pellets in terms of size change per unit drum revolution has been found to be independent of the speed of rotation below about 35% of the critical speed of the drum [2,4,19–24, 27–35]. Of course, longer times are required to grow agglomerates of a given size with a slower speed of rotation. Very high speeds can produce very rapid and uncontrolled growth, due to the cascading and pulverizing action of the larger agglomerates.

The critical speed is defined as the speed at which material can be just carried completely around the drum pelletizer by centrifugal action. The critical speed applies only to dry materials in a clean drum. Wet material is carried right around the drum at much lower speeds than the critical speed. Oyama [36] has shown that in drums of different diameters (D), equivalent motion is obtained if the drum speed is scaled as $D^{-1/2}$. Brook [37] suggested that for granulating drums, the best rotational speed is one-half the critical speed. Other researchers [38–40] reported that the critical speed corresponds to the speed at which the Froudes number (NF$_r$) for the drum is equal to unity. The Froude number is a dimensionless group describing the ratio of inertial to gravitational forces. For a rotating drum it is n^2D/g. In SI units the rotational speed at which NF$_r$ = 1.0 is $n = 42.3D^{-1/2}$, where n is the rotational speed in rpm and D is the drum diameter in meters. In practice, good granulation can be achieved at speeds such that NF$_4 \sim$ 0.3 to 0.5.

The best speed for rotation for a drum is influenced by the properties of the material being granulated. For example, in iron-ore granulation too much cascading in the drum can lead to extensive granule breakdown. A drum granulator in which the drum is rotated above the critical speed has been patented [40]. Linkson [41] reported that with increasing drum speeds, greater compaction of the granules occurred.

There is an appreciable divergence of opinion both in theory and practice as to what constitutes the best rotational speed of a dish pelletizer. The derivation of the critical speed at which NF$_r$ = 1 is complicated by the need to include the angle of inclination θ (between the base of the dish and the horizontal) of the dish. The requisite formula as derived by Pietsch [26] is

$$n = \frac{42.3}{D^{1/2}} (\sin \theta)^{1/2} \tag{4}$$

where n is the rotational speed in rpm and D is the dish diameter in meters. Macavei [42] derived much more complicated expressions for the rotational speeds in terms of dish radius.

Papadaski and Bombled [43] derived the equation

$$n = (\sin \theta - \sin \phi)^{1/2} \frac{42.3}{D} \tag{5}$$

where φ is the angle of repose of the material being granulated. They recommended that the dish be rotated at 75% of the speed given by the equation. Bazilevich [44] derived the expression

$$n = \frac{\sin(\theta - \phi)^{1/2}}{\cos \phi} \frac{42.3}{D} \tag{6}$$

Hence there are different expressions for the variation of speed and angle of inclination of the pelletizer. A graphical comparison of various expressions for a dish pelletizer 2.5 m in diameter shows very pronounced dependencies of speed of rotation on the angle of inclination. Ball [45] granulated iron-ore concentrates in a 5-ft dish. He found that as the dish speed was increased, the proportion of unpelletized material passing through the spray region decreased linearly. Also, at higher speeds, unpelletized materials appeared in the discharge region, with particles of maximum size moving in from the edge, an effect described as vortex formation. Vortex formation becomes more pronounced as the speed of the dish pelletizer is increased and the angle of inclination is reduced. Thus the speed of rotation should be high enough to give a good rolling action for pelletization and efficient use of the pan area, but not so high that the particle size segregating effect is lost. The angle of inclination of the dish is correlated primarily with the size of the granule [45].

D. Amount of Material Fed to the Pelletizer

In a rotating cylinder pelletizer a solid particle is taken up the wall to a particular height that depends on the wall friction, specific density, and shape. A collection of pellets is taken up higher as a result of the interaction between the particles and the restricted relative movements of the individual particles within the mass. The center of gravity of the entire collection of particles is displaced to a position eccentric to the axis of the cylinder. The eccentricity decreases with increasing loading. With greater loading, the rate of mixing decreases [46]. Some feed material may rotate through a greater angle in the bulk before approaching the rolling layers again. This results in an increased rate of growth.

V. BINDERS

In many a pelletization process, the inclusion of a binder is often necessary. The choice of a binder can determine the success or failure of the entire process. Binders can contribute significantly to agglomerate strength. Binders are additives that impart cohesive properties to the powdered material through particle–particle bonding. It may also be desirable to incorporate lubricants, as they decrease the coefficient of friction between individual particles in the pellet or between the surfaces of the pellets and the rolls that form them.

　　Apart from imparting strength to pellets, binders also improve the flow properties by appropriate formulation of the pellets with the desired pellet size and hardness. Binders can be divided generally into three groups [47]:

1. *Matrix binders.* In these binders, the particles are embedded in an essentially continuous network of the binding material. A rather substantial quantity of the binder is required because its film strength tends to be low. The amount required thus depends largely on the void fraction of the pellet (e.g., dry starch and dry sugars).

2. *Film binders.* These binders are not necessarily used as solutions or dispersions. The amount of binder used is important, for it has considerable effect

on the product. The quantity required depends on the extent of dilution. Satisfactory wetting of the surface of the material can be achieved with 0.5 to 2% of a liquid binder (i.e., pure liquid without any additive present) or a binder solution. When used by itself, water can be considered as a special case of a film binder and can produce a number of effects. Water has a solvent action in soluble material. It can dissolve the surface of the particles and when dried cause recrystallization across the particle boundaries, resulting in solid bridges. By its surface tension effects, it can effect coalescence between insoluble materials. Water can soften and plasticize the surface of particles or agglomerates, resulting in plastic surface, which can then effect further growth by coalescence. Water can promote bonding by van der Waals forces by increasing the true area of contact of the particles. The surface effects of water are so pronounced that the success or failure of the pelletization process rests solely on holding the moisture content of the particles within narrow limits. Some examples of film binders are poly(vinylpyrrolidone), starch paste, celluloses, bentonite, and sucrose.

3. *Chemical binders*. These binders develop their strength through a chemical reaction between the components of the binder or between the binder and the powdered material. They may be either a matrix or a film, depending on the strength and characteristics of the reaction products. For example, when mixed with water, portland cement makes a matrix type of binder, whereas a solution of sodium silicate hardened with carbon dioxide is a film type of binder.

Hundreds of patents have been granted for binders and hundreds more have been tried over the years. But to choose a binder that meets all specifications is a major task. The appropriate binder for a given application remains a matter of experience. Binders tend to be specific to a material. They must be capable of wetting the surface of the particle being pelletized or granulated. Binders must have sufficient wet strength to allow agglomerates to be handled and must have adequate dry strength to make them suitable for their intended purpose. Each process, however, makes use of a different system of forces and may require a different agglomerate strength. The final selection of the binder should therefore be made on the basis of the type of agglomeration equipment that is used. Many reviews of binder systems used in fuel briquetting and in pharmaceutical [48–51], and ceramics [52] applications are available.

VI. EVALUATION OF PELLET CHARACTERISTICS

The objective of all agglomeration processes is to produce a product with the properties desired. It is therefore important to quantify the physical properties of the agglomerates produced and to relate them to the initial feed properties and the operating conditions. The most common physical characteristics evaluated are

(1) pellet size and the size distribution, (2) bulk density, (3) pellet strength, and (4) flow property. To quantify the above, techniques applied for their measurement should be indicative of the nature and magnitude of the properties desired.

A. Size and Size Distribution of Pellets

Size and size distribution can be measured by an enormous number of methods and can supply information concerning practically any dimension. The aim of sizing procedures is the quantitative determination of a dimension that can be related to the particle-forming mechanism. In any agglomeration process, the initial particle size distribution is transferred into a product size distribution. By far the simplest and most widely used method for the determination of size and size distribution of pellets, granules, and powders is by sieving [51,52]. A sieving curve can be obtained by plotting the cumulative percent by weight retained by sieve. Plots may be constructed on log-log paper or log-probability paper. The more important factors influencing the performance of the sieving operation [53] are sieve load and sieving time, screen movement, aperture size, particle orientation, particle shape, and sampling of material.

The size distribution data can be expressed in numerous ways for the establishment of valid frequency distribution of size. The observed size distribution serves as a basis for establishing descriptive characteristics or constants, such as median diameter by weight or number, percent by weight greater or less than the stated size, or standard deviation. For quantitative analysis there is an ever-growing choice of techniques that provide varying information on the measured size or size distribution. There are excellent reports [54–56] giving detailed descriptions of various techniques and their relevance to several physical processes. Davies and Gloor [57–59] have noted that variables such as binders and particle size of powders had an effect on granule size for granulations prepared in a fluidized-bed dryer.

B. Bulk Density

The packing characteristics of particles play an important role in determining the physical properties associated with the product of the material. Two simple methods are often used to quantify the bulk packing characteristics of an assemblance. One is to fill a container by pouring a measured weight of the material and noting its initial volume. This provides the poured packing density (when the particle density is known). If the vessel is then tapped until no further reduction in volume is noted, this corresponds to dense random packing density or tapped density [60,61]. Considerable variations exist in the literature as to the length of tapping or number of taps used [62–64]. However, the important point to note is that the number of taps required to achieve the tightest packing will be dependent on the material under study. Therefore, a sufficient number of taps should be employed to ensure reproducibility of the material in question.

From a knowledge of the bulk packing density, the available void space P_b can be calculated for the given distribution, since

$$P_b = P_s(1 - E)$$

where P_s is the particle density and E is the voidage or porosity. Packing density, which is $P_b = P_s(1 - E)$, has been shown to influence the mechanisms of granule growth and hence has a direct effect on granule size distribution. This, in turn, plays an important part in influencing the bulk flow properties of the product. Many workers [57,59,62,65–67] have used the concept of bulk density to relate particle packing characteristics to the strength of the product material, shape of particles and granules, process changes, and flowability of granules.

C. Strength of Pellets

Destructive test methods and equipment used to assess quality of bonding in agglomerates or in powder compacts can conveniently be considered in two categories, tensile testing and compressive testing. Equipment employed for this purpose contains at least three elements, one for gripping the specimen, a second for deforming it, and a third for measuring the load required in performing the deformation. In applying the tensile test to an agglomerate, the main problem is to find the best way to mount and grip the specimen without changing its properties during the test [10]. An inherent weakness in this procedure is that great care is needed to avoid the disruption of the specimen. Ashton et al. [68] designed an apparatus (the Warren spring tensile tester) for measuring the tensile strength of loosely coherent assemblies of particles. Although not suited to measuring the tensile strength of individual agglomerates, this tester can be used to determine powder strength in a bed of particles held together by relatively weak bonds: for example, those due to intermolecular forces and low-viscosity liquid bridges. Tensile strength can also be measured indirectly by means of a number of compression, bending, and indentation tests [69,70]. The diametral compression test is one form of indirect tension test that has been applied to tablets [71].

Compressive strength is widely used as an index of granule strength. In the laboratory, a compressive strength test consists of crushing the agglomerate between flat parallel plates in which the lower plate is the pan of a balance that registers the load at failure [72]. In another variation the lower plate is the pan of an equilibrated overhead beam balance. A load is applied to the agglomerate by allowing water to run from a graduated burette into a beaker placed on the other balance pan. The burette reading gives the equivalent weight of water required to cause failure [73]. A more sophisticated variation of the compression strength test is that employed by Gold et al. [74] and Ganderton and Hunter [65] using electrical signals. A simple apparatus to measure the crushing strength of pellets has been devised [75].

In evaluating the performance of four types of commercially available "crushing strength" instruments, Brooks and Marshall [76] found that varying methods of load application leads to variation in crushing strength values. Cahn and Karpinski [77] showed that the fracture load of pliable agglomerates was highly dependent on the rate of loading.

Impaction tests can be carried out by dropping granules repeatedly from a constant height and counting the number of times it is necessary to drop the granule before it breaks [78,79]. The abrasion resistance of dry iron-ore balls has been measured by Fitton et al. [80] by tumbling the balls in a cylinder constructed from wire mesh. The amount of material passing through the screen as a fraction of the initial weight of balls gives the abrasion index.

Other methods of studying granule strength are those that relate to friability measurements [57,62,81,82]. These methods are a variation of the American Society for Testing and Materials (ASTM) tumbler for the friability of coal. A common method is one that involves taking granules of certain mesh size and packing them in a container which is then tumbled or shaken for a predetermined time. The percentage of the material passing through a screen is taken as a measure of granule strength or friability. The method provides a means of measuring the propensity of granules to break into smaller pieces when subjected to disruptive forces.

Granule strength and friability are important, as they affect changes in particle size distribution of granulations and consequently compressibility of the granules into tablets. The measurement of granule hardness and friability can be used as a characterizing tool, and this together with the granule size, density, and porosity can be used to characterize a granulation quantitatively to the extent that highly reproducible granulation can be produced.

D. Flow Property of Pellets

Several methods are available to measure the extent of interparticle forces. Such measurements are often employed as an index of flow. The more common methods are the angle of repose [83–85] and hopper flow measurements. The angle of repose is best suited for particles greater than or equal to 150 μm [86]. In this size range, cohesive effects will be minimal. In the fixed-funnel and freestanding cone method, a funnel is secured with its tip at a given height above a graph paper placed on a flat horizontal surface. Pellets or granulation is carefully poured through the funnel until the apex of the conical pile just touches the tip of funnel; thus

$$\tan \theta = \frac{H}{R}$$

where θ is the angle or repose, R the radius of the base of cone, and H the distance between the tip of the funnel and the base. In the fixed cone method, the diameter of the base is fixed by using a circular dish with sharp edges. The angle of repose is calculated as noted previously. In the revolving cylinder method, a cylinder half-filled with pellets is rotated. The maximum angle that the plane of pellets makes with the horizontal on rotation is taken as the angle of repose. In the tilting box method, a rectangular box is filled with pellets and tipped until the content begins to slide. Other methods of determining the angle of repose are given by Pathirama and Gupta [87] and Pilpel [88].

Hopper flow rates have been employed as a method of assessing flowability. Most of the instrumentation employed is a variation of that described by Gold et al. [89–92] and involves basically a recording balance. From their studies it was found that the quantity of fines, the amount and type of granulating agent, particle size distribution, and type of glidants used all had a measurable effect on flow rate.

CONCLUSION

Core pellets can be produced by balling using pelletizers. An understanding of the principles of operation of pelletizers and the effects of such operation on the physical properties of pellets so formed is important in pellet production. In addition, the mechanism of pellet formation, the kinetics of pellet growth, and the role of additives incorporated in the formulation are equally important in determining the quality of the pellets. Such qualities include size, strength, and flowability of pellets. With this information on hand, many a successful formulation can be achieved.

REFERENCES

1. C. W. Lyne and H. G. Johnston, *Powder Technol. 29*:211 (1981).
2. D. M. Newitt and J. M. Conway-Jones, *Trans. Inst. Chem. Eng. 36*:422 (1958).
3. R. Fogel, *J. Appl. Chem. 10*(3):139 (1960).
4. C. E. Capes and P. V. Dankwerts, *Trans. Inst. Chem. Eng. 43*:T116 (1965).
5. H. J. Koch, *Agric. Food Chem. 7*(11):748 (1959).
6. C. H. Chilton, *Chem. Eng.*, October, p. 169 (1951).
7. J. J. Fisher, *Chem. Eng. 69*, 5, February p. 83, (1962).
8. S. A. Kut, *J. Am. Oil Chem. Soc. 55*(1):141 (1978).
9. W. A. Knepper, ed., *Agglomeration*, Wiley, New York, 1962.
10. H. Rumof, *Chem. Eng. Technol. 30*:144 (1958).
11. H. C. Hamaker, *Physics 4*:1058 (1937).
12. M. Corn, in *Aerosol Science* (C. N. Davies, ed.), Academic Press, New York, 1966.
13. H. Krupp, *Adv. Colloid Interface Sci. 1*:111 (1967).
14. H. Rumpf, *Chem. Eng. Technol. 46*:1 (1974).
15. S. A. Turner, M. Balasubramaniam, and L. Offen, *Powder Technol. 15*:97 (1976).
16. N. Pilpel, S. S. Jayasingle, and C. F. Harwood, *Mater. Sci. Eng. 5*:287 (1069/70).
17. E. A. Guggenheim, *Trans. Faraday Soc. 36*:422 (1958).
18. P. J. Sherrington, *Chem. Eng. (London) 220*:CE 201 (1968).

19. P. C. Kapur and D. W. Furstenau, *Trans. AIME* 229:348 (1964).
20. P. C. Kapur and D. W. Furstenau, *Ind. Eng. Chem. Process Des. Dev.* 5(1):5 (1966).
21. I. Sekiaguchi and H. Tohata, *Kagaku Kogaku* 32(10:1012 (1968).
22. P. J. Sherrington, *Can. J. Chem. Eng.* 47:308 (1969).
23. J. D. Hardesty, A. Szabo, and J. G. Cummings, *J. Agric. Food Chem.* 4:60 (1956).
24. S. M. Janikowski, *Chem. Eng.* 246:51 (1971).
25. C. E. Capes, R. L. Germain, and R. D. Coleman, *Ind. Eng. Chem. Process Des. Dev.* 16:517 (1977).
26. W. Pietsch, *Aufbereit. Tech.* 7(4):177 (1966).
27. P. C. Kapur and D. W. Furstenau, *Ind. Eng. Chem. Process Des. Dev.* 16:517 (1977).
28. D. Reynolds, *Philos. Mag. Ser. 5* 20:469 (1885).
29. M. Butensky and D. Human, *Ind. Eng. Chem. Fundam.* 10(2):212 (1971).
30. N. N. Bharny, R. T. Johnson, T. L. Muron, and E. A. Pelezarski, in *Agglomeration* (W. A. Knepper, ed.), Wiley-Interscience, New York, 1962, p. 229.
31. K. Kayatz, *Zement-Kalk-Gips* 17(5):183 (1964).
32. P. C. Kapur, *Chem. Eng. Sci.* 27:1863 (1972).
33. P. C. Kapur, S. C. D. Arora, and S. V. B. Subbarao, *Chem. Eng. Sci.* 28:1535 (1973).
34. V. V. Kanetkar, S. K. Gupta, and R. P. Krishna, *Trans. Indian Inst. Met.* 30(1):50 (1977).
35. S. K. Nicoll and Z. P. Adamiak, *Inst. Min. Metall. Trans. Sec. C* 82:C26 (1973).
36. Y. Oyama, *J. Phys. Chem. Res. Inst. (Tokyo)* 12(570):770 (1933).
37. A. T. Brook, *Proc. Fert. Soc.* 47:1 (1957).
38. R. Rutgers, *Chem. Eng. Sci.* 20:1079 (1965).
39. R. E. Johnstone and M. W. Thring, *Pilot Plants and Scale Up Methods in Chemical Engineering*, McGraw-Hill, New York, 1957, p. 227.
40. H. W. Haines and F. Large, *Ind. Eng. Chem.* 48:996 (1956).
41. P. B. Linkson, *Can. J. Chem. Eng.* 47(5):519 (1969).
42. G. Macavei, *Br. Chem. Eng.* 10(9):867 (1965).
43. M. Papadaski and J. P. Bombled, *Rev. Mater. Constr. Trav. Publics* 549:289 (1961).
44. S. V. Bazilevich, *Stal. Engl. (USSR)*, 8:551 (1960).
45. P. D. Ball, *J. Iron Steel. Inst. London*, May, p. 40 (1959).
46. R. Rutgers, *Chem. Eng. Sci.* 20:1079 (1965).
47. K. R. Komarek, *Chem. Eng.* 74(25):154 (1967).
48. L. Lachman, H. A. Lieberman, and J. L. Kanig, eds., *The Theory and Practise of Industrial Pharmacy*, Lea & Febiger, Philadelphia, 1970.
49. R. E. King, *Remingtons Pharmaceutical Science*, 14th ed., Mack Publishing, Easton, Pa., 1970, p. 1649.
50. S. Levine, *Ceram. Age*, January, p. 39 (1960).
51. D. E. Fonner, G. S. Banker, and J. J. Swarbrick, *J. Pharm. Sci.* 55:576 (1967/68).
52. K. T. Whithby, *ASTM Spec. Tech. Publ.* 234 (1958).
53. M. L. Jansen and J. R. Glastonbury, *Powder Technol.* 1:334 (1967/68).
54. R. R. Irani and C. F. Callis, *Particle Size: Measurement, Interpretations and Application*, Wiley, New York, 1963.

55. C. Orr, *Particulate Technology*, Macmillan, London, 1966.
56. G. Herdan, *Small Particle Statistics*, Butterworth, London, 1966.
57. W. L. Davies and W. T. Gloor, *J. Pharm. Sci. 60*:1869 (1971).
58. W. L. Davies and W. T. Gloor, *J. Pharm. Sci. 61*:618 (1972).
59. W. L. Davies and W. T. Gloor, *J. Pharm. Sci. 62*:170 (1973).
60. B. S. Neumann, *Flow Properties of Disperse Systems*, North-Holland, Amsterdam, 1953, Chapter 10.
61. Z. Ormes and K. Pataki, *Hung. J. Ind. Chem. 1*:207 (1973).
62. D. E. Fonner, G. S. Banker, and T. Swarbrick, *J. Pharm. Sci. 55*:181 (1966).
63. A. M. Marks and J. J. Sciara, *J. Pharm. Sci. 57*:497 (1968).
64. Q. A. Butler and J. C. Ramsey, *Drug Stand. 20*:217 (1952).
65. D. Ganderton and B. M. Hunter, *J. Pharm. Pharmacol. 23*:15 (1971).
66. A. A. Chalmers and P. H. Elworthy, *J. Pharm. Pharmacol. 28*:234 (1976).
67. H. J. Heywood, *Imp. Coll. Chem. Eng. Soc. 2*:9 (1946).
68. M. D. Ashton, R. Farley, and F. H. H. Valentin, *Rheol. Acta, 4*:206 (1968).
69. K. Ridgway and J. B. Shotton, *J. Pharm. Pharmacol. 22*:245 (1970).
70. K. Ridgway and M. H. Rubinstein, *J. Pharm. Pharmacol. 23*:115 (1971).
71. J. T. Fell and J. M. Newton, *J. Pharm. Pharmacol. 20*:657 (1968).
72. C. E. Capes, *Powder Technol.*, *4*:77 (1970/71).
73. H. P. Meissner, A. S. Micheals, and R. Kaiser, *Ind. Eng. Chem. Process Des. Dev. 3*:202 (1964).
74. G. Gold, R. N. Duvall, B. T. Palmero, and R. H. Hurtle, *J. Pharm. Sci. 60*:922 (1971).
75. L. S. C. Wan and T. Jeyabalan, *Acta Pharm. Technol. 32*:197 (1986).
76. D. B. Brooks and K. Marshall, *J. Pharm. Sci. 57*:481 (1968).
77. D. S. Cahn and J. M. Karpinski, *Trans. Am. Inst. Mining Eng. 241*:475 (1968).
78. D. F. Ball, J. Dartnell, J. Davison, A. Grieve, and R. Wild, *Agglomeration of Iron Ores*, Heinemann, London, 1973.
79. J. S. Wakeman, *Trans. Am. Inst. Mining Eng. 252*:83 (1972).
80. J. T. Fitton, W. D. Williams, and A. Grieve, *Proc. 9th* Commonw. *Mining Met. Congr. 3*:297 (1969).
81. A. M. Marks and J. J. Sciara, *J. Pharm. Sci. 62*:1215 (1973).
82. A. Mehta, M. A. Zoglio, and J. T. Cartensen, *J. Pharm. Sci. 67*:905 (1978).
83. R. L. Carr, *Chem. Eng. 72*:163 (1965).
84. N. Pilpel, *Br. Chem. Eng. 11*:699 (1966).
85. D. Train, *J. Pharm. Pharmacol. 11*:699 (1966).
86. N. Pilpel, *Advances in Pharmaceutical Science*, Vol. 3, Academic Press, New York, 1971, p. 173.
87. W. K. Pathirama and B. K. Gupta, *Can. J. Pharm. Sci. 11*:30 (1976).
88. N. Pilpel, *J. Pharm. Pharmacol. 16*:705 (1964).
89. G. Gold, R. N. Duvall, and B. T. Palermo, *J. Pharm. Sci. 55*:1133 (1966).
90. G. Gold, R. N. Duvall, B. T. Palermo, and J. Slater, *J. Pharm. Sci. 55*:1291 (1966).
91. G. Gold, R. N. Duvall, B. T. Palermo, and J. Slater, *J. Pharm. Sci. 57*:2153 (1968).
92. G. Gold, R. N. Duvall, B. T. Palermo, and J. Slater, *J. Pharm. Sci. 57*:667 (1968).

2

Preparation of Micropellets by Spray Congealing

A. Atillâ Hincal and H. Süheyla Kaş

Hacettepe University
Ankara, Turkey

I. INTRODUCTION

Spray congealing has been used for many years as a unit operation in the microencapsulation of pharmaceuticals and flavors. This method can also be used in changing the structure of the materials, to obtain free-flowing powders from liquids, and to provide sustained-release pellets ranging in size between 0.25 and 2.0 mm. Spray congealing is a process in which a drug is allowed to melt, disperse, or dissolve in hot melts of gums, waxes, fatty acids, and so on, and is sprayed into an air chamber where the temperature is below the melting points of the formulation components, to provide spherical congealed pellets [1]. The air removes the latent heat of fusion. The temperature of the cooled air used depends on the freezing point of the product. The critical requirement in this process is that substances should have well-defined melting points or small melting zones. The particles are held together by solid bonds formed from the congealed melts. Due to the absence of solvent evaporation during most spray congealing processes, the particles are generally nonporous and strong and remain intact upon agitation [1].

In other words, spray congealing is the transformation of a feed from a fluid state into a dried solid particulate form by spraying the molten feed into an ambient temperature. Since ambient temperature is used, this method of pelletization can be applied to heat-stable substances [2]. Moisture-sensitive drugs can also be encapsulated by using this nonaqueous coating system. This method is similar to spray drying except that no solvent is used for the coating material, which has the property of melting at elevated temperature when atomized and congealing while the droplets formed are air cooled in a spray dryer [3] Spray congealing is a uniform and rapid process and is completed before the product comes into contact with any equipment surface.

II. PRINCIPLES OF SPRAY CONGEALING TECHNIQUE

A. Theoretical Basis

Spray congealing techniques may modify the physical and chemical behavior of pharmaceutical materials. The characteristics of the final congealed product depend in part on the properties of the particular matrix used. The selection of a suitable coating or matrix material is based on such properties as solubility, hydrophobicity, and permeability [4].

The coating material is melted by heating up to a suitable temperature; then the active ingredient is suspended in the hot melt. Then the hot mixture is atomized into a cooling chamber through an atomizer. When the molten feed is atomized, a very large surface area is exposed. Uniform mixing of cooling air is assured by an air dispenser. Air used under atmospheric pressure cools the droplets and conveys them to the collection unit.

Tha rate of feeding as well as inlet and outlet temperatures are adjusted to ensure congealing of the atomized liquid droplets. The powder collected consists of tiny particles, each of which contains active ingredient dispersed in a matrix of coating agent. The coating material should have an adequate viscosity to ensure that it contains a uniform dispersion of core substance at elevated temperatures. In spray congealing, the air supply to the turbine can be preheated to prevent the feed from congealing in the atomizer and blocking it [3].

When spray congealing using a spray dryer, the coating is applied as a hot melt and solidification of the coating occurs upon spraying into cold air. Much higher concentrations of coatings are required for spray congealing than for spray drying, because only the molten coating forms the coat of the liquid phase [3]. The conversion of molten feed into powder is a single, continuous step. The fast cooling of the droplets prevents the formation of uneven or nonhomogeneous particles. Each droplet represents a uniform mixture of the feed [4]. Proper atomization is a key to the spray congealing process, as it results in a very large surface area. This high ratio of surface area to mass permits rapid congealing and gives uniform particles [4].

B. Equipment

Conventional spray dryers operating with cool inlet air have been used for spray congealing purposes [4]. The spray congealing apparatus has two main parts: the cooling chamber and the atomizer. The atomizer is a tool inside the cooling chamber that breaks the feed into small particles. There are three methods of atomization: pressure (single-fluid nozzle), pneumatic (two-fluid nozzle), and centrifugal atomization. Centrifugal atomizers are of different types, such as curved vanes, slotted wheel, impact pin, radial hole, and multiple-tier design.

Scott et al. used a Niro laboratory spray dryer operating with inlet air at ambient temperature (25 to 27°C) for spray congealing studies [5]. The Niro unit employs an air-driven rotating wheel for atomization of the liquid feed. Wheel speeds are controlled by regulating the turbine air pressure through a pressure reducer and needle value assembly, which is standard equipment on the Niro unit. Feed temperature at the wheel is obtained from a thermocouple that had been introduced down the atomizer feeding tube terminated at the liquid entrance to the wheel (Fig. 1). For these studies the wheel speeds ranged from 11,900 to 43,100 rpm.

Javaid [4] used the Nerco Niro spray dryer model 1422 for preparing spray-congealed micropellets. He kept the temperature of the slurry as 98°C and the temperature of the separatory funnel at 110°C. The centrifugal wheel atomizer was preheated to 120°C. Compressed air at a pressure of 6 kg/cm^2 rotated the atomizer at 35,000 rpm. Lerk and Sasburg have described an operation for the

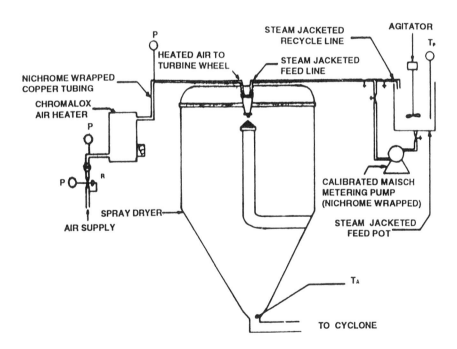

Fig. 1 Schematic diagram of equipment used for spray congealing. Instruments and controls include: T_p, pot thermometer; P, pressure gauge; R, reducer valve; T_A, outlet air thermocouple; jacketed gate valve. (From Ref. 5.)

production of uniform spray-congealed granules based on the principle of the interrupted liquid jet. In this method an axial wave is launched onto the liquid jet by forced vibration of the nozzle (Fig. 2). Eldem et al. used a Büchi model 190 spray dryer with a prototype spray nozzle (0.5-mm nozzle cap) in the spray congealing process [2,16]. The molten feed is spray congealed by using micro-filtered pressure air into the spray cylinder, which is at room temperature.

III. FACTORS AFFECTING MICROPELLETIZATION BY SPRAY CONGEALING

In this section formulation aspects such as carrier (matrix materials), viscosity, and composition, and processing aspects such as temperature, cooling rate, and atomization pressure are discussed. These factors are very important for achieving good physical properties of micropellets.

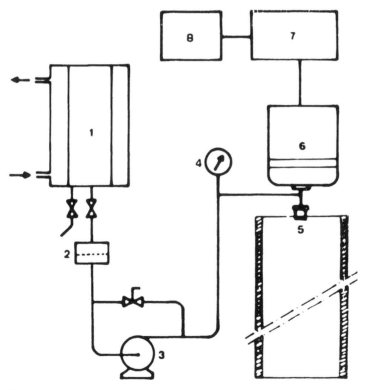

Fig. 2 Apparatus for spray congealing. 1, Supply tank; 2, filter; 3, pump; 4, manometer; 5, spray nozzle; 6, vibration generator; 7, amplifier; 8, sound generator. (From Ref. 6.)

A. Formulation Aspects

Carriers (Matrix Materials)

Waxes such as white wax [7–10], synthetic wax–like ester [7,9], and carnauba wax [8], fatty acids [11,12], stearic acid [5], 12-hydroxy stearic acid [8,13], glyceryl tristearate [2,8,14–16], glyceryl monostearate [8,15], glycerol tripalmitate [2,16], glycerol monodistearate [2,16], stearyl alcohol [13,14], liquid paraffin [7–19], hydrogenated castor oil [10,20], lecithin [2,16], gelatin [21,22], monodiglyceride [23], ethyl cellulose [5], alcohols, sugars, plastics, and other materials that are solids at room temperature and melt without decomposition can be used as matrix agents. Coatings of ion-exchange resins are also employed to alter the taste of the drug product [24]. The behavior of the final congealing product depends on the properties (solubility, hydrophobility, permeability, etc.) of the matrix materials selected.

Viscosity

It is observed that the viscosity of the formulation at the processing temperature plays an important role in the physical characteristics of the micropellets. High viscosity produced a higher percentage of large micropellets. The optimum viscosity of 24 cP at 55°C is observed to be best for obtaining acceptable particle size distribution [17].

The viscosity of the drug–matrix mixture at the processing temperature influenced pellet diameter. The viscosity is again influenced by the total amount of the drug dispersed into the coating solution and its water content. At high viscosities much larger micropellets are formed; at lower viscosities much smaller micropellets are produced. High viscosity also causes blockage of the nozzle and results in nonhomogeneous particle size. Scott et al. [5] examined feeds with viscosities ranging from 0.098 to 1.195 P. Eldem et al. [2,16] observed clogging in the nozzle when high-viscosity slurries were used.

Composition

Formulations are dependent on the physical properties of the wax, drug–wax particles, solution composition, and dissolution medium. Modifiers in spray-congealed wax formulations, such as surfactants and hydrogenated castor oil, altered the release rate significantly [7–9]. Hamid and Becker [9] studied the effect of sorbitan monooleate on the release of sulfaethidole (SETD) from spray-congealed products made into compressed tablets. The concentrations of surfactant used in the formulations were 0, 1, 4, and 10%. They observed a gradual decrease in the amount of SETD released as the concentration of sorbitan monooleate increased in the formulation using acid pepsin medium. In alkaline pancreatin medium, an increase in the percentage of SETD released was noted as the concentration of surfactant increased.

Cusimano and Becker [8] also studied the effect of a surfactant, sorbitan monostearate, on the release of SETD from a spray-congealing formulation. They showed that the dissolution rate increased with surfactant concentration. This was noted with waxes such as white wax, glyceryl tristearate, carnauba wax, and hydrogenated castor oil. An unusual surfactant effect was noted with cetyl alcohol and glyceryl monostearate, because the amount of drug released decreased as the surfactant concentration increased. This was related to the hydrophilicity of the waxes.

John and Becker [7] studied the effect of the nonionic lipophilic surfactant sorbitan monooleate over the range of concentrations 0 to 10 wt %. They observed that sorbitan monooleate in concentrations up to 4% increased the dissolution rate and depressed the dissolution rate markedly at 10% when 0.1 N HCl was used. However, the surfactant consistently promoted faster release rates in alkaline pancreatin solution.

B. Processing Factors

Temperature

Temperature changes in the medium have a critical influence over the size of the micropellets. Smaller micropellets are obtained if during solidification there is a delay in lowering the temperature. To obtain micropellets with the optimum size distribution, it is necessary to siphon out hot air from the outer jacket of the manufacturing vessel immediately. The inlet and outlet temperatures have to be adjusted to ensure cooling of the atomized droplets.

Atomization Speed

Atomization speed is found to be one of the most important factors influencing micropellet size distribution. An increase in atomization speed increases the percentage of small micropellets, and visa versa. The hot mixture is atomized into a cooling chamber through an atomizer running at a velocity of 30,000 to 40,000 rpm by compressed air. The air and the cooled product are exhausted from the bottom of the chamber through a duct to the cyclone collector, where the product is separated from the air and collected in a container. To collect the small micropellets it is necessary to use compressed air with a low velocity; otherwise, all the small particles are lost in the system.

Cooling Rate

A melt—the liquid form of a fat—can crystallize in different polymorphic forms when different cooling rates are applied [2,16]. Especially rapid cooling rates crystallize many of the triglycerides in this unstable α form. Because of the rapid cooling rate in spray congealing, lipids may be transformed into their unstable polymorphic structures.

IV. PROPERTIES OF MICROPELLETS PREPARED BY SPRAY CONGEALING

In this section the properties of micropellets, including morphology, particle size distribution, polymorphism, and dissolution characteristics, are discussed.

A. Morphology

Spherical micropellets with smooth surfaces are obtained by a spray-congealing method (Fig. 3). The addition of lecithin improves the surface morphology as shown. Apart from the solvent effect, the chain length of the glycerides also plays an important role in the surface morphology of micropellets. Eldem et al. [2] showed the effect of the chain length of glycerides on the surface morphology of formulations containing Dynasan 116 and Tripalmitin. The authors also observed smooth surface morphology with Compritol 888, which is a mixture of mono, di, and tribehenate. Due to the absence of solvent evaporation, the

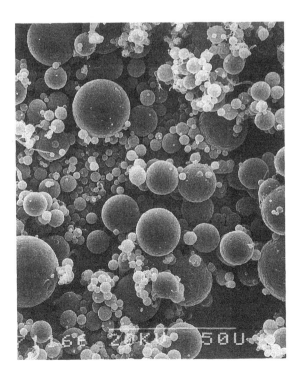

Fig. 3   SEM of freshly spray-congealing micropellets. (From Ref. 2.)

micropellets are generally nonporous and remain intact upon agitation and after dissolution studies.

Eldem et al. [2,16] prepared micro- and nanometer-sized spray-congealed lipid pellets to examine the role of the process on the surface morphology of the lipids. Since the structures are the result of rapid crystallization from melts, the spray-congealed lipid pellets obtained have unstable polymorphic structures with a smooth surface morphology. These spherical smooth surfaces of sprayed lipid micropellets corresponded to a less stable form.

B. Particle Size Distribution

The particle size of micropellets ranging from 0.5 to 1.5 mm are important in pharmaceutical technology. The characteristics of congealed micropellets are also governed by the particle size of the product. This factor is particularly important in controlling the distribution and average particle size of the active substance in the matrix. Scott et al. were the first to conduct an extensive study on the factors influencing the size and size distribution of spray-congealed particles [5]. They showed that variables such as atomizer wheel speed, feed rate, and feed velocity influenced the size of the pellets. This study showed that the average particle size of spray-congealed stearic acid varied directly with the 0.17 power of the feed rate, inversely with the 0.54 power of the peripheral wheel velocity, and inversely with the 0.02 power of the feed velocity. The correlation developed in this study is useful in arriving at first estimates of the wheel speed and feed rate required to obtain a desired particle size. When specifications for the product have been established in laboratory runs, a correlating equation developed by Scott et al. [5] may be helpful in approximating scale-up factors for the pilot- and production-level operations. The results of this experimental system have particular application to spray-congealing procedures of interest in pharmacy.

The viscosity of the oil and of the drug–carrier mixture played an important role in micropellet size, high viscosity producing a high percentage of large particles. Cusimano and Becker [8] showed that the particle size of the spray-congealed SETD–wax particles are affected primarily by the composition of the wax and the nozzle size used. In this study the SETD–cetyl alcohol formulations gave the highest values for the volume-surface diameter. These formulations were observed to be the least viscous of all the waxes used. Temperature changes in the medium have a critical influence on micropellet size. An increase in stirring speed increased the percentage of small micropellets.

C. Polymorphism

Eldem et al. studied the polymorphic behavior of spray-congealed lipid micropellets by differential scanning calorimetry and scanning electron microscopy

[2,16]. The results obtained showed that the spraying process exerted an important effect on the polymorphic and crystallization properties of micropellets. The lipid micropellets obtained possessed an unstable polymorphic form. This unstable form was gradually transformed toward a stable form by storage at elevated temperatures. The type of glyceride (composition, chain length) and the presence of a stabilizing agent such as lecithin were shown to affect the polymorphic transition and its rate (Fig. 3). The authors illustrated that the polymorphic properties are more severe with drug incorporation than without it [18]. They showed that incorporation of estradiol cypionate and medroxyprogesterone acetate caused an adverse effect in the crystalline structure of lipid micropellets, resulting in suppression of the stabilizing effects of lecithin.

D. Dissolution Characteristics

The in vitro release rates of spray-congealed formulations were affected by the initial amount of the drug, type of nozzle [8], type of wax [9], addition of surfactant [7,8], and micropellet size [17]. It is clear from Fig. 4 that particles of 20/32-mesh size failed to release the total drug within the scheduled period of experimentation [17]. The drug release rate increased with decreased particle size because the diffusional path length is less for smaller particles.

Hamid and Becker [9] evaluated the release of SETD from tablets prepared by spray-congealed micropellets and showed that the release was through erosion and leaching of the drug through the tablet. John and Becker [7] demonstrated that the surfactant used—sorbitan monooleate—in concentrations up to 4% increased the dissolution rate. However, a decrease in dissolution rate was observed when the surfactant concentration reached 10%. Cusimano and Becker [8] observed that the composition of the wax and the dissolution medium affected dissolution behavior. The surfactant used promoted wettability and increased the amount released. Raghunathan and Becker [10] showed that the addition of glyceryl ester of hydrogenated rosin decreased the dissolution rate of the active drug (Fig. 5).

V. SPECIFIC EXAMPLES

In this section, micropellets of sulfamethizole, sulfaethidole, sulfasomidone, indomethacin, trimethoprim, estradiol 17β-cypionate, and medroxyprogesterone acetate will be given as examples of micropellets prepared by spray congealing (Table 1).

A. Sulfamethizole

Javaid et al. prepared lipase–lipid–sulfamethizole granules by spray congealing [15]. Spray congealing was used in the formulation of lipase–lipid–drug granules

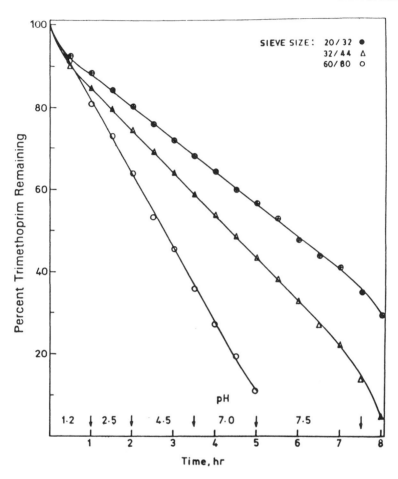

Fig. 4 Effect of particle size of micropellets on in vitro dissolution profile of tri-methoprim. (From Ref. 17.)

to minimize the possible thermal inactivation of lipase. In this study, compressed air at a pressure of 6 kg · cm^2 rotated the atomizer at about 35,000 rpm and the rate of atomization by centrifugal disk atomizer wax approximately 100 g/min. Micropellets of 23/35- and 35/50-mesh particle size were used in the experiments. This study showed that additives such as calcium carbonate and glyceryl monostearate could affect the rate of drug release (Fig. 6).

Scott et al. used molten stearic acid thickened with dissolved ethyl cellulose at 70°C in a modified laboratory Niro spray dryer to produce sulfamethizole micropellets [5]. The particle size of the spray-congealing material was 17 to 40 μm and was found to increase within this range with decreasing wheel speed,

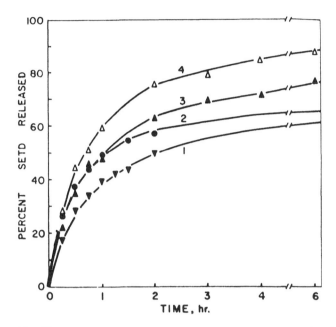

Fig. 5 Plot of cumulative percent of SETD released versus time in acid pepsin medium (1,2) and alkaline pancreating medium (3,4) from SETD–hydrogenated castor oil (2,4) and SETD–hydrogenated castor oil with 2.5% ethyl cellulose (1,3). (From Ref. 10.)

increasing feed rate, and decreasing viscosity of the feed. Scott et al. developed a relationship between the average particle size of the spray-congealed products and many important variables, including liquid feed rate, feed viscosity, and atomizer wheel speed.

B. Sulfaethidole

Robinson and Swintosky [20] encapsulated sulfaethidole (SETD) by mixing it with molten hydrogenated castor oil at 110°C. The produce was shown to have sustained-release properties when tested in vitro and in vivo.

John and Becker [7] studied the effect of the surfactant sorbitan monooleate on the particle size and dissolution behavior of spray-congealed SETD particles. The effect of these treatment variables, wax matrix material, atomizing nozzle orifice size, and surfactant concentration on the rate of production, resulting particle size, density, and porosity were also investigated. Nozzle orifice size was found to determine almost entirely the rate of production. As the nozzle orifice size increased from 0.02 in. to 0.04 in. and to 0.06 in., the rate of output increased from 3.9 g/min to 54.0 g/min and to 127.4 g/min.

Table 1 Some Active Principles Microencapsulated by Spray Congealing

Active principles	Coating material	Ref.
Sulfamethizole	Stearic acid, ethyl cellulose	5
	Glyceryl trilaurate, glyceryl tristearate, glyceryl monostearate, lipase	15
Sulfaethidole (SETD)	Hydrogenated castor oil	20
	White wax, synthetic wax–like ester	7
	White wax, hydrogenated castor oil, glyceryl tristearate, glyceryl monostearate, cetyl alcohol, carnauba wax	8
	Wax, hydrogenated castor oil	10
	White wax and synthetic wax–like ester	9
Ferric salts	Fatty acids	11
Vitamins B_1, B_2, B_{12}	Fatty acids	12
Vitamins B_1, B_2, B_6	Mono- and diglyceride	23
Vitamin A palmitate	Gelatin	22
Sulfanilamide	Gelatin, mineral oil	21
Riboflavin	Gelatin, mineral oil	21
Amobarbital	12-Hydroxystearic acid, stearyl alcohol	13
Phenobarbital	12-Hydroxystearic acid, stearyl alcohol	13
Amphetamin sulfate	12-Hydroxystearic acid, stearyl alcohol	13
Dextromethorphane	Glyceryl tristearate and 12-hydroxystearic acid	14
Sulfasomidine	Gelatin + liquid paraffin + light paraffin − lysine	19
Indomethacin	Gelatin + liquid paraffin + light paraffin	18
Trimethoprim	Gelatin + liquid paraffin + light paraffin	17
Estradiol 17β-cypionate	Glycerol tripalmitate, glycerol tristearate, glycerol monodistearate, lecithin	2
Medroxyprogesterone acetate	Glycerol palmitate, glycerol tristearate, glycerol monodistearate, lecithin	16

Source: Ref. 25.

The size of the nozzle used to atomize the melted wax formulations was found to largely influence the volume-surface mean diameter (DVS) of the product. The average DVS value was approximately linearly related to the nozzle size: 14.18 μm for the 0.02-in. nozzle, 21.33 μm for the 0.04-in. nozzle, and 29.24 μm for the 0.06-in. nozzle. The authors found that the synthetic wax–like ester formulation produced the smallest particles and that a 1:1 combination of

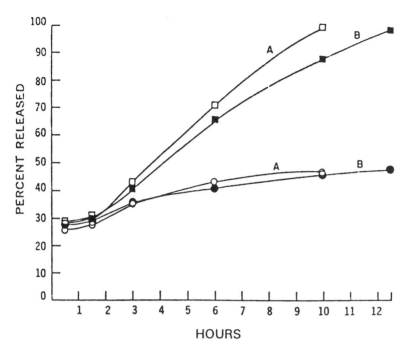

Fig. 6 Percent of sulfamethizole released from double-layered tablets made from drug–lipid granules containing: A, 2% glyceryl monostearate; and B, 5% glyceryl monostearate and initial-release granules with 25% sulfamethizole in tablets during in vitro dissolution. □, Tablet Y; ■, tablet Z; ○, tablet YF; ●, tablet ZF. (From Ref. 15.)

white wax and synthetic wax–like ester resulted in particles approximately 50% larger. Surfactant concentration was found not to affect the particle size. John and Becker [7] also showed that in concentrations up to 4%, the surfactant sorbitan monooleate increased the dissolution rate, while at 10% a remarkable depression was observed.

Cusimano and Becker [8] prepared spray-congealed SETD–wax products by pneumatic atomization using six different waxes, white wax USP, glyceryl tristearate, carnauba wax, hydrogenated castor oil, cetyl alcohol, and glyceryl monostearate; two nozzle sizes, 0.2 and 0.25 cm; and three concentrations of the surfactant sorbitan monostearate, 0, 1 and 5%. They observed that the dissolution behavior depended on the chemical composition of the wax and the composition of the dissolution medium. The authors showed that the ease with which the products were sprayed depended on the wax and the nozzle size used. The production rates were found to be greatest when white wax, glyceryl tristearate, and cetyl alcohol were used as the drug carrier. Formulations containing glyceryl monostearate were the most difficult to spray, and this difficulty increased with increasing concentrations of sorbitan monostearate. This was attributed to the increase in viscosity. The largest nozzle size gave higher production rates for all the wax formulations.

In this work, the authors found that the volume-surface diameter of the spray-congealed particles varied inversely with the feed viscosity. They also showed that the type of wax and the nozzle size used had a significant effect on the bulk density, porosity, and specific surface (Table 2). SETD–carnauba wax products were found to be least porous, and SET–white wax products were the most porous formulations, with high values of specific surface. In general, they showed that the products atomized through small nozzles gave higher porosities.

Increasing surfactant concentration and decreasing nozzle size tended to increase drug release, and the type of the waxy coating materials used affected the dissolution behavior. The authors showed that all of the main effects (i.e., wax, nozzle size, and concentration of sorbitan monostearate) had a significant effect on the percent of SETD released in both media (Table 3). In general, it was found that the increase in concentration of sorbitan monostearate increased the dissolution rate of the products. The surfactant used promoted the wettability of the drug–wax particles and increased the amount released with increase in surfactant concentration.

Raghunathan and Becker investigated the effect on the dissolution characteristics of spray-congealed pellets of SETD of the addition of low-molecular-weight polyethylene, ethyl cellulose, glyceryl ester of hydrogenated rosin, carnauba wax, hydrogenated castor oil, and synthetic wax–like ester [10]. The addition of low-molecular-weight polyethylene was observed to retard the dissolution rates of SETD from hydrogenated castor oil and synthetic wax–like ester formulations. The addition of ethyl cellulose tended to increase the dissolution

Table 2 Analysis of Variance for Specific Surface (S_W)

Source of variation	df	SS	MS	F value
Main effect				
W (wax)	5	1,534,170	306,834	19.35[a]
N (nozzle size)	1	1,486,774	1,486,774	93.76[a]
S (sorbitan monostearate)	2	39,947	19,947	1.26
Interaction				
W × S	10	137,458	13,746	0.87
W × N	5	359,716	71,943	4.54[a]
N × S	2	6,383	3,192	0.20
Error	10	158,567	15,587	
Total	35	3,723,015		

Source: Ref. 8.
[a]Significant at $p < 0.01$.

Table 3 Analysis of Variance for Percent SETD Released During Initial 15-Min Period in Alkaline Pancreatin Medium

Source of variation	df	SS	MS	F value
Main effect				
W (wax)	5	12,679,684	2,535,937	107.99[a]
N (nozzle size)	1	391,314	391,314	16.66[a]
S (sorbitan monostearate)	2	612,111	306,056	13.03[a]
Interaction				
$W \times S$	10	833,060	83,306	3.55[b]
$W \times N$	5	175,688	35,138	1.50[a]
$N \times S$	2	44,206	22,103	0.94
Error	10	234,832	23,483	
Total	35	14,970,895		

Source: Ref. 8.
[a]Significant at $p < 0.01$.
[b]The wax–sorbitan monostearate interaction was not significant at the 2.5% level.

rate from carnauba wax and synthetic wax–like formulations. The addition of glyceryl ester of hydrogenated rosin tended to decrease the dissolution rate of the drug from carnauba wax and hydrogenated castor oil formulations (Fig. 5).

Hamid and Becker [9] studied the in vitro dissolution patterns of spray-congealed products: granules and tablets of SETD–wax. The rate of release of SETD from the tablets was evaluated in acid and alkaline media using a rotating bottle method. The mechanism of release of SETD appeared to be due to erosion, solubilization, and leaching of the drug from the tablet. Tableting caused a decrease in the rate of SETD release as shown in Fig. 7.

C. Sulfasomidine

A controlled-release oral drug delivery system of sulfasomidine was developed by a spray-congealing technique using gelatin as the matrix [19]. The pellets were hardened by treating with a formalin–isopropanol mixture. The stirring rate employed was 200 to 250 rpm, and the cooling rate was between 5 and 10°C.

D. Indomethacin

A controlled-release oral drug delivery system of indomethacin was developed using as the matrix system gelatin that was rigidized using various formalin concentrations without using alcohol by a spray-congealing method. Sa et al. have shown that an increase in the drug/gelatin ratio retarded drug release from the matrix. Experiments have also shown that an increase in the formalin concentration significantly decreased the release of indomethacin from gelatin

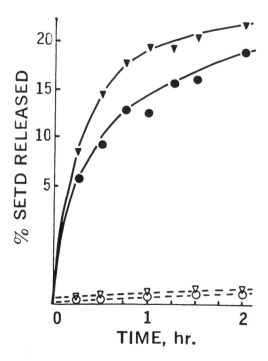

Fig. 7   Plot of percent SETD released as a function of time for dissolution from SETD/ synthetic wax–like ester/white wax (1 : 1) tablet and powder in acid pepsin medium. ▼, ▽, $W_3N_6S_{10}$; ●, ○, $W_3N_6S_4$. Solid line, powder; dashed line, tablets. (From Ref. 9.)

matrix systems. It could be concluded that the release rate of drug from the matrix system is dependent on both drug/gelatin ratio and formalin concentration [18].

E.   Trimethoprim

To produce optimum controlled-release properties, Das and Gupta investigated the design of trimethoprim micropellets of cross-linked gelatin matrix and studied the drug release profile [17]. The formulations were prepared in the proportions of 3:10, 1:2, and 1:1 as the drug/gelatin ratio. The stirring rate and cooling temperatures employed were 300 to 350 rpm and 5 to 10°C, respectively.

F.   Estradiol 17β-Cypionate and Medroxyprogesterone Acetate

Eldem et al. [2,16] have shown that lipids that have a sufficiently high melting point can be spray congealed to yield fine-powdered fat pellets. However, the authors have also demonstrated that the liquid-fat form can crystallize in various polymorphic forms when different cooling rates are used. Due to the rapid cooling rate in spray congealing, the lipids investigated in this study are transformed into their unstable polymorphic structures. Spherical micropellets having smooth surfaces are obtained with tristearin and Compritol 888. The most favorable triglyceride concentrations for the formulations containing Dynasan 116, Dynasan 118, tristearin, tripalmitin, and GTS-33 are found to be 20 to 25%. The authors are not able to use concentrations higher than 20 to 25%, due to the high viscosity of the triglyceride solutions. The investigators were unsuccessful in their attempts to prepare spray-congealed micropellets of preciral. Incorporation of medroxyprogesterone acetate and estradiol 17β-cypionate showed the same effect on the thermal behavior of liquid micropellets. The structures of these micropellets have been proved by scanning electron microscopy of the microparticles. These results were in agreement with thermal analysis microscopy results, in which the round spherulitic appearance of the unstable α form of tristearin is illustrated.

VI. CONCLUSIONS

The use of spray-congealing techniques to modify the physical, chemical, and/or physiological behavior of pharmaceutical materials has gained increasing attention in recent years [5]. Several researchers have extended the conventional spray-congealing unit operation to pharmaceutical formulations [5,7,9–23]. The spray-congealed particles may be used in tablet granulation form, encapsulated form, or be incorporated into a liquid suspension.

The general technique was applied in the preparation of tasteless forms of several water-soluble vitamins, iron salts [11], and in formulating stabilized, sugar-based flavoring agents. Several congealing methods have been applied to citric acid, sodium bicarbonate, vitamin A [22], vitamins B_1, B_2, B_6, and B_{12} [12,21,23], and other pharmaceutical materials [2,5,7–10,14–18,20,21]. Spray congealing has also been used to enhance the stability and palatability of iron salts and vitamins A and B [11,12,22,23].

Although spray congealing produces uniform particles, it is expensive and there is nearly a 40% loss in the process, due to the fact that the slurry sticks to several parts of the equipment. This loss of particles can be observed especially during collection. The small micropellets can easily stick to the filter if the aspirator rate is very high. To obtain uniform micropellets with a high yield value it is necessary to control the feed rate, inlet and outlet temperatures, composition of materials used, slurry viscosity, and atomization speed. Spray congealing has the advantage over many other microencapsulation procedures of being a rapid single-stage operation suitable for batch or continuous production of large quantities of products.

REFERENCES

1. I. Ghebre-Sellassie, in *Pharmaceutical Pelletization Technology*, (I. Ghebre-Sellassie, ed.), Marcel Dekker, New York, 1989, p. i.
2. T. Eldem, P. Speiser, and A. A. Hincal, *Pharm. Res.* 8(1):47 (1991).
3. P. B. Deasy, in *Microencapsulation and Related Drug Processes* (P. B. Deasy, ed.), Marcel Dekker, New York, 1984, p. 181.
4. A. K. Javaid, Ph.D. thesis, The University of Mississippi (1970).
5. M. W. Scott, M. J. Robinson, J. F. Pauls, and R. J. Lantz, *J. Pharm. Sci.* 53(6):670 (1964).
6. C. F. Lerk and R. H. Sasburg, *Pharm. Weekbl.* 106:149 (1971).
7. P. M. John and C. H. Becker, *J. Pharm. Sci.* 57(4):584 (1968).
8. A. G. Cusimano and C. H. Becker, *J. Pharm. Sci.* 57(7):1104 (1968).
9. I. S. Hamid and C. H. Becker, *J. Pharm. Sci.* 59(4):511 (1970).
10. Y. Raghunathan and C. H. Becker, *J. Pharm. Sci.* 57(10):1748 (1968).
11. L. Stoyle, P. Quellette, and E. Hanus, U.S. Patent, 3,035,985 (1962).
12. L. Stoyle, P. Quellette, and E. Hanus, U.S. Patent, 3,037,911 (1962).
13. R. J. Lantz and M. J. Robinson, U.S. Patent 3,146,167 (1964).
14. W. E. Smith, J. D. Buehler, and M. J. Robinson, *J. Pharm. Sci.* 59(6):776 (1970).
15. K. A. Javaid, J. H. Fincher, and C. W. Hartman, *J. Pharm. Sci.* 60(11):1709 (1971).
16. T. Eldem, P. Speiser, and H. Altorfer, *Pharm. Res.* 8(2):178 (1991).
17. S. K. Das and B. K. Gupta, *Drug Dev. Ind. Pharm.* 14(12):1673 (1988).
18. B. Sa, S. Roy, and S. K. Das, *Drug Dev. Ind. Pharm.* 13(7):1267 (1987).
19. S. K. Das and B. K. Gupta, *Drug Dev. Ind. Pharm.* 11(8):1621 (1985).
20. M. J. Robinson and J. V. Swintosky, *J. Am. Pharm. Assoc. Sci. Ed.* 48:473 (1959).
21. N. Tanaka, S. Takino, and I. Utsumi, *J. Pharm. Sci.* 52(7):664 (1963).
22. J. C. Hecker and O. D. Hawks, U.S. Patent 3,137,630 (1964).
23. A. Koff, U.S. Patent 3,080,292 (1963).
24. S. Motycka and J. G. Nairn, *J. Pharm. Sci.* 67(4):500 (1978).
25. E. Doelker and P. Buri, *Pharm. Acta Helv.* 50(4):73 (1975).

3

Cryopelletization

Axel Knoch

Parke-Davis Pharmaceutical Research
Gödecke AG
Freiburg, Germany

I. INTRODUCTION

Multiparticulate dosage forms are gaining importance in the pharmaceutical industry. This applies especially to pellets, beadlike granules 0.8 to 2.0 mm in diameter. Pellets can be classified in the following groups, based on their structure. On the one hand, there are pellets with a homogeneous distribution of active ingredient, and on the other hand, pellets in which the active substance is applied to a starter core. Pellets with modified release of active substance are usually film coated. With the selection of suitable polymers, both enteric pellets and pellets with sustained-release characteristics can be produced. Pellets that are not film coated, on the other hand, are usually immediate release unless they are conceived as matrix pellets. In matrix pellets, the extent of retardation depends on the type of matrix and is limited because of pellet geometry. Pellets with a starter core are usually produced by conventional coating procedures such as pan coating. If only a thin layer of material with active ingredients is to be used, fluidized-bed procedures are more suitable. Pellets with homogeneous structure can be produced by means of extrusion/spheronization procedures, in rotor process equipment, or in high-shear mixers. In addition, pellets with homogeneous structure can be produced using spray congealing or dropping methods. In spray congealing, carriers with a low melting point and the narrowest possible congealing range are melted. After the active ingredient is incorporated, the melted product is sprayed through nozzles or rotating disks to form drops that are congealed by cooling in air baths or cooling towers.

Figure 1 gives an overview of dropping procedures in general. In producing drop pellets—starting with aqueous solutions or suspensions—there are

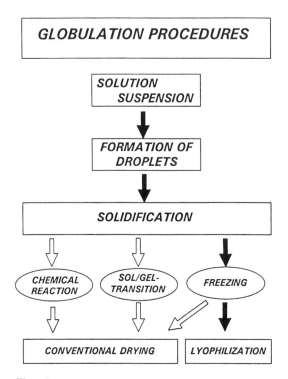

Fig. 1   Globulation procedures to create spherical particles.

several possibilities of obtaining the solid form. These possibilities differ mainly in the way in which the liquid droplets are congealed. One of the possibilities is given in chemical precipitation reactions: for example, the formation of hard-to-dissolve salts and complexes, or hard-to-dissolve, undissociated forms of the polymers used (such as Alginate). Suitable electrolyte solutions are used as the congealing baths [1]. Moreover, temperature-dependent solution/gel transformation of polymer solutions (gelatin, agar-agar) can be used. This is done by dripping the warm polymer solution into hydrocarbon cooling baths. Congealing the dripped liquid by freezing is another method.

European Patent EP 0 081 913 describes a method where drops are formed and frozen in a liquid coolant that is immiscible with the solution or suspension to be formed [2]. The coolants mentioned are trichloroethane, trichloroethylene, dichloromethane, diethyl ether, or fluorotrichloromethane. In the Cryopel method described in somewhat more detail below, liquid nitrogen at − 196°C is used as the coolant.

II. CRYOPEL TECHNOLOGY AND EQUIPMENT

The Cryopel procedure is a new freeze congealing pelletization technology, patented by Buse Gase GmbH & Co., Messer-Griesheim Group, 47805 Krefeld, Germany, for the conversion of aqueous solutions or suspensions to solid, beadlike particles (German Patent DE 37 11 169 [3]). It was developed for the nutrition industry for lyophilization of viscous bacteria suspensions, whereby a pourable, easily dosed product is obtained. Solutions or suspensions are dripped by means of an appropriately designed device into liquid nitrogen at −196°C. The use of liquid nitrogen offers several advantages over the precipitation and cooling media that are used in other procedures:

1. Nitrogen is inert.
2. No extraction of the active ingredient occurs.
3. Rinsing steps are not necessary.
4. There is no problem with residual solvent.
5. There is no solvent to be disposed of.

    The production of small drops and direct heat exchange between the solution or suspension and the liquid nitrogen coolant permit very rapid and even freezing of the material processed. Pellets are dried in conventional freeze-dryers. Thereby the trays can be charged with pellet layers up to 5 cm thick. Due to the low bulk density of pellets, there is enough interspace for sublimation of water and water vapor removal. This allows for optimal utilization of freeze-dryer capacity. The drying process is accelerated by the large surface and the small diameter of the pellets.

    Figure 2 shows a schematic drawing of the unit and a detail of its bottom plates. The Cryopel arrangement (1, Fig. 2) consists of a supply container (21, Fig. 2) in which the solution or suspension (22, Fig. 2) to be dripped is kept at a set level with the help of a sensor (28, Fig. 2). The container, which can be pressurized if needed, can be heated by means of a heating belt (30, Fig. 2). The bottom plates of the container (23, 24, Fig. 2) have holes (25, Fig. 2) with a shearing edge (27, Fig. 2). Within a short falling distance, the droplets (26, Fig. 2) become round due to surface tension, and these beads are fixed in a bath of liquid nitrogen (6, Fig. 2). The frozen pellets (20, Fig. 2) which are thus formed are removed continuously from the nitrogen bath (16, Fig. 2) by means of a conveyer belt (17, Fig. 2). The transport baffles (18, Fig. 2) of this conveyer belt are immersed in liquid nitrogen so that the frozen pellets are transported beneath the liquid nitrogen surface (19, Fig. 2) until they are solidified. They are then transported out of the nitrogen bath and transferred to a storage container, where they can be stored at −60°C. The residence time of the pellets in the cooling bath can be controlled by setting the speed of the conveyer belt so that complete

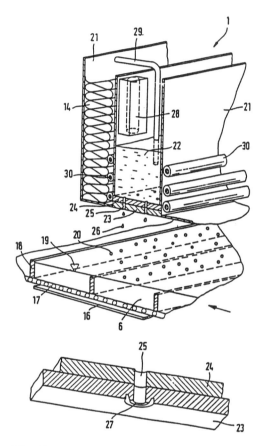

Fig. 2 Schematic drawing of the Cryopel equipment, pilot and production scale. 1, solution/suspension dropping device; 6, liquid nitrogen; 14, isolation jacket; 16, nitrogen bath; 17, conveyor belt; 18, transport baffle; 19, surface of liquid nitrogen; 20, frozen pellet; 21, reservoir; 22, solution/suspension; 23, 24, bottom plates; 25, hole; 26, droplet; 27, shearing edge; 28, filling height sensor; 29, liquid nitrogen supply; 30, heating belt. (From Ref. 3.)

freezing is guaranteed. The unit that produces the droplets and the nitrogen bath are contained in a closed chamber that is well isolated by a jacket (14, Fig. 2).

The technology described above is available from Buse Gase. Five types of Cryopel equipment are on the market, starting with a laboratory unit, Cryopel Lab, which is designed to freeze 0.5 to 2 kg/h, and going up to a production-type machine, Cryopel 1000, which allows freezing up to 250 kg/h. The sizes in between are Cryopel 200, 400, and 600, with corresponding dropping and

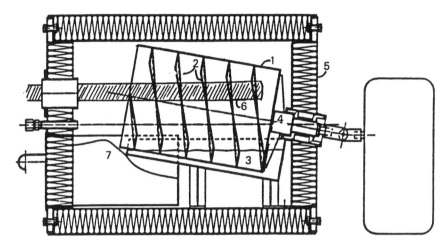

Fig. 3 Schematic drawing of Cryopel equipment, laboratory scale. 1, rotating drum; 2, helical transport baffle; 3, liquid nitrogen; 4, tilted axis; 5, isolated jacket; 6, nozzle; 7, frozen pellets.

freezing capacities of 5 to 15, 10 to 50, and 50 to 120 kg/h. The construction of the laboratory equipment differs from that of the pilot- to production-scale devices, where the conveyer belt moves horizontally in the liquid nitrogen bath.

A schematic of the laboratory-scale device is presented in Fig. 3. Typically, it has a rotating drum (1, Fig. 3) that rotates around a tilted axis (4, Fig. 3) and is equipped with a helical baffle (2, Fig. 3). All materials in contact with the product and liquid nitrogen are made of stainless steel. The equipment can be set up to be sterilized by heat or with vapor, which means that sterile pellet products can be produced under aseptic conditions. In case of aseptic processing, sterile liquid nitrogen is required, which can be produced using Cryobran, a low-temperature filter equipment of Buse Gase, which also allows conventional filter integrity testing at −196°C.

III. PROCESSING VARIABLES

The most critical step in producing spherical particles by globulation is the droplet formation process itself. The formation of drops is influenced on the one hand by formulation-related variables such as viscosity, surface tension, and the solids content. These variables are discussed in detail in Section IV. On the other hand, the quality of drops is influenced by processing variables and equipment design. Some of the critical factors are given in Fig. 4. These are, for instance, the design of the holes, which need to have a shearing edge, and the diameter of

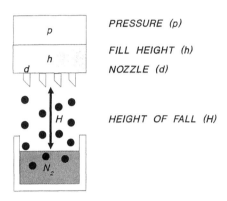

Fig. 4 Processing variables.

the holes. The diameter of the holes influences the flow rate, which must be optimized so that individual droplets are formed. The diameter of the holes also has to be adjusted to the viscosity of the formulation. Moreover, the diameter of the holes influences not only the size of the pellets but also their shape. The smaller the nozzle diameter, the smaller the resulting particles. The flow rate can be regulated via the filling level of the supply container (i.e., via hydrostatic pressure) by pressurizing the reservoir or by varying the position of the two bottom plates and thereby varying the area of the openings.

To get pellets of a narrow particle size distribution, process parameters have to be adjusted so that individual drops are sheared off. The so-called "Leidenfrost phenomenon" contributes to keep droplets in their spherical shape during solidification. However, the shape of the droplets depends among other things on the distance the droplets travel before contacting the solidification fluid. This distance has to be sufficient to allow the drops to become round, but it may not be too long, since the drops would then enter the cooling bath with too great a speed, which would have a detrimental influence on the shape of the resulting pellets. When all processing parameters are optimized, very nice, round particles can be produced with a very smooth surface, as can be seen in Fig. 5. If particles less than 2 mm in diameter are to be produced, care must be taken to prevent agglomeration of pellets in the liquid nitrogen bath. Agglomeration can be avoided by circulation or stirring of the coolant.

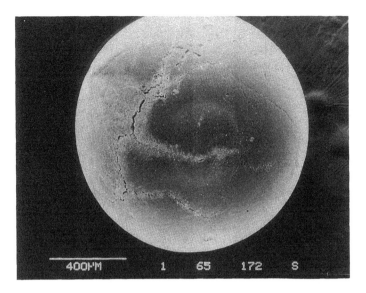

Fig. 5   Electron microscopic image of a lyophilized pellet (Collagel, mannitol, placebo formulation).

IV. FORMULATION VARIABLES

There are some requirements that solutions or suspensions must meet in order to be processed using a Cryopel device. For instance, the viscosity may not exceed a critical limit, if droplets that produce spherical pellets are to be formed. It is also important to develop a recipe that has the highest possible solid materials content in order to have the highest possible concentration of active ingredient in the finished product and to minimize the drying time. The solid materials content of solutions can vary within a wide range (10 to 30%), depending on the formulation. In suspensions, an even higher solid materials content can be achieved. However, when the solid materials content is increased, the viscosity of liquid preparations increases as well.

As mentioned above, either solutions of an active drug substance or suspensions can be processed. Special attention has to be paid to suspension formulations, since the formula must ensure that the active can be dispersed homogeneously and does not tend to settle. If this requirement is met, problems with content uniformity in the final product can be avoided. The surface tension of the medium is another factor that influences the pellet diameter to some extent. A reduction of the surface tension of the formulation by addition of a surfactant results in smaller particle size. The particle size also depends upon the

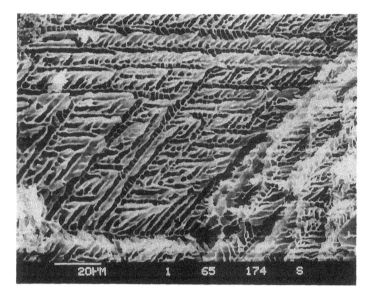

Fig. 6 Porous structure of pellets produced by Cryopel technology (Collagel, mannitol, placebo formulation).

surface activity of the drug itself. Since pellets have to be processed after drying in either a capsule- or pouch-filling machine, a certain mechanical strength of the lyophilized particles is required. The hardness of the particles depends on the fillers, binders, or matrix-forming excipients used. Pellets produced by freeze-drying show a low relative density due to the very porous structure seen in electron microscopic images (Fig. 6).

The selection of excipients is related to the biopharmaceutical design of the dosage form. There are several strategies in the formulation development of Cryopel pellets, depending on the release behavior. First, preparations that dissolve as quickly and as completely as possible have to be considered: that is, immediate-release formulations similar in dissolution behavior to ''melting tablets'' or Expidet formulations [4]. Expidets are solid dosage forms that do not have to be taken with water and are useful where swallowing is difficult or esophageal clearance is impaired. This type of formulation is also a promising approach for buccal delivery and for local delivery of drugs to the gastric mucosa [5].

Another goal in developing formulations for the Cryopel technique is to produce multiparticulate controlled-release dosage forms. This includes enteric pellets and sustained-release matrix pellets as well. An additional coating step can be avoided, as indicated in preliminary experimental trails [6] (unpublished data from Alfatec-Pharma). These trials showed that controlled-release behavior can be achieved in a single-step procedure using tailormade formulations. Since an additional coating step is not necessary, Cryopel technology combined with appropriate formula development offers real advantages over conventional pelletization procedures. Above all, a variety of release patterns can be realized in a simple process without any heat stress for the product, which is unique in pelletization technology.

V. PRACTICAL EXAMPLES

In this section examples of immediate-release, enteric, and sustained-release formulations that are suitable for manufacture by the Cryopel technology are discussed. The formulas presented are derived from preliminary experimental trials. To meet special product requirements, investigation has to be thorough, taking the properties of the active drug substance into consideration.

A. Immediate-Release Formulations

Immediate-release formulations usually consist of a soluble filler (mannitol, lactose), a binder [gelatin, gelatin hydrolysates, poly(vinylpyrrolidone)], and the active drug substance. Mannitol is superior to lactose as a filler. Poly(vinylpyrrolidone) (PVP 90) and a gelatin hydrolysate (Gelita-Collagel, Deutsche Gelatine-Fabriken Stoess & Co. GmbH, 69412 Eberbach, Germany) proved to be superior to other binders, such as agar, gelatin, hydroxypropyl methylcellulose, or carboxymethylcellulose. As could be expected, a higher concentration of a binder (2% w/w) and a high portion of filler (20% w/w) in the formulation resulted in harder pellets than those for a formulation that has only 1% of binder and 10% of filler. A typical formulation without an active is given in Table 1.

B. Enteric Formulations

In one study, use was made of the incompatibility of polyacrylic acid with high-molecular-weight poly(ethylene glycol) described in the literature [7,8] to produce enteric matrix pellets. While polyacrylic acids form very highly viscous

Table 1 Immediate-Release Formulation

Excipient	Function	Content (%)
Mannitol	Filler	20
Poly(vinylpyrrolidone) (PVP 90)	Binder	2
Water	Solvent	78

Table 2 Enteric Pellet Formulation

Excipient	Function	Content (%)
Polyacrylic acid (Carbopol 940)	Complexing agent	0.5
Poly(ethylene glycol) 20 000	Complexing agent	0.5
Poly(vinylpyrrolidone) (PVP 90)	Binder	1.0
Mannitol	Filler	8.0
Water	Solvent	90.0

gels in even low concentrations, the viscosity decreases dramatically in the presence of high-molecular-weight poly(ethylene glycols) under defined pH conditions. Thus it was possible to process the recipe shown in Table 2, even though it has a Carbopol content of 0.5% [6]. In this context, *enteric* means that the pellets did not dissolve or disintegrate in the paddle apparatus (USP) for 120 min when exposed to the following trial conditions: pH of the aqueous medium, 1.2; temperature, 37°C; and rotation speed, 100 rpm. Initially, the pellets floated on the surface of the test fluid. Later, the pellets wetted and volume increased due to hydration. The volume change, however, was little over a period of 120 min. After a 2-h residence time in an aqueous medium of pH 1.2 and a switch to pH 6.8, the pellets had disintegrated after 15 min.

In vitro release tests were then performed to determine whether the matrix could be shown to be resistant to gastric juice, as far as drug release is concerned. Two model formulations containing 4% of active ingredients of different solubilities were tested. The active drug substances were incorporated in the formula described in Table 2 to form a suspension. The in vitro release profiles are compared in Fig. 7. The results using model compound A seemed at first to indicate that the pellets are enteric. As can be seen in the left graph, release does not begin until after the pH is changed.

Pellets with a different drug model, B, by contrast, behaved completely differently. Here the active ingredient is released from the matrix within the first 30 min. The different behavior of the two test drugs can be explained on the basis of their saturation solubility. In the first case, the active substance is of extremely low water solubility with a saturation solubility of about 1 mg/L, whereas the saturation solubility of the second model is 6 mg/mL. In this study a matrix was produced which did not dissolve in an acidic test medium within 120 min. These pellets, however, did not prove to be enteric in the technological sense.

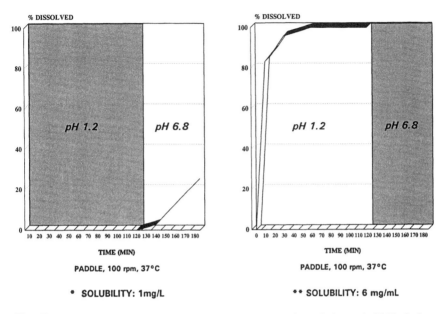

Fig. 7 In vitro release from enteric matrix formulations (formulation as in Table 2 plus 4% of model compound A or B).

C. Sustained-Release Formulations

Alfatec-Pharma GmbH, 69120 Heidelberg, Germany, prepared sustained-release formulations using the Cryopel technology. The galenical principle is a cross-linked biopolymer matrix system based on collagen derivatives as matrix-forming excipients. The collagen derivatives are patented by Alfatec. They are well-characterized polymer fractions that can be manufactured reproducibly. By varying the polymer properties, several in vitro release profiles can be realized, as can be seen in Figs. 8 and 9. Figure 8 shows a diffusion-controlled release of ibuprofen from an Alfatec ibuprofen sustained-release formulation. The in vitro release behavior in this case follows square-root-of-time kinetics according to Higuchi, which is typical for diffusion-controlled drug delivery from matrix-type dosage forms. About 40% of the active (dose strength: 400 mg) is delivered within the first 2 h. Release is completed at 8 h, using a rotating basket in vitro dissolution apparatus (USP) at 100 rpm, filled with 900 mL of phosphate buffer pH 7.2.

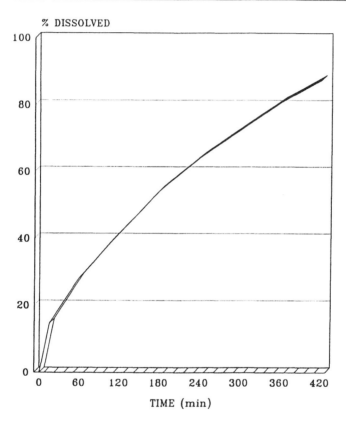

IN VITRO DISSOLUTION
Ibuprofen SR Cryopellets
(dose: 400 mg)

**Basket, 100 rpm, 37°C,
Phosphate buffer, pH 7.2**

Fig. 8   In vitro release of ibuprofen from Alfatec sustained-release cryopellets. Dose, 400 mg. (Unpublished data from Alfatec-Pharma.)

In another formulation, a zero-order release profile was achieved for flurbiprofen (Fig. 9). In this matrix, a non-Fickian, swelling-controlled diffusion mechanism [9] is postulated, with a transition of the glassy matrix to a swollen matrix and a progressively increasing diffusion coefficient in the matrix during the swelling process. One hundred percent of the active drug (dose strength: 150 mg) is delivered within 5 h in the rotating basket in vitro dissolution apparatus (USP) at 100 rpm and 900 mL of phosphate buffer at pH 7.2. An average release rate of 34 mg of flurbiprofen per hour was observed. It has to be pointed out here that both release profiles are generated only by proper matrix design and that no coatings are applied to the pellets. Tailormade matrix design is based on the unique properties of the active drug substance and on the well-standardized properties of the matrix-forming excipient. As depicted in Fig. 10, the drug's physicochemical properties, its stability in solid state and in solution, the dose strength, and specific biopharmaceutical aspects of its administration have to be taken into consideration in selection of the appropriate polymer. The polymer itself, on the other hand, is selected according to its well-defined physicochemical properties. Identity, purity, and safety requirements of these collagen derivatives meet general compendial requirements and are approved by several health authorities. The properties of the formulation can be influenced additionally by other excipients added to the formulation.

D. Cost Considerations

The quantity of liquid nitrogen used depends on the solid content of the solution or suspension and on the temperature of the medium being dripped. It is 3 to 5 kg per kilogram of pelletized formulation. With respect to production costs, one has to take into consideration that the price for 1 kg of liquid nitrogen is about $0.15 (price in Germany in December 1992). As far as energy costs of the manufacturing process are concerned, the following calculation can be made: To pelletize— to form 100 kg of solution or suspension—about 400 kg of liquid nitrogen is needed. Additional cost will also be incurred for the subsequent freeze-drying step. Equipment costs range total $42,000 to $300,000 for Cryopel machines producing 15 kg/h up to 250 kg/h frozen pellets, respectively.

VI. FUTURE TRENDS

The Cryopel technology is versatile. Its most promising advantages are that spherical solid particles of a wide range of particle sizes and of almost any release behavior can be produced in a solvent-free procedure without any heat stress for the product. This opens new paths to peptide formulation technology. For example, Cryopel can be an alternative to wet spherical agglomeration techniques for peptides that use organic solvents. But the technology is not

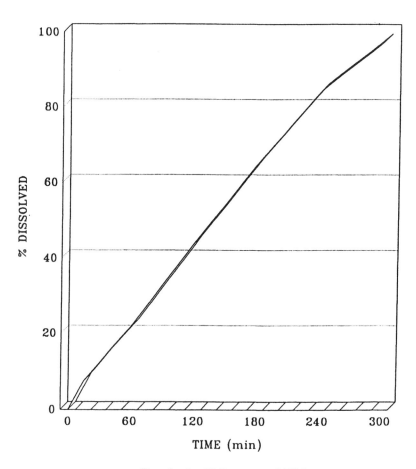

Fig. 9 In vitro release of flurbiprofen from Alfatec sustained-release cryopellets. Dose, 150 mg. (Unpublished data from Alfatec-Pharma.)

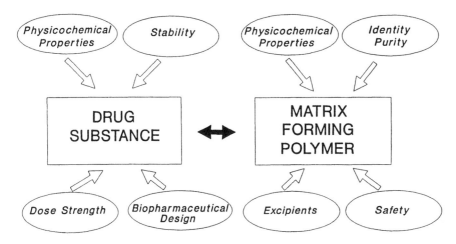

Fig. 10   Factors influencing formulation design of cryopellets.

restricted to peptides; all thermolabile drug substances are good candidates for processing using the Cryopel technology.

The variety of release patterns established so far makes it possible to generate immediate-, sustained-, timed-, or pulsed-release systems by combining various pellet formulations. Floating dosage forms can potentially be made in the Cryopel process as well. The particle size of pellets made using the technology described can be varied, ranging from 0.5 to 6 mm. This means that the spherical particles obtained can be packed into capsules or pouches to form multiparticulate dosage forms. Moreover, with the larger spheres, single units can also be manufactured in a very simple way. These can be filled in dispensing systems or in conventional blister packages or bottles. So many attractive, innovative, and unique dosage forms can be realized.

In combination with Cryodry, freeze-drying equipment of Buse Gase, which uses only liquid nitrogen as coolant, in contrast to conventional lyophilizers that require fluorocarbons, the Cryopel technology meets all environmental requirements. Alfatec-Pharma GmbH is investigating matrix formulations that can be dried conventionally in fluidized-bed dryers to minimize energy costs where thermal stress is not a problem. Moreover, formulations are under development that do not need to be dried, which includes only those formed using the Cryopel technology.

ACKNOWLEDGMENTS

The author thanks J. Wunderlich from Alfatec-Pharma GmbH, 69120 Heidelberg, Germany, and J. Buchmüller and G. Weyermanns from Buse Gase GmbH

& Co., 47805 Krefeld, Germany, for their support. This chapter is dedicated to Dr. Ruth Dillmann, Heidelberg, Germany, on the occasion of her sixtieth birthday.

REFERENCES

1. R. Bodmeier and O. Paeratakul, Spherical agglomerates of water-insoluble drugs, *J. Pharm. Sci. 78*(11):964 (1989).
2. European Patent EP 0 081 913 (1985).
3. German Patent DE 37 11 169 (1988).
4. C. G. Wilson and N. Washington, Assessment of disintegration and dissolution of dosage forms in vivo using gamma scintigraphy, *Drug Dev. Ind. Pharm. 14*(2–3): 211 (1988).
5. N. Washington, C. G. Wilson, J. L. Greaves, S. Norman, J. M. Peach, and K. Pugh, A gamma scintigraphic study of gastric coating by Expidet, tablet and liquid formulations, *Int. J. Pharm. 57*:17 (1989).
6. A. Knoch, G. Weyermanns, and F. Stanislaus, Cryopel: evaluation of a new procedure for the manufacture of pellets containing active ingredient, *Eur. J. Pharm. Biopharm. 38*(1):15S (1992).
7. K. L. Smith, A. E. Winslow, and D. E. Petersen, Association reactions for poly(alkylene oxides) and polymeric poly(carboxylic acids), *Ind. Eng. Chem. 51*(11):1361 (1959).
8. F. E. Bailey, R. D. Lundberg, and R. W. Callard, Some factors affecting the molecular association of poly(ethylene)oxide and poly(acrylic acid) in aqueous solution, *J. Polym. Sci. Part A 2*:845 (1964).
9. R. W. Korsmeyer and N. A. Peppas, Effect of the morphology of hydrophilic polymeric matrices on the diffusion and release of water soluble drugs, *J. Membr. Sci. 9*:211 (1981).

4

Coating of Multiparticulates Using Polymeric Solutions
Formulation and Process Considerations

Klaus Lehmann

Röhm GmbH
Darmstadt, Germany

I. INTRODUCTION

When pellet formulations as multiparticulate dosage forms for controlled drug release were introduced in the pharmaceutical market around 1950, organic solvents were the preferred media for natural waxes or polymers that were available to make the membranes. Due to the relatively small consumption of organic solvents for this special application, a broader selection of such solvents were acceptable. But increasing problems of air pollution from industrial processes let to severe restrictions from the governments of highly industrialized countries. So it is now necessary to prevent the release of organic solvents into the atmosphere, and the highly toxic or even cancerogenic chlorinated hydrocarbons especially have to be eliminated from modern coating formulations. Acetone has a high risk of explosion and is difficult to condense due to its low boiling point. So mainly alcohols that are also suitable as mixtures with water remain as acceptable solvents. Aqueous solutions of water-soluble polymers are suitable only for isolating layers or fast-disintegrating coatings with limited use in controlled-release multiparticulate systems. Aqueous dispersions of water-insoluble polymers are described in Chapter 5. Solvent-free systems using molten materials are described in Chapter 6. It is thought that there will be an urgent but limited need in the future for organic solvent coating systems for special drugs, perhaps those whose sensitivity to water is very high or where very special release profiles must be prepared.

II. ASPECTS OF SOLVENTS

A. Water

Some hydrophilic cellulose derivatives, such as methylcellulose (MC) and hydroxypropyl methylcellulose (HPMC), are soluble in water. Therefore, only water-soluble film coatings can be prepared, which have very limited application in controlled-release formulations. They are of some value in coloring layers and to some extent in taste masking (see Section VII.A). Due to the high viscosity of polymeric solutions, special low-molecular-weight grades (Methocel types E3 and E5 and Pharmacoat 603 and 606) were developed so that a concentration of approximately 5 to 10% can be handled in the film coating process. The sensitivity of cores to water as well as the process conditions are very similar, as described in Chapter 5. Also, the process considerations given there for aqueous systems are more relevant.

B. Organic Solvents

The physical properties of some organic solvents are compared with those of water in Table 1. The preferred organic solvents have lower boiling points than those of water and much higher evaporation numbers, which means that they evaporate much faster than water. Also, the heat of evaporation is lower, which means that the coating process can be conducted with a minimum of heating. On the other hand, with an excess of low-temperature air, the rate of evaporation of the solvent can be so high that the surface temperature of the cores is cooled below room temperature or even below the freezing point of water. If the relative humidity of the introduced air is high, condensation of water can take place at the surface of the cooled cores. Table 1 also contains data on the toxicity, the minimum odor detection concentration, and the ignition temperature as well as the flash points, which are very important for flameproof installation and safe handling of the materials in production.

   An important aspect is the relatively high viscosity of polymer solutions, which depends on the molecular weight and affinity of the polymer to the solvents. If the solvent has a high affinity to the polymer chains, the apparent molecular sphere of action of the polymer is very high, due to the spreading of chain segments, resulting in high viscosity. If the solvent has a lower affinity to the polymer, some polymer chain aggregation and shrinkage of the polymer molecule results in lower viscosity. Sometimes, solvent mixtures give better dissolution properties of the polymers as well as a lower solution viscosity, in correlation with the concentration of the solid polymer. The most common solvents for coating polymers are listed in Tables 2 and 3.

Table 1 Physical Data and MAK/TLV Values of Selected Solvents

Solvent	Boiling point (°C) at 1013 mbar	Evaporation number[a]	Heat of evapo- ration (J/g)	Vapor pressure at 20°C (mbar)	MAK/ TLV ppm (mL/m³)	Minimum odor detection (mg/m³)	Ignition temper- ature (°C)	Flash point (°C)	Flammability range at 760 torr (vol %)
Ethanol	78.3	8.3	855	60	1000	93	425	+16	3.5–15.0
Methanol	64.7	6.3	1102	128	200	7800	508	+6.5	5.5–26.5
Isopropanol	82.3	11.0	667	40	400	90	634	+15	2.0–12.0
Acetone	56.2	2.0	520	240	1000	770	540	−19	2.5–13.0
Dichloromethane	40.2	2.0	321	475	100	550	605	No	13.0–22.0
Trichloromethane	61.2	2.5	247	210	10	1000	—	No	No
Water	100.0	60.0	2264	17.5	No	Very high	Very high	No	No

Source: Modified from Ref. 22.
[a]Diethyl ether = 1.

Table 2 Solubility of Cellulose Derivatives in Organic Solvents and Water

	MC	HPMC	CAP	HPMCP	EC
Water	+	+	>pH 6	>pH 5–6	−
Methanol	+		−	sw	+
Ethanol	−	+	−	sw	+
Isopropanol			−	sw	+
Acetone	+		+	+	+
Dichloromethane		−	−	sw	+
Ethyl acetate			−		+
Mixtures	Ethanol–water 80:20 to 40:60		Acetone–ethanol or Acetone-isopropanol or Acetone-dichloromethane 50:50 Dichloromethane–methanol or Dichloromethane–ethanol or Dichloromethane–isopropanol 50:50		Toluene–ethanol 70:30 Ethanol– dichloromethane 50:50

[a] +, soluble up to 10% and more; sw, swelling; −, insoluble.

Table 3 Solubility of Polymethacrylates in Organic Solvents and Water

	Polyamino-methacrylate E100	Methacrylic acid copolymers		Ammoniomethacrylate copolymers	
		L 100–55	L100/S100	RL100	RS100
Water	<pH 5	>pH 5.5	>pH 6–7	sw	sw
Methanol	+	+	+	+	+
Ethanol	+	+	+	(+)	(+)
Isopropanol	+	+	+	(+)	(+)
Acetone	+	+	+	+	+
Dichloromethane	+	−	−	+	+
Ethyl acetate	+	−	−	+	
Mixtures	Ethanol–water 60:40 Isopropanol–acetone 60:40			Ethanol–water 50:50 Isopropanol–acetone 60:40	

+, soluble up to 10% or more; (+), soluble at lower concentration; sw, swelling; −, insoluble.

C. Solvent Mixtures

Cellulose derivatives were used primarily in solvent mixtures of alcohols and chlorinated hydrocarbons or acetone. For polymethacrylates, solvent mixtures of alcohols and acetone are recommended. A few percent water will improve the solubility of polymers, especially polymethacrylates, when added at the 3 to 5% level. These traces of water will reduce the time for dissolution of the solid polymer and will also reduce the turbidity of the polymer solution. The molecular weight of the polymers is responsible for the viscosity in a given solvent. To get good, homogeneous spray with small droplets in the range of about 20 μm with airborne spray guns, a relatively low viscosity in the range 150 to 500 cP in the formulation is necessary. If the molecular weight of polymers increased above a critical level, polymer fibers like cobwebs are formed in the spray, so such products are not suitable for film coating.

Mixed-solvent systems containing a good solvent of high evaporation speed under the given coating conditions are sometimes very useful in the film-forming process. When a droplet of polymer solution reaches the surface of the core, it has a high spreading tendency due to the low viscosity of the diluted polymer solution; when the more volatile solvent is a better solvent for the polymer and evaporates first, the polymer solution in the other solvent becomes less sticky and gelation occurs at higher polymer concentration. There is a higher tendency of the polymer to retain the solvent of higher affinity. For more details, see Section IV.A.

D. Residual Solvents

In all coating processes with organic solvents, it is easy to eliminate residual solvents from the film and also from the core as long as the film is very thin and does not cover the entire surface area of the core. If solvent is entrapped in the core or the first layers of the coating, the elimination of the residual solvent is strongly reduced by the following coating layers. To prevent high residual solvents in coated multiparticulate dosage forms, spraying should be done slowly at the beginning of the process and stopped for intermediate drying before an entire isolating layer covers the core. The following spray process is then controlled such that very thin layers of coatings are applied and dried immediately before the next layer is applied. So the spray rate is correlated directly with the drying capacity and movement of the core in the spraying zones.

III. ASPECTS OF CORES

A. Surface Area

The surface area of particles in multiparticulate dosage forms is much higher than in tablets. When a normal tablet approximately 10 mm in diameter has a surface area of approximately 3 cm^2, some 3% of the wall material is sufficient for coating. The equivalent amount of 1000 pellets of 1-mm diameter have a surface area of 30 cm^2. The surface area of spherical particles and the amount of coating that is needed to form a membrane of a given thickness can be calculated from the median diameter as follows:

$$A = \pi d^2$$

where A is the surface area (mm^2) and d is the diameter of particles (mm), and

$$c = \frac{AT}{w}$$

where c is the coating material (%), T the thickness of the coating (μm), and w the weight of the core (mg).

If the particle size is reduced to one-tenth of the diameter, the surface area increases 10-fold. So the amount of coating material to encapsulate particles around 1 mm in diameter comes up to approximately 30% of the weight of the cores. But when film thickness is really the same as that of tablets, the diffusion rate of the microparticulate system is still higher, due to the higher surface area of the coated particles. When the diffusion rate is proportional to the surface area, it is 10- or 100-fold with 1.0- or 0.1-mm particles compared to the tablets. Therefore, coating layers of high quality and polymers of very low permeability are needed. Films for enteric coating normally have a very low diffusion rate, so that it is possible to get enteric multiparticulate dosage form with a release rate of less than 5% per hour in gastric fluid, and also other coatings for sustained release are low enough in their permeability to get the desired release rate with approximately 10 to 30% wall material with particles in the diameter range 0.1 to 1.0 mm.

From the calculation above it is clear that a particle size below 0.1 mm is normally very uneconomical with regard to the consumption of wall-forming encapsulating material. If, on the other hand, particles are larger than 1.0 mm, the number of particles per dose may not be high enough to give acceptable content uniformity and release characteristics. A rough calculation gives a minimum of several hundred particles per dosage form to meet the usually requirements [2]. From these limitations the coating of small particles for multiparticulate dosage forms is preferably conducted in the particle diameter size range 0.1 to 1.0 mm.

B. Particle Shape

Particle shape is a very important factor. Nearly globular particles have, of course, the smallest surface area with a given particle diameter. But in most cases a special spherionization process is needed to form such particles. Drugs used in very large amounts are to some extent available from the producer in the form of more-or-less spherical particles. In such particles produced using the spherical crystallization process (e.g., acetylsalicylic acid and potassium chloride) the quality of the surface is very near that of spheronized particles and the surface area is very close to the theoretical value, which can be calculated from the median particle size and an ideal sphere.

Many drugs are available in the form of crystals or compact granules. The quality of the coating of such irregular particles is more problematic using rotating pans, due to the high agglomeration tendency, but can be very good using fluidized-bed systems [1]. If the coating formulation is one of low viscosity with good spreading and poor sticking effects, encapsulation can be achieved with acceptable amounts of raw material [22].

Crystals should be as compact in shape as possible, as cubic and columnlike structures give the best results. Friable particles may break and can be agglomerated during the coating process, resulting in an increased surface area and poor encapsulation. Compact granules from dry compaction or extrusion processes may be similar to crystals and thus are often suitable cores for fluidized-bed encapsulation. To keep agglomeration to a minimum, it is very useful to have a narrow particle size distribution. Otherwise, smaller particle pieces will adhere to larger pieces and stimulate agglomeration, causing increasing amounts of oversized particles to be formed during the coating process.

C. Particle Porosity and Stability

The porosity of particles is more critical in aqueous coating processes with latexes, where the film-forming process is disturbed [4]. With an organic coating solution there is a good chance of filling the pores and stabilizing the surface area in the first phase of the coating process. Intermediate drying may be useful to eliminate residual solvent and dry the film for optimal stabilization of the core. If the particles are not stable under the conditions of movement in coating pans or in the fluidized bed and some abrasive material is formed continuously during the coating process, the encapsulation quality is often questionable. The rate of agglomeration may increase considerably or abrasive material may be incorporated in the film layer, resulting in a high-porosity coating. Methods to determine relevant quality data for core materials are given in the literature [5].

IV. ASPECTS OF COATING MATERIALS

A. Solubility of Polymers in Organic Solvents

Tables 2 and 3 provide a summary of solubility data of the most common coating materials. The solvents listed were limited to those that can be used in large-scale production. If solvent recovery systems are installed to prevent air pollution, it may be possible to use other, more expensive solvents, provided that the recovery system is good enough and the purity sufficient after simple redistillation. The general tendency is to use cheap alcohols or acetone, which can be burned off. Water as a component of organic solvents is important to reduce costs and inflammability as long as the presence of water does not cause stability problems with the drug cores. If the sensitivity of drug cores against water is not important, pure aqueous systems are preferred (see Chapter 5).

Solvents used in organic coating systems fall into three categories [6,7]:

1. Active or good solvents have a high degree of affinity with the structural elements of polymer molecules and lower their intermolecular forces.
2. Intermediate and poor solvents as described by Deasy [6] correspond to the latent solvents and diluters described below.
3. Latent or intermediate solvents also have some affinity to macromolecules but not enough to overcome the intermolecular forces between polymeric molecules. In a mixture with an active solvent, a latent solvent can increase polymer solubility.
4. Diluters or poor solvents are low-boiling liquids used to lower the viscosity of polymeric solutions and to improve spraying conditions. The vapor pressure of the diluter should be higher than that of the solvent to prevent the enrichment of a poor solvent during the film-forming process and cause precipitation of the polymer.

B. Preparation of Polymer Solutions

During the dissolution process, the polymer swells first in the solvent and a viscous layer is formed around the polymer particles. Therefore, the polymer should be added to the solvent under vigorous stirring, and stirring should be continued in a manner that does not allow sticky polymer particles to sediment or layer down at the bottom or walls of the kettle. The dissolution process can be facilitated by heating, which reduces the viscosity of the polymeric solution.

Some cellulose derivatives show a lower degree of solubility at higher temperatures. This effect can be used to get good dispersion of the polymer in the range of lower solubility and faster dissolution when the temperature is lowered. It is recommended that methylcellulose be dispersed in one-third to one-half of the final volume of water at 80°C, soaked for 5 to 10 min and mixed for 5 to 10 min, then cold water added and the solution cooled to 20°C.

If the polymer contains structural units that form hydrogen bonds, water is a good cosolvent. For example, 3 to 5% water is recommended for dissolution of polymethacrylate in alcohol–acetone [8]. Plasticizers may be added during the dissolution process to stimulate dissolution of the polymer. Most plasticizers show good solubility in alcohols. An exception is poly(ethylene glycol) (PEG), which is much more soluble in water and precipitates out when an aqueous solution of PEG is added to an alcoholic solution of the polymer.

C. Use of Excipients

Table 4 lists some solubility data of commonly used plasticizers and the types of plasticizers that are recommended for certain coating polymers. Other excipients are normally added to organic polymer solutions during the coating process. Talc reduces the stickiness of coating formulations by forming latice structures, and these particles are very easily embedded in the polymer layers, thus significantly reducing sticking during the film-forming process. Talc also reduces the porosity of film coatings and lowers their water permeability. A minimum of 20% talc is recommended for use with polymethacrylates, and up to 150% can be incorporated without disturbing the dissolution properties of these materials in the digestive fluids. In formulations with methylcellulose, 5 to 20% talc is used. Magnesium stearate, which functions similarly, is normally used in a mixture of a similar amount of talc. Although fumed silica (Carbosil, Aerosil) is recommended as an antisticking agent in coating formulations, due to its bulkiness it is more suitable if it is added in dry form at the end of the coating process. Micronized silica (Syloid No. FP 244, Grace USA) is easy to suspend in coating formulations and has very good antisticking effects in amounts of 5 to 15%, but the permeability of the coating is somewhat increased and large amounts will give a mat surface.

Pigments are widely used for colored film coatings. The preferred white pigment is titanium dioxide. All pigments will reduce the stickiness of coating formulations to some extent, but they are not as active as talc or the other excipients mentioned earlier. The addition of pigments and other solid additives is limited by the critical pigment volume concentration of the polymers [9]. For hydroxypropyl methylcellulose films this value is approximately 25 to 30%; polymethacrylates have a much higher pigment binding capacity: up to 200% of such materials can be incorporated. This is useful for covering bad or irregular colored cores without increasing the amount of film formers.

V. ASPECTS OF MACHINERY

A. Safety

To work with a range of formulations of organic solutions, flameproof equipment is necessary. There are several categories of such installations. It may be useful to exclude or limit the most harmful components and utilize less dangerous formulations. Acetone has a very low flash point and very high vapor pressure, resulting in a high evaporation rate, so any details of installation and handling that could cause inflammation must be done very carefully. Di- and trichloromethane are not flammable but are highly toxic and for air pollution abatement purposes are the most highly restricted solvents. Alcohols are solvents of limited risk, methanol being the most toxic and flammable. The industrial use of ethanol is restricted in certain countries due to customs regulations and to its

Table 4 Solubility and Recommended Use of Plasticizer in Organic Coatings

| | Solubility in: (g/L) | | | | Recommended for: | | |
	Water	Ethanol	Acetone	Dichloromethane	Cell. ether	Cell. ester	Polymethacrylates
Diethyl phthalate	1.0	×		×	×	×	
Dibutyl phthalate	0.4	×		×	×	×	×
Acetyltriethyl citrate	7.2					×	×
Acetyltributyl citrate	Insoluble					×	×
Triethyl citrate	65	×				×	×
Glycerol triacetate	71	×				×	×
Glycerol	Miscible	×			×		
Castor oil	Insoluble	×					×
1,2-Propylene glycol	Miscible	×	×		×	×	×
PEG 6000	Miscible	×	×	×	×	×	×
Dibutyl sebaccate	Insoluble	×			×	×	×

high price. The addition of water to alcohols is recommended to reduce their flammability.

Isopropanol is a very acceptable compromise. It has a high flash point, low toxicity, and a medium-level evaporation rate; its boiling point and heat of evaporation are significantly lower than those of water. To prevent air pollution, it can be burned or, alternatively, condensed using acceptable technical installation.

The traditional open coating pans are normally not acceptable but the pans can be enclosed in cabinets. Although such pans are not absolutely airtight, the air supply can be controlled such that the drum is always under reduced pressure, thus preventing the escape of organic solvent vapor. Fluid-bed systems are very airtight and are normally operated under reduced pressure and sometimes at slightly elevated pressure.

The formation of dust from the core material during the coating process can be a severe problem, since the dust can be incorporated into the film coating layer. Before starting the coating process, dust has to be eliminated as much as possible, preferably by the use of stable core materials.

B. Movement of Cores and Drying Air Capacity

Important in the construction of coating machinery are devices that facilitate movement of the cores and enhance the drying air capacity. In rotating drums, continuous movement of the particles should be achieved by the introduction of buffles that reduce the agglomeration tendency in the sticky phase of the film-forming process and prevent sliding of the cores when the polymeric film layer dries. Continuous movement of the cores must be achieved to prevent adhesion of the cores to each other and to the wall.

There is often a considerable tendency for cores to slide on the walls of coating pans and to pick up metal abrasive particles, which can result in black spots on the cores. To prevent this, it is recommended that a layer of film-coating material be sprayed onto the wall before the cores are placed in the pan. A suitable amount of drying air is necessary to dry the sprayed coating solution as fast as possible and to restrict sticking to very low levels.

Due to the high evaporation rate of organic solvents, the heating of drying air is controlled so as to keep the core around room temperature. Higher temperatures stimulate sticking due to the softening tendencies of polymer at higher temperatures; too low core temperatures may cause water condensation if the air introduced is humid and below the dew point.

The greater agglomeration tendency of small particles limits their use in rotating drums to nearly spherical pellets of particle size around 1 to 2 mm. Smaller, irregular particles are much more effectively coated in fluidized-bed systems. The working principles and optimization of processes in these devices are described in more detail elsewhere [22].

C. Electrostatic Effects

Electrostatic effects are observed in both rotating drums and fluidized-bed systems, especially under very dry conditions. The effects occur normally at the end of the drying cycle, when spraying has been stopped and the film dried to a very low concentration of residual solvent, and then disappear when spraying is continued. Electrostatic effects can be lowered or eliminated by adding water to the solvent system, by using magnesium stearate as an antisticking agent, or by using PEG as a plasticizer.

D. Special Equipment to Handle Flammability and Air Pollution

Fluidized-bed systems can be operated with nitrogen as carrier gas and equipped with a solvent recovery system. If the fluidized-bed process is conducted in a vacuum, there is no need to use nitrogen as the inert carrier gas. The vacuum fluidized bed process works at a solvent vapor pressure of 100 to 300 bar. Part of the solvent gas evaporated is used as a carrier medium for heat transfer and fluidization; another part is condensed continuously in the recycled gas stream. The loss of density of the gas stream is compensated by a higher gas velocity, so that fluidization is comparable to working conditions under normal air pressure [12].

When using the vacuum coating pan developed by Glatt GmbH, Binzen, Germany, the pan is heated from the outside by hot water and the solvent is condensed in a cooled trap. The drying capacity is limited by the heat transfer through the pan wall, for the only energy carrier is very low inert gas [10,11]. Thus spray rates are low, so that especially in pellet coating, a long processing time results. The equipment is compact and may be useful for medium-scale production of sensitive specialties.

There are several possibilities for combining standard coating machinery with modern technology to prevent air pollution. A relatively simple process is thermic postcombustion, where the waste air is transported through a firebox in which a supporting flame maintains the temperature required to burn off the solvent substances. Such pieces of equipment operate at low cost since the energy can be used for heating purposes. Thermic combustion is most acceptable for solvent free of chlorine, such as alcohols, acetone, and hydrocarbons.

Solvent absorption on charcoal is possible on large absorption towers, which can subsequently be regenerated by water vapor. The solvents–water mixture recovered needs further processing by condensation and/or redistillation. Solvent condensation in a closed circulation system appears to be very effective in a film-coating process [13]. Coupled with catalytic oxidation of traces of organic material in the remaining gas stream, solvent condensation is becoming more popular in the control of air pollution. Biodegradation of solvents in wet filter towers is a modern method of cleaning an airstream of several 100 m³/h. Halogen-containing solvents are even more difficult to handle, so alcohols and ketones should generally be used as coating solvents if aqueous systems are not feasible.

VI. PROCESS CONSIDERATIONS

A. Air Supply

Fast solvent evaporation is essential for the formation of a stable film on the core surface as soon as possible after the spray droplets have reached the core and spread on the surface. Especially during the coating of small particles, there is a strong tendency toward agglomeration when the core surface is sprayed with polymer solution and the drying film layer is in a highly sticky phase. High levels, even an excess of drying air is thus very important for effective coating. In fluidized-bed systems a strong stream of air is in place to keep the particles fluidizing, so that interparticle contact is kept to a minimum and excess drying capacity is available. In rotating pans the amount of drying air is based primarily on the level acceptable for drying and so is normally much lower than that in fluidized-bed systems. If small particles are to be coated using such equipment, it is important to increase the air supply to a maximum and to introduce the incoming air directly into the core bed, to optimize the drying effectiveness and to stimulate more intensive movement of the cores. Such a system is known as an immersing tube process [14]. The drying capacity of the inlet air is also high in perforated drums.

B. Temperature

The temperature of drying air can be relatively low, in the range 20 to 40°C, with organic coating solutions containing highly volatile solvents such as acetone and methylene chloride. Formulations with isopropyl alcohol may require temperatures around 30 to 50°C. The product temperature should be kept near room temperature or slightly above, normally not higher tan 30°C, which normally requires the temperature of incoming drying air to be between 30 and 50°C; to some extent, this depends on the spray rate (Table 10).

Solvent evaporation cools the surface of the cores. If the spray rate is high and the energy transported with the drying air not high enough to compensate for the heat of evaporation, the temperature at the surface of the cores may fall below room temperature. If the incoming air is taken directly from a highly humid atmosphere without dehumidification, the dew point may be reached, leading to water condensation. If the temperature of the incoming drying air is too high, the product temperature goes up and film stickiness increases. The solvent is normally a very active plasticizing agent. The percent of residual solvent retained in the polymer is therefore critical. A low level of residual solvent in the film is retained when spraying and drying are conducted continuously during the coating process and the very thin film layers are dried immediately. This occurs if the concentration of solvent in the airstream is low. As a result, it is recommended that the temperature of the incoming air be kept as low as possible to keep the temperature of the core around room temperature and to increase the amount of drying air to the maximum consistent with the capability of the apparatus. Under such conditions, stickiness is reduced. If cores are porous and solvents tend to diffuse into the core, additional intermittent drying may be necessary as long as the coating is thin enough to allow diffusion of residual solvent from the core to the surface. At the end of the process, the coating normally acts as a tight barrier for traces of solvents entrapped in the core, and very long final drying times are necessary to attain low levels of residual solvents.

C. Spray-Rate/Spray Systems

The spray rate of a coating solution depends on several parameters: (1) the drying air capacity of the machinery, (2) the mixing intensity of the cores, and (3) the spray area. To get fine spray droplets of approximately 20 μm, an atomization air pressure of about 2 to 4 bar is sufficient, and the spray rate can be regulated with spray nozzles approximately 0.8 to 1.5 mm in diameter. The coating solution can be fed to the nozzle by a peristaltic pump. The spray rate must be reduced if the level of stickiness is so high that more agglomerates are formed than destroyed in the normal cycle of movement of the particles in the machinery. If it is not possible to overcome the sticking effects by increasing the drying air capacity or by intensifying the movement of the particles in the machinery, the coating solution should be diluted. Diluted coating solutions can be sprayed at a much higher rate, so that often more solid polymer material can be applied in the same amount of time. Diluted coating solutions also give smoother, more homogeneous film layers.

Airless spray systems are operated at high hydraulic pressures, 50 to 150 bar. This is achieved with compressed air, but there is no air in the spray stream. This prevents loss of atomized mist. The spray rate is determined by the choice of the nozzle diameter and spray pressure. Here it is more difficult to regulate the relevant spray rate for the film-coating process, as the minimal spray pressure that is necessary to form sufficiently fine droplets in the spray gives a very high output of material, approximately 250 mL/min at 100 bar. Therefore, normally the spray has to be interrupted every few seconds, followed by a drying phase, so airless spraying is usually done in cycles of spraying/drying. A continuous airless spray is difficult to manage and needs big batches, a large spray area, and fast movement of cores through the spray.

D. Adding Powder

In coating processes, it is often desirable to incorporate solid materials as a dry powder to the cores, primarily to overcome phases of stickiness or to incorporate drug substances into the film. The first choice is to suspend this powder material into the coating solution as is normally done with pigments, but if this is not possible, a powder feeding system may be attached to the machinery [15].

E. Attrition

If the stability of the particles is not high enough, attrition of the core material may occur during the coating process and powder particles get embedded in the coating layer. If the core material is water soluble, pores may be formed in the coating, altering the release profile. In addition, abrasion of the polymer coating material can occur during the process or, even more likely, when drying is done under fluidization following coating. To prevent such effects, it is helpful first to warm the cores under very mild fluidization or a low rotation speed and to begin applying the coating at a low spray rate. The first layers of coating material will stabilize the surface of the cores and intermediate drying can be introduced at this early stage of the process to eliminate residual solvent. Similarly, another stabilizing layer can be applied with intermediate drying done under reduced fluidization. Also, at the end of the coating process, postcoating drying must be conducted under reduced movement of the particles to prevent abrasion or even damage to the coating shell.

VII. FORMULATION CONSIDERATIONS

A. Isolating Layers

Thin layers of 5 to 20 μm are applied to reduce the porosity of particles, to prevent dust formation during handling, and more important, to mask taste, which may also prevent irritation of the esophagus [16]. Film coating is also sometimes used to reduce odor or to stabilize moisture-sensitive products [21]. The isolating effect is proportional to the film thickness, although thick films can also cause delayed release of the drug.

   The water-soluble cellulose ethers are preferred for all applications where the requirements as to the isolating effects are low and release in an aqueous environment should be as rapid as possible. A formulation is given in Table 5 (formulation 1). To get better taste masking, a combination of HPMC and ethyl cellulose is recommended (formulation 2). A film-forming material that is insoluble in the mouth but dissolves rapidly in gastric juice and in the stomach in a slightly acidic environment (up to pH 5) is dimethylaminoethyl methacrylate

Table 5 Formulations for Taste Masking

	Formulation number				
	1	2	3	4	5
HPMC	5.0	4.3	—	—	—
EC	—	1.1	—	—	—
E100	—	—	2.0	—	—
RL100	—	—	—	5.0	—
L100	—	—	—	—	5.0
Glycerol triacetate	—	0.6	—	—	—
PEG 6000	1.1	—	0.4	0.5	0.7
Talc	—	—	2.8	—	6.0
TiO_2/pigments	3.9	—	2.8	—	3.3
Methanol	—	32.0	—	—	—
Isopropanol	—	—	45.0	45.5	41.0
Acetone	—	—	45.0	45.5	41.0
Dichloromethane	—	62.0	—	—	—
Water	90.0	—	2.0	1.5	3.0

copolymer (Eudragit® E100, formulation 3). This material gives very specific taste-masking effects with layers approximately 10 μm thick. Due to its good solubility in gastric fluid, thicker film layers can also be applied without greatly affecting the release. In a neutral or slightly alkaline environment, this polymer swells and within a few minutes forms a very permeable film. If the pH in the stomach is above 5, due to a low secretion of gastric fluid by the patient or the high buffering capacity of food ingested, it is possible that the coated particles reach the intestine without dissolution of the coating; in this case release will take place by permeation of the drug and mechanical disintegration of the coating membrane. This has to be verified by testing the coated particles in a neutral buffer solution.

Hydrophilic coatings of very high permeability containing trimethylam-moniomethyl methacrylate chloride copolymer (Eudragit RL100, formulation 4) can be used for isolating layers. The polymer is not soluble in digestive fluids but is very permeable, independent of pH. The films swell within a few minutes and the drug permeates quickly out of the coating. The taste-masking effect is lower than with Eudragit E100.

Another possibility for forming isolating layers is to use enteric coating materials (formulation 5). The water permeability of such materials is lower than that of the materials described above, so they give very good taste-masking effects and also very low water permeability in a humid atmosphere. As long as film thickness is in the range 5 to 10 μm, these films are not resistant to gastric juice but disintegrate within 10 to 30 min. So isolating effects together with the release pattern should be investigated very carefully with increasing film thickness to find the best compromise. If such coated particles will pass the stomach without disintegration, the coatings will be dissolved at the latest in intestinal fluid.

B. Enteric Layers

To get resistance to gastric juice as required in the USP and other pharmacopoeias, it is necessary to apply films of approximately 20 to 30 μm thickness to have enough mechanical stability and to reduce the permeation rate of the encapsulated drug below the given limit of 5% per hour. Enteric coating materials normally contain in their polymer structure carboxylic groups which are susceptible to salt formation at a nearly neutral pH of 5 to 7 to give water-soluble polymeric salts. Gastric fluid has a pH of 1 to 2 in an empty stomach, but after ingestions of food, the pH increases and can reach pH values of 3 to 5, especially with a food rich in protein. The secretion of gastric fluid is stimulated by a meal and lowers the pH during the following hours. Patients with less gastric fluid secretion will have less acidic or even nearly neutral stomach contents. Transport of drug particles through the stomach depends on particle size [17]. Enteric-coated tablets that do not disintegrate in the stomach are transported into the intestine mainly by "housekeeper waves" when the stomach is empty. Smaller particles are transported continuously with the food through the pylorus.

The formulation of enteric coatings should be consistent with the biopharmaceutical aspects of the drug under the following considerations:

1.   How much drug release into the stomach is acceptable?
2.   Is the drug sensitive to gastric fluid and/or low pH?
3.   Where in the intestine should the drug be released?
4.   What is the desired transit time of the dosage form through stomach and intestine?
5.   When is drug expected to be administered relative to the intake of food (i.e., before, during, or after a meal)? What is the dosing interval?

To prevent release of coated drug in the stomach, the film-forming material should be insoluble in the range pH 1 to 5. The permeability of the coating layer should also be as low as possible if the drug is sensitive to low pH or if the drug irritates the stomach. The disintegration and dissolution of enteric coatings can be controlled in the range pH 5 to 7.5 by selection of several polymer materials. Table 6 describes the solubilities and permeabilities of polymers in the various sections of the digestive tract.

Although normally, absorption of drugs in the distal parts of the intestine is low, absorption in the colon is still possible. Sometimes drug release in the colon

Table 6 Behavior of Polymeric Films in the Digestive Tract

	Mouth > 1 min pH 5–8.5	Stomach ~ 1–2 h pH 2–5	Duodenum ~ 0.5 h pH ~ 6	Jejunum ~ 2–4 h pH ~ 6.5	Ileum ~ 2–4 h pH ~ 7	Colon ~ 10 h pH ~ 7
MC/HPMC	Swelling	→ soluble				
EC	Swelling	→ swelling/permeable —— (pH independent) ——————————————————————————————→				
HPMCP	Insoluble	resistant	→ soluble			
CAT	Insoluble	resistant	→ soluble			
CAP	Insoluble	resistant	→ swelling	→ soluble		
E100	Permeable	→ soluble	permeable			
L100-55	Insoluble	resistant	→ soluble			
L100	Insoluble	resistant	→ swelling	→ soluble		
S100	Insoluble	resistant	→ resistant	resistant →	swelling →	soluble
RL100	Permeable	→ disintegrating				
RS100	Insoluble	→ permeable (pH independent) ———————————————————————————————————→				

68

is desired when the drug is degraded in the upper parts of the intestine or applied there for local therapy. Since the pH in the colon is around pH 7, Eudragit S or mixtures of Eudragit L/S are used for delivery in this lower part of the intestine. Due to the fact that bacterial growth starts in the colon, a remarkable activity of polysaccharide-dependent enzymes (glucosidases) is found [18]. Galactomannans such as guar or locust bean gum can be dissolved in alcohol–water 1:1 and mixed with polymethacrylate solutions in the same solvents; such coatings are susceptible to degradation by colonic enzymes [19] (Table 7, formulation 5).

All substances used in enteric coating are relatively brittle polymers and need the addition of plasticizer in coating formulations in the range 10 to 20%. Higher amounts of hydrophilic plasticizer may increase the permeability of coatings in digestive fluids. The recommended plasticizers for several coating materials are given in Table 4. The optimal amount is to be developed in correlation with the most important requirements for enteric dosage forms: permeability of the coating, mechanical stability, and release pattern. Table 7 summarizes some characteristic formulations for enteric coating of drug particles.

Table 7   Formulations for Enteric Coating

	Formulation number				
	1	2	3	4	5
CAP	5.0	—	—	—	—
HPMCP	—	10.0	—	—	—
HP50	—	—	5.0	—	—
L100/S100	—	—	—	7.3	—
L100-55	—	—	—	—	6.0
DEP	2.0	—	—	—	—
DBP	—	—	0.5	1.5	—
GTA	—	1.0	—	—	—
TEC	—	—	—	—	1.2
Talc	0.25	—	—	1.8	—
Mg stearate	—	—	—	—	1.5
TiO$_2$/pigments	0.75	—	—	—	—
Isopropanol	46.0	44.5	—	89.4	92.8
Acetone	—	44.5	—	—	—
Dichloromethane	—	—	47.25	—	—
Ethanol	—	—	47.25	—	—

C. Sustained-Release Coatings

The release profiles for sustained-release dosage forms are the result of multiple considerations. The main purpose is to reduce side effects and prolong drug intake intervals. Permeable coating membranes are widely used, and in this case drug release is controlled by permeation of water into the core and diffusion of dissolved drug through the membrane into the lumen of the digestive tract. As long as a reservoir of undissolved drug is in the core surrounded by a saturated drug solution and enveloped by an intact membrane, the release rate is zero order, or constant in time. When all drug in the core is dissolved and concentration inside the membrane is lowered by further diffusion, the mechanism turns to first-order reaction. If the core is a matrix and additionally coated, a more complicated release mechanism is observed.

Coating materials that form water-insoluble, more-or-less permeable membranes are ethyl cellulose and some neutral cellulose esters, such as cellulose acetate, cellulose triacetate, and acetate/butyrate. To alter the permeability of these coatings, they are often combined with hydrophilic, water-soluble substances such as methyl- and hydroxypropyl cellulose, polyethylene glycole, poly(vinylpyrrolidone), or water-soluble solids as pore-forming materials.

Polymethacrylates containing hydrophilic quarternary ammonium groups as functional units in the polymer chain (Eudragit RL100 and RS100) can be used without other release-controlling excipients as permeable membranes. The permeability depends directly on the content of the hdyrophilic units (trimethylammonioethyl methacrylate chloride). RL100 contains 10% w/w of this unit, resulting in very permeable films that tend to disintegrate quickly in water; therefore, this type is also used for fast-disintegrating cores. RS100 contains only 5% w/w hydrophilic units and exhibits very low permeability, so that with increasing film thickness up to about 100 μm, the release rate can be dropped down to very low rates, to give sustained release over 24 h or even more. Sometimes the release can be near zero and even incomplete if the permeability is too low. To get the desired release profile, both types can be mixed in any proportion, so that a wide range of permeability can be established with approximately 10 to 30 μm thickness and adapted to dissolution and diffusion properties of the actives as well as to the pharmacological and phamakokinetic requirements of the therapy.

If the drug is definitely transported through the membrane via diffusion, the partition coefficient of the drug in the polymer layer and the surrounding media determines the release rate. Pore-free membranes normally show more effective retardation and a more reproducible release pattern. Pores may always be present as long as very thin polymer layers are applied to rough surfaces and not all irregularities are sufficiently covered by the polymer. So it is essential to have smooth core surfaces to get good retardation with acceptable amounts of wall material. Drug release is also modified by substances in the core that influence osmotic pressure or are responsible for a distinct pH inside the diffusion cell to influence the solubility of the drug during release. Osmotic pressure is the driving force for release from Oros® products, which are formulated with semipermeable membranes and a hole of definite diameter generated by a laser beam.

It is very difficult to predict the release profile of a drug from a coated core through a very thin membrane. The thickness of effective sustained-release coatings is in the range 5 to 50 μm. Due to the application technique and processing variables, which include the influence of the core surface, attrition during spraying, the droplet size in the spray, the drying temperature under fluidization conditions, drying air supply, and others, it is nearly impossible to predict release rates from preliminary laboratory trials with free films, which are normally dried in one layer and must be more than 10 μm thick for practical handling. It is normally true that such films are much less permeable than film layers on cores.

The usual way in which sustained-release dosage forms are developed is to apply a selection of coating formulations with increasing or decreasing permeability or porosity to drug cores in increasing amounts and to investigate coated cores of various coating thicknesses for their release pattern. A selection of coating formulations is given in Table 8. In their simplest form they contain

Table 8   Formulations for Sustained Release

	Formulation number					
	1	2	3	4	5	6
HPMC	—	2.5	—	—	—	—
EC	5.0	2.5	—	—	—	—
RL100	—	—	2.5	—	3.5	—
RS100	—	—	2.5	2.6	—	—
S100	—	—	—	1.6	—	—
NE30D	—	0.5	—	—	—	5.0[a]
Diethyl phtalate	—	0.5	—	—	—	—
Dibutyl phthalate	—	—	—	0.9	—	—
PEG 6000	1.0	—	0.5	—	—	—
Talc	—	2.5	2.5	6.3	2.0	5.0
Galactomannan	—	—	—	—	10.5	—
Ethanol	94.0	46.0	—	—	—	—
Isopropanol	—	—	45.5	43.0	43.5	31.0
Acetone	—	—	45.5	43.0	—	47.5
Dichloromethane	—	46.0	—	—	—	—
Water	—	—	1.0	2.6	40.5	11.5
Reference	(7)	(21)	(7)	(22)	(19)	—

[a]Dry substance from 30% aqueous dispersion.

only polymers as film-forming agents, a plasticizer, and an antisticking agent. If the release of the drug is too slow, it is normally relatively easy to increase the permeability of the coating by adding more hydrophilic polymers or pore-forming substances.

If the release rate is too high, even with a very high film thickness (e.g., around 50 μm), the coating may be porous or mechanically unstable. In many cases it may be difficult to obtain a complete, homogeneous shell around the core. From experience it can be stated that a stable drug core surrounded by a polymer layer of very low permeability, such as Ethocel or Eudragit RS100, will normally show significant retardation of drug release over 8 to 24 h. Talc or other antisticking agents, such as magnesium stearate, together with a hdyrophobic plasticizer such as dibutyl phthalate and dibutyl sebaccate, often improve the application technique and make the film layer more hydrophobic, resulting in slower release.

D. Mixing of Polymers and Combination of Layers

Mixing of polymers is one possibility to alter the release pattern. When polymers with different functional groups are used, they may be incompatible and form insoluble complexes. For example, Eudragit L100/S100, which contain carboxylic groups, and E100, which has amino groups, may react to form polymeric salts. On the other hand, Eudragit RL100/RS100 can be combined with Eudragit L100/S100 to increase the permeability of such films in a neutral or weakly alkaline medium. This is often necessary with drug substances that become insoluble at higher pH (i.e., salts of weakly drug molecules) [20].

A huge number of sustained-release formulations have been developed by combining different polymer layers, which may be porous, nonporous, enteric, or gastrosoluble; more or less permeable and swellable; and some containing additional drug to be released in the stomach or, later, in defined amounts in the intestine to give repeat action. There is a possibility for adsorption of drugs on the coating material, especially if the drug is in a low dose and incorporated in the film layer, where the amount of coating material is much higher than the amount of the drug. Adsorption may be caused by ion exchange or hydrophobic interactions. Adsorption is normally reversible and the drug is released when it comes in contact with an excess of digestive fluids.

VIII. EXAMPLES OF COATED DRUG PARTICLES

Some examples of coated controlled-release drug particles are given in Table 9.

IX. SUMMARY AND OUTLOOK

Coating of drug particles with a polymer solution has a broad range of applications. Even in the future, organic solvents will have important advantages over aqueous dispersions (latexes). The first choice are water–alcohol mixtures of low flammability potential. Many polymers are soluble in alcohols and acetone, which have low toxicity and are easy to handle in solvent recovery and waste air purification. Chlorinated hydrocarbons and others require much higher levels of investment and more precautions. The possibility of using a melt coating process with polymers dissolved in a low-molecular-weight melting wax is an interesting concept for future development.

Coating processes with organic solvents can use a much broader selection of polymers and polymer mixtures than aqueous systems offer today (Table 10). With water-sensitive drugs, it is difficult or sometimes impossible to use aqueous systems directly, so there is a substantial demand for isolating layers in the form of organic solutions. A good selection of special machinery in the form of fluidized-bed or -pan coating systems is available as flameproof equipment with condensation or other means to recover or eliminate the organic solvents. Organic solvents evaporate very easily at low temperature, so heat-sensitive formulations may also need such coating processes. In the future it may be useful or even necessary to have special processes and equipment for the application of organic solvents where aqueous coating cannot be applied.

APPENDIX: ABBREVIATIONS OF COATING POLYMERS WITH THEIR CHEMICAL NAMES, MONOGRAPHS, AND COMMERCIAL PRODUCTS

MC   methylcellulose, Methocel® A (Dow)
HPMC   hydroxypropyl methylcellulose (USP/NF), Methocel E (Dow), Phar-
          macoat® (Shin Etsu)
EC   ethyl cellulose NF
CA   cellulose acetate
CTA   cellulose triacetate
CAB   cellulose acetate butyrate
CAP   cellulose acetate phthalate USP-NF
CAT   cellulose acetate trimellitate (Eastman Kodak)
CMEC   carboxymethylethyl cellulose (Duodcel®/Freund)
HPMCP   hydroxypropyl methylcellulose phthalate NF, HP 50, HP 55 (Shin
          Etsu)
          polymethacrylic acid–methacrylic acid copolymer (USP/NF)
              Type A: poly(methacrylic acid, methyl methacrylate) 1:1, solid
                  polymer (powder), Eudragit L100
              Type B: poly(methacrylic acid, methyl methacrylate) 1:2, solid
                  polymer (powder), Eudragit S100
              Type C: poly(methacrylic acid, ethyl acrylate) 1:1—solid polymer
                  (powder), Eudragit L100-55; aqueous dispersion 30% solids,
                  Eudragit L30D
          poly(meth)acryl ester: poly(ethyl acrylate, methyl methacrylate 2:1),
              Eudragit NE30D aqueous dispersion 30% solids

Table 9 Coated Drug Particles

Active ingredient	Particle shape	Particle size (mm)	% Dry lacquer substance	Type	Release rate [time (h)/% release]			
Amobarbital (sedative)	Pellets	0.8–1.2	6	E/RL 1:1	0.5/25	2/50	4.5/75	7/85
Chlorpheniramine maleate (antihistaminic)	Pellets	0.8–1.2	10	RL/RS/S 2:2:1	1/25	2/40	4/65	6/90
Indomethacin (antihistaminic)	Pellets	0.8–1.2	10	S		Enteric coated		
Potassium chloride (K-supplement)	Pellets	0.8–1.2	10	RS	1/6	2/20	4/40	6/65
Lithium carbonate (antidepressive)	Pellets	0.8–1.2	15	RL/L	1/20	2/55	3/75	4/90
Phenylpropanolamine HCl (sympathomimetic)	Pellets	0.8–1.2	8	RS/S 5:3	1/20	2/30	4/50	6/75
Trifluoperazine HCl (tranquilizer)	Pellets	0.8–1.2	15	RS	1/35	2/50	4/70	6/85
Vitamin C (vitamin supplement)	Pellets	0.8–1.2	2	RL	1/100 (stabilized)			
Xanthinol nicotinate (vasodilator)	Pellets	0.8–1.2	7	RS/RL 3:1	1/45	2/91		

Drug (indication)	Form			Polymer		Application	
Isopropamide iodide	Pellets	0.8–1.2	1.2	RL	1/4	Insulation	2/16 4/43 6/73
Phenylephrine HCl	Pellets	0.8–1.2	17.5	RL/S	1/15		2/27 4/51 6/78
Acetylsalicylic acid (antiarthritic)	Crystals	0.3–0.8	2	L/S 1:1	1/1		2/3 3/81 4/94
Chinidine sulfate (antiarrhythmic)	Crystals	0.3–1.5	10	L		Insulation	
	Crystals	0.1–0.3	10–20	E			
Isoniazid (tuberculostatic)	Crystals	0.1–0.3	20	RS	1/45		2/65 4/88 6/98
Lithium citrate (antidepressive)	Granules	0.5–1.0	25	RL/RS	1/45		2/70 4/86
Methaqualone (hypnotic)	Crystals	0.3–0.8	5	RL/RS 1:1	0.5/35		1/50 2/75 3/85
N-Acetylmethionine	Crystals	0.3	8	L		Taste masking	
Nitrofurantoin (urinary tract antibacterial)	Crystals	0.08–0.2	10	L		Enteric coated	
Pancreatin (enzyme substitute)	Granules	0.2–1.0	6–10	L		Enteric coated	
Paracetamol (antipyretic)	Granules	0.3–0.8	6	L	1/12	2/18 4/94	
Piracetam (antiemitic)	Crystals	0.1–0.4	10	E		Taste masking	
Potassium chloride (K-supplement)	Crystals	0.5	12, 5	RS	1/8	2/20 4/50	6/70
Propicillin (antibiotic)	Granules	0.1	10	E		Insulation	
Propyphenazone (antipyretic)	Crystals	0.2–0.5	10	E		Taste masking	
Vitamin B_1 complex	Granules	<0.1	1.3	L		O_2 stabilization	
Vitamin C	Crystals	0.05–1.25	20	L	2/21	(partially enteric coated)	

Table 10 Process Data of Fluidized Bed Coating on a Laboratory Scale (up to 8 kg)[a]

	Particle structure	Spray time (min)	Solvent mixing ratio	Polymer type/ % solids	Spray nozzle position	Inlet air temp. (°C)	Outlet air temp. (°C)	Applied polymer (% w/w)	Applied solids (% w/w)	Spray rate (g/min · kg)
Amitriptyline	Pellets	117	Ac-Ip 44–56	EU-RS/8.5	Top	36	35	10.0	17.0	17.1
Acetylsalicylic acid	Crystals	40	Ip	EU-L-S 1:18.5	Top	40	23	2.0	3.4	10
Nicotinic acid	Powder	114	Ac-Ip 45–55	EU-RS/9	Top	40	26	10.0	13.5	13.2
Placebo/Sucrose	Pellets	54	Ac	EU-S/10	Top	38	26	2.0	6.0	14.8
Methaqualone	Crystals	75	Ac-Ip 45–55	EU-RL-RS/8.6	Top	38	23	7.5	12.0	18.7
Acetylsalicylic acid	Granules	153	Ac-Ip 27–73	EU-L/10.2	Top	52	36	8.0	13.6	8.7
Isoniazid	Crystals	125	Ac-Ip 71–29	EU-RS/10	Top	50	28	10.0	16.0	12.8
Bisacodyl	Pellets	160	Et-Me 50–50	CMEC/9	Bottom	68	40	15.4	17.4	6–13
	Pellets	230	Et-Me 50–50	HP-50/5.5	Bottom	53	27	16.0	17.6	10–16
	Pellets	225	Ip	EU-L/7.3	Bottom	46	26	9.1	10.0	8–14
Theophylline	Granules	200	Ac-Ip-W 53–35–12	EU-NE30D/5	Top	30	28	6.0	12.0	6

[a]Ac, acetone; Et, ethanol; Ip, isopropanol; W, water; Me, methylene chloride; pellets 0.5–1.2 mm; granules/crystals 0.3–0.8 mm, powder 0.1–0.3 mm.

polyaminomethacrylate (DAB in preparation: poly[butyl methacrylate, (2-dimethylaminoethyl)methacrylate, methyl methacrylate] 2:1, solid polymer (granules), Eudragit E 100

poly(trimethylammonioethyl methacrylate chloride)–ammoniomethacrylate copolymer (USP/NF)

Type A: poly(ethyl acrylate, methyl methacrylate, trimethylammonioethyl methacrylate chloride) 1:2:0.2—solid polymer (granules), Eudragit RL100; aqueous dispersion 30% solids, Eudragit RL30D

Type B: poly(ethyl acrylate, methyl methacrylate, trimethylammonioethyl methacrylate chloride) 1:2:0.1—solid polymer (granules), Eudragit RS100; aqueous dispersion 30% solid Eudragit RS30D

Eudragit® (Röhm)

Plasticizer

DEP   diethyl phthalate
DBP   dibutyl phthalate
TEC   triethyl citrate
GTA   glycerol triacetate
DBS   dibutyl sebaccate

REFERENCES

1. K. Lehmann and D. Dreher, Coating small particles with acrylic resins, *Pharm. Technol, 3*:53 (1979).

2. R. Goldman, Sustained release capsules, *Drug Cosmet. Ind. 107*:52 (1970).

3. M. D. Coffin, R. Bodmeier, K. Chang, and J. W. McGinity, The preparation, characterization and evaluation of poly(*dl*-lactide) aqueous latices, *J. Pharm. Sci. 76*:261 (1987).

4. K. Lehmann, Chemistry and application properties of polymethacrylate coating systems, Chapter 4, in *Aqueous Polymeric Coatings for Pharmaceutical Dosage Forms*, (J. W. McGinity, ed.), Marcel Dekker, New York, 1989.

5. A. M. Metha, Evaluation and characterization of pellets, Chapter 11 in *Pharmaceutical Pelletization Technology*, (I. Ghebre-Sellassie, ed.), Marcel Dekker, New York, 1989.

6. P. B. Deasy, *Microencapsulation and Related Drug Processes*, Marcel Dekker, New York, 1984, p. 34ff.

7. K. H. Bauer, K. Lehmann, H. P. Osterwald, and G. Rothgang, *Überzogene Arzneiformen*, Wissenschaftliche Verlagsgesellschaft, Stuttgart, 1988, p. 78ff.

8. K. Lehmann, G. Rothgang, H. M. Bössler, D. Dreher, H. U. Petereit, C. Liddiard, W. Weisbrod, *Practical Course in Lacquer Coating*, Röhm Pharma, Darmstadt/ Weiterstadt, 1989, p. 24ff.

9. R. C. Rowe., Materials used in the film coating of oral dosage forms, *Crit. Rep. Appl. Chem. 6*:1 (1984).

10. W. Rölz, *Acta Pharm. Technol. 26*:287 (1980).
11. K. Wehrle, *Pharm. Ind. 44*:83 (1984).
12. B. Luy, Vakuum-Wirbelschicht, Dissertation, Universität Basel, Switzerland (1991).
13. Th. Köblitz; R. Bergauer, G. Körblein and L. Ehrhardt, The use of a solvent recovery plant in connection with filmcoating, *Drugs Made Ger. 27*:164 (1984).
14. W. Rothe and G. Groppenbächer, *Pharm. Ind. 35*:11 (1973).
15. F. W. Goodhart and S. Jan, Dry powder layering, Chapter 8 in *Pharmaceutical Pelletization Technology*, (I. Ghebre-Sellassie, ed.), Marcel Dekker, New York, 1989.
16. M. Marvola, M. Rajaniemi, E. Marttila, K. Vahervuo and A. Sothmann, *J. Pharm. Sci. 72*:1034 (1983).
17. J. T. Fell, Gastrointestinal transit of microcapsules, Chapter 8 in *Microcapsules and Nanoparticles in Medicine and Pharmacy*, (M. Donbrow, ed.), CRC Press, Boca Raton, Fla., 1992.
18. C. M. Lancaster and M. A. Wheatley, Drug delivery to the colon, *Poly. Prepr. 30*(1):480 (1989).
19. K. O. R. Lehmann and K. D. Dreher, Methacrylate–galactomannan coating for colon-specific drug delivery, *Proceedings of the 18th International Symposium on Controlled Release of Bioactive Materials*, Amsterdam, July 8–11, 1991.
20. K. Lehmann, Programmed drug release from oral dosage forms, *Pharma. Int. 3*:34 (1971).
21. E. Horváth and Z. Ormós, Film coating of dragée seeds by fluidized bed spraying methods, *Acta Pharm. Technol. 35*(2):90 (1989).
22. K. Lehmann, Fluid bed spray coating, Chapter 4 in *Microcapsules and Nanoparticles in Medicine and Pharmacy*, (M. Donbrow, ed.), CRC Press, Boca Raton, Fla., 1992.

5

Coating of Multiparticulates Using Polymeric Dispersions
Formulation and Process Considerations

Yoshinobu Fukumori

Kobe Gakuin University
Kobe, Japan

I. INTRODUCTION

Organic solvent–based polymeric solutions have traditionally been used for coating. By appropriate selection of organic solvents or their mixtures, a variety of compounds can be dissolved in solutions to make the products display the functions desired [1]. The amount of coating material that is required to display a specific function can also be reduced by additives [1,2]. This has a great economic advantage in coating, which usually requires lengthy processing times. However, the high cost of solvents, high price of solvent recovery systems, strict air quality controls, and potential toxicity and explosiveness of solvents have motivated pharmaceutical and food supplement processors to remove organic solvents from the coating process [3].

As a result, water-based systems have been developed for pharmaceutical dosage forms instead of organic solvent–based polymeric solutions because of their environmental and economic advantages. Because water has a high heat of vaporization, aqueous systems that might require lengthy processing times seemed initially to have a serious economic disadvantage despite their environmental advantages. However, in addition to the progress of fluidized-bed technology, the aqueous polymeric latexes and pseudolatexes that have been developed since the 1970s have overcome the disadvantage by their low viscosities and high polymer content.

There are still many problems to be solved or overcome for wider applications of aqueous dispersions to pharmaceutical multiparticulate dosage forms.

While film formation in polymeric solution systems (either aqueous or organic) is easily achieved by drying solutions, the polymeric particles in dispersions have to be fused for the formation of continuous films. Some complicated processes are involved in the formation of continuous films. Practically, this brings about some additional difficulties in coating using aqueous dispersion systems. For their efficient application to pharmaceutical dosage forms, a clear understanding of the mechanisms of film formation may be required.

There is one other problem with regard to multiparticulate dosage forms, which arises from their relatively small particle sizes. Because they have low inertia and momentum, small-core particles easily agglomerate in the coating process. The coating of such small particles as those a few hundreds or less micrometers in size has recently been tried using a spouted bed coater (the Wurster process); even the coating of particles on the order of 10 μm in size has been reported [2,4]. Although attention so far has focused on film formation, complete film formation in the fluidized bed can lead to core particle agglomeration. Hence there can be some cases where polymeric particles adhesive to the core surface, but not film formable, may be required. In fact, when coating is done under poor film-forming conditions, the agglomeration of core particles can be kept to 1% or less, even in coating of particles smaller than 100 μm [5].

In this chapter, after the available aqueous dispersions are explained, the mechanisms of film formation from aqueous polymeric dispersions are reviewed. Then the formulation and processing factors related to the coating of multiparticulates are discussed.

II. MARKETED AQUEOUS POLYMERIC DISPERSIONS

The commercially available polymeric dispersions are classified into three types based on the preparation methods: (1) latexes synthesized by emulsion polymerization; (2) pseudolatexes prepared by emulsion processes, such as emulsion–solvent evaporation, phase inversion, and solvent change; and (3) dispersions of micronized polymeric powder (Table 1).

Eudragit L30D and NE30D are acrylic copolymer latexes synthesized by emulsion polymerization [6]. They have particle sizes in the submicron order. L30D is a copolymer of ethyl acrylate (EA) as ester component with methacrylic acid (MA) (MA/EA 1:1). It is used as an enteric coating because it contains carboxyl groups. NE30D is a copolymer of ester components only, EA and MMA (2:1). The films formed from NE30D have a very low softening temperature and hence are flexible and expandable even under room conditions.

Cellulose derivatives can not be synthesized directly in latexes; therefore, they are prepared as pseudolatexes (Aquacoat, Aquateric [7], Surelease [8]), or micronized powders (Aqoat (HPMCAS) [9] and EC N-10F [10]). While the pseudolatexes can be prepared as submicron particles, the micronized powders

Table 1 Commercially Available Aqueous Polymeric Dispersions[a]

Brand	Type	Polymer component	Dispersion type	Additives
Eudragit	L30D	Copoly(MA-EA)	Latex	Tween 80 (2.1%), SDS (0.9%)
	RS/RL30D	Copoly(EA-MMA-TAMCl)	Pseudolatex (solvent change)	Sorbic acid, no surfactant
	NE30D	Copoly(EA-MMA)	Latex	PNP
Aquacoat		Ethyl cellulose	Pseudolatex (solvent evaporation)	Cetyl alcohol (9%), SDS (4%)
Surelease		Ethyl cellulose	Pseudolatex (phase inversion)	Dibutyl sebacate, oleic acid, ammonia, fumed silica
EC	N-10F	Ethyl cellulose	Powder (2.6 μm)	
Aquateric		CAP	Pseudolatex	Pluronic F-68, Myvacet 9-40, Tween 80
Coateric		Poly(VAP)	Micronized powder	Plasticizer, pigments
Aqoat		HPMCAS	Micromized powder (3 μm)	

Source: Ref. 3.

[a]MA, methacrylic acid; EA, ethyl acrylate; MMA, methyl methacrylate; TAMCl, trimethylammonioethyl methacrylate chloride; CAP, cellulose acetate phthalate; VAP, vinyl acetate phthalate; HPMCAS, hydroxypropyl methylcellulose acetate succinate; SDS, sodium lauryl sulfate; PNP, polyoxyethylene nonyl phenyl ether; Myvacet 9-40, acetylated monoglyceride.

have minimum mean sizes of a few micrometers. Poly (VAP) is also supplied as a micronized powder (Coateric) [8]. Eudragit RS and RL are copolymers of EA and MMA as ester components with trimethylammonioethyl methacrylate chloride (TAMC1) as hydrophilic quaternary ammonium groups; RS and RL are 1:2:0.1 and 1:2:0.2 copoly(EA-MMA-TAMC1), respectively. Because Eudragit RS and RL contain MMA-rich ester components (EA/MMA 1:2), their softening temperatures are higher than those of NE30D (EA/MMA 2:1) and form hard films under room conditions. Eudragit RS and RL powders are easily transformed into pseudolatexes by emulsifying their powders in hot water without additives [6]. It is costly to ship aqueous dispersions around the world; therefore, the Aquateric pseudolatex is supplied as a spray-dried powder; it is redispersed just before use [7].

A variety of additives are incorporated in the dispersions as surfactants (Tween 80, sodium lauryl sulfate, polyoxyethylene nonyl phenyl ether, cetyl alcohol, Pluronic F-68), plasticizers (dibutyl sebacate, oleic acid, Myvacet 9-40), pigments, antiadherents (fumed silica), anticoagulant (Myvacet 9-40), preservatives (sorbic acid), and stabilizers (ammonia).

III. MECHANISM OF FILM FORMATION

The coating is achieved by a spraying and drying process of the dispersions, where the system is composed of three phases: environmental gas phase, aqueous solution, and polymeric particles. Water is evaporated, leaving the polymeric solid. The subject to be discussed here is how the residual solid, which is originally composed of discrete particles, becomes a homogeneous film.

The mechanisms of film formation from aqueous polymeric dispersions have been discussed for a long time and many theories have been proposed [11–16]. Muroi [16] reviewed them in detail from a basic point of view. Film formation in pharmaceutical applications was discussed by Lehmann [6] and Steuernagel [7]. Fusion and film formation of polymeric particles during the coating process can be explained by the wet sintering theory [15] for particles suspended in water, the capillary pressure theory [13] for particle layers containing water in various degrees of saturation, and the dry sintering theory [11,12] for dry particle layers.

A. Wet Sintering Theory

In dispersion, pressure acts on the small polymeric particles, depending on their radius, r, and the water–polymer interfacial tension, $\gamma_{p,w}$. Let the pressures on the inner and outer surfaces of particles be represented by P_1 and P_2, respectively. Then the pressure difference can be expressed by Laplace's equation:

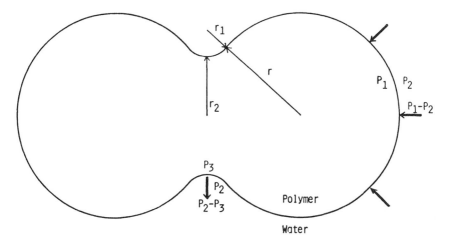

Fig. 1   Fusion of polymeric particles.

$$P_1 - P_2 = \frac{2\gamma_{p,w}}{r} \tag{1}$$

When $\gamma_{p,w}$ is 10 dyn/cm [15], this pressure difference is calculated to be 2 kg/cm^2 for polymeric particles of 100 nm.

When two particles are in contact, as shown in Fig. 1, the pressure difference at the neck in the contact region can again be expressed by Laplace's equation:

$$P_2 - P_3 = \gamma_{p,w} \left(\frac{1}{r_1} - \frac{1}{r_2} \right) \tag{2}$$

Because r_1 can be very small, especially at the beginning of deformation, it is possible for very high pressure difference to be induced on the contact region. Consequently, the pressure gradient is induced within the polymeric particle:

$$P_1 - P_3 = \gamma_{p,w} \left(\frac{1}{r_1} - \frac{1}{r_2} + \frac{2}{r} \right) \tag{3}$$

The polymer is forced to flow by the pressure gradient, leading to expansion of the contact region if it will yield mechanically. The pressure gradient is decreased with increase in the curvatures r_1 and r_2 during deformation. The two particles should finally be fused into a sphere if the polymer continues to yield.

When particle fusion at this stage can proceed easily, dispersion stability is lost by particle coalescence or aggregation. To facilitate film formation in subsequent processes, the polymeric particles usually have to be made deform-able. Therefore, the polymeric phases are prevented from direct contact by adding surfactants or by other procedures.

B. Capillary Pressure Theory

By water evaporation from droplets of a dispersion sprayed on and adhering to core particles, the dispersion is condensed and the polymeric particles are closely packed on the surfaces of core particles. When the surface layer of polymeric particles is exposed to air, the air–water surface tension generates capillary pressure on the surface of a water-saturated layer of polymeric particles (Fig. 2). The capillary pressure compresses the polymeric particles, the compression acting to squeeze out the water. Since the curvatures of the capillary surface are thus enlarged, capillary pressures are decreased. This process is repeated until water resides only among the interparticulate contact points. Then the polymeric particles can be fused by capillary pressure. When water is located only at inter-particulate contact points, capillary pressure is exerted in the contact regions of polymeric particles (Fig. 3). Although precise prediction is difficult, the capil-lary pressure is qualitatively increased with water–air surface tension and is inversely proportional to the polymeric particle radius [13,16].

How far particle fusion can proceed at this stage depends on the rate of disappearance of the aqueous phase in addition to the surface tension and mechanical properties of the polymeric phase. The aqueous phase can disappear through evaporation and by permeation of the water into the core or underlying layer. The evaporation rate depends on the temperature, humidity, and flow rate of air. When coating is done under high humidity conditions, the capillary pressure can act over a much longer period, leading to enhanced film formation. When the underlying surface is water absorbable or porous, the aqueous phase

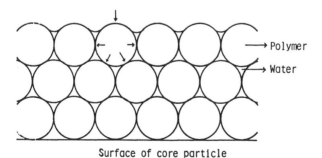

Fig. 2 Capillary pressure acting on the layer of polymeric particles saturated with water.

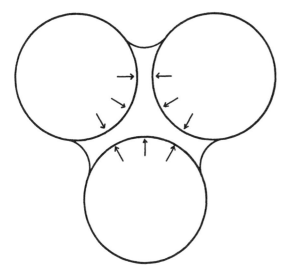

Fig. 3 Capillary pressure acting on polymeric particles holding interparticulate bridge water.

will be shorter and there will not be enough time for film formation to be completed [14].

It is useful in film formation that particles be under capillary pressure for a long period. But, polymeric particles can have the best chance to act as a adhesive among core particles. This can lead to the agglomeration of core particles, which is not preferable in multiparticulate dosage forms.

C. Dry Sintering Theory

After the water is completely evaporated, the pressures caused by replacing the water–polymer interfacial tension in Eqs. 1 to 3 by air–polymer surface tension are exerted on the polymeric particles. By the induced pressure gradient [Eq. 3], dry sintering can proceed if the polymer is deformable and the fusion of polymeric particles is not completed.

If dry sintering proceeds at a significantly high rate during the coating process, agglomeration of core particles cannot be avoided. However, the rate of dry sintering is usually low; therefore, further gradual coalescence [15,17] can proceed for a long period after coating. During that time, membrane properties such as drug permeability continue to change. This should cause serious problems. To complete film formation, the products are usually cured by heating (aging) at temperatures higher than the softening temperature of the membrane material; then, to avoid aggregation of the particles with softened membrane, powder is dusted on the particle surface.

IV. FORMULATION AND PROCESSING FACTORS

A. Polymeric Particle Size

As discussed briefly above, the driving force of film formation becomes stronger with decreased polymeric particle size. Hence there is an essential difficulty in the film formation from dispersions of micronized polymeric powders of large particle size. Film formation is easier with latexes and pseudolatexes because of their submicron particle size.

The effect of particle size on the film formation of 2-methyl-5-vinyl-pyridine–methylacrylate–methacrylic acid copolymer (MPM-47, Tanabe Pharmaceutical Co. Ltd.) suspension was reported by Nakagami et al. [18]. Although free films were obtained from polymeric particles smaller than 150 μm by casting suspensions, film formation was not observed with coarser particles. This showed polymer particle size to be an important formulation factor. They compared EC latex (Aquacoat, FMC), with a particle size of 0.171 μm, to micronized EC (N-10F, Shin-Etsu Chemical Co. Ltd.), with a particle size of 4.54 μm (Table 2). Aqueous coating dispersions containing 10 to 20% w/v polymer and various amounts of plasticizer were placed in a petri dish and dried in a thermostated chamber at constant temperature for 24 to 48 h. The micronized EC powder required more plasticizer to form a continuous transparent film. In this case, the large particle size of micronized EC can account for the poor film formation, but the 4% sodium lauryl sulfate and 9% cetyl alcohol contained as

Table 2 Effect of Plasticizer and Temperature on Film Formation from, Ethul Cellulose Dispersions[a]

Temp. (°C)	Plasticizer level							
	Aquacoat				EC N-10F			
	0%	10%	25%	40%	0%	10%	25%	40%
25	−	+	+ +	+ +	−	−	+	+ +
30	−	+	+ +	+ +	−	−	+	+ +
40	−	+	+ +	+ +	−	−	+	+ +
50	−	+ +	+ +	+ +	−	−	+ +	+ +
60	−	+ +	+ +	+ +	−	−	+ +	+ +

Source: Ref. 18.
[a]−, No film formation; +, formation of cracked or partially transparent film; + +, formation of continuous, transparent film.

stabilizers in the Aquacoat formulation can improve the film formability of EC through their plasticizing effect.

B. Film-Forming Temperature

Polymers have to be mechanically deformable to form films under specified conditions. Polymers are deformable above the softening temperature (T_s) of their cast films. The softening of polymer films is related to the glass transition of the polymer; both glass transition and softening correspond to a sharp increase in polymer chain mobility [19]. The glass transition temperature (T_g) is usually determined by the differential scanning calorimetry and T_s, for example, by a penetration test in thermomechanical analysis.

Amer et al. [20] reported on the relation between T_g and bed temperature in the Wurster process (Table 3). When coating was performed at bed temperatures of about 40°C, the acrylic latex, whose T_g of 35°C was lower than the bed temperature, formed a membrane with few imperfections. As the T_g value rises above the bed temperature, the film quality is reduced. They concluded that coating at 10°C or less above the glass transition temperature resulted in a high degree of efficiency, reflective of good film formation. If the inlet air temperature were elevated above the T_g value, film formation would be certain. However, such an operation can lead to the particle agglomeration or aggregation, especially in the coating of small multiparticulates. This type of agglomeration can be avoided to some extent by introducing a dusting powder directly into the coating chamber. The excellent performance reported by Amer et al. [20] was achievable due to the large (800 μm) cores used in their work.

Fortunately, film formation from aqueous dispersions usually occurs below the softening temperature of the cast films because the interfacial or surface tension can force the small polymeric particles to be deformed, and the polymer

Table 3   Effect of Process Parameters on Capsule Quality in the Coating of Particulate Oxidizing Material with Acrylic Latex by the Wurster Process

T_g (°C)	Bed temperature (°C)	Quality[a]
35	38	Excellent
53	41	Fair
60	40	Poor

Source: Ref. 20.
[a]Excellent, few imperfections; fair, surface cracks, good adhesion to core, holes within coating; poor, cracks extending to core, poor adhesion to core.

yield strength is lowered by hydration, which can even cause the polymer to gel. The minimum film-forming temperature (MFT) has been determined at slow water-evaporation rates by casting dispersions. Film formation is, however, a dynamic property of dispersion, dependent on the water evaporation rate. Therefore, the film formation is sensitive to coating conditions. With coarse cores, coating can be performed under wet conditions, which can enhance film formation. However, the coating of finer-core particles requires drier operating conditions to produce discrete or unagglomerated multiparticulates. In this case, the film formation can be achieved only near the softening temperature. In practice, the film-forming temperature should vary with the coating conditions and range between the softening temperature and the MFT as determined by slow drying of the dispersion.

Amer et al. [20] demonstrated the relation of process conditions to coating efficiency, which reflected film formation (Table 4). In run 1 the low efficiency (42%) and yield (88%) clearly resulted from a T_g value (67°C) higher than the bed temperature (45°C). In run 2 the improved efficiency of 60% was achieved by the almost twofold higher spray rate, despite using latex with a T_g value higher than 100°C. This suggests how much the humidification can enhance film formation. The more improved efficiency (69%) in run 3 clearly resulted from a T_g value (35°C) lower than the bed temperature. It is interesting that bottom spraying in the Wurster configuration led to a remarkably high efficiency (90%), as shown in run 4. In this case the conditions were made less wet because the

Table 4 Effect of Process Conditions on Efficiency in the Coating of Particulate Oxidizing Material with Methacrylic Acid–Based Latex

	Run number			
	1	2	3	4
Conditions				
Percent solid in spray dispersion	30	30	30	46
Spray type	Top	Top	Top	Bottom
Charge of cores (g)	600	600	800	800
Spray rate (mL/min)	12	23	13	10
T_g of latex	67	>100	35	35
Bed temperature (°C)	43	45	45	42
Performance				
Coating efficiency (%)	42	60	69	90
Total yield (%)	88	93	94	98

Source: Ref. 20.

spray rate was reduced somewhat and the percent of solid in the spray solution was enlarged to 46%.

In the Wurster apparatus, the coating zone is partitioned by a draft tube in which particles and spray droplets are cocurrent and therefore may have many chance to come into contact. However, one disadvantage in this process is that the spray rate must be reduced because too many droplets cause particles to adhere to the inner surface of the partition. This is the reason the spray rate had to be reduced and, consequently, the percent of solid in the spray solution had to be increased to avoid a time-consuming operation in run 4 (Table 4). It is a distinct advantage that spray dispersions can be used at high concentrations in aqueous dispersion systems. In general, the high concentration leads to core particle agglomeration in the coating of fine cores, but the collision of particles with the partition eliminates agglomerates in the Wurster process. On the other hand, the fluidizing airflow and spraying are countercurrent in top-spray coating; therefore, droplets are more easily spray dried, leading to a low coating efficiency.

The effect on film formation of bed temperatures in fluidized coating was also demonstrated by Yang and Ghebre-Sellassie using an Aquacoat formulation (Table 5) applied to water-soluble diphenhydramine HCl pellets [21]. The dissolution rates followed biexponential first-order kinetics; the rate constants are shown in Table 6. When the bed temperature ranged between 22 and 50°C under constant spray conditions, optimal film formation, as evaluated by the drug release rate, was achieved between 30 and 40°C. The poor film formation at higher bed temperatures could be explained by the high water evaporation rate, which prevented development of the requisite capillary pressure. At lower temperatures, the migration of solute (diphenhydramine HCl) from the drug

Table 5   Example of Aquacoat Formulation in the Coating of Water-Soluble Diphenhydramine HCl Pellets[a]

	Ingredients (% w/w)	
	Sustaining coat	Overcoat
Aquacoat ECD-30	38	—
HPC NF	—	6
Triethyl citrate	3.6	—
Mistron talc	—	1
Distilled water	58.4	93

Source: Ref. 21.
[a]Fluidized-bed coating machine; Stera I (Aeromatic); charge of pellets (12–18 mesh) with diphenhydramine HCl layer on non-pareils, 300 g; spray pressure; 0.5–0.6 bar; spray rate; 1 g/min for the first 10 min in sustaining coating and 3–4 g/min thereafter.

Table 6 Release-Rate Constant for Diphenhydramine HCl Pellets Coated with Aquacoat Formulation (Table 5) at Various Bed Temperatures[a]

Bed temperature[b] (°C)	k_1[c] (h^{-1})	k_2[d] (h^{-1})
22	1.045	—
25	0.402	0.201
30	0.300	0.109
35	0.262	0.101
40	0.312	0.123
45	0.335	0.264
50	0.450	0.381

[a]Pellets were cured at 45°C for 14 h.
[b]Allowed to vary as a function of the inlet air temperature.
[c]First-order rate constant for the initial phase (up to 2 h).
[d]First-order rate constant for the final phase (after 2 h).

layer into the film layer during the coating process and hardening of the polymeric particles prevented the formation of continuous film.

C. Plasticizer

Plasticizers are used to lower T_s and T_g values. The degree of decrease in T_s is dependent on the amount used and the physicochemical properties of the plasticizers. The plasticizer has to be compatible with the polymer to promote polymer chain mobility and flexibility. Toyoshima [22] evaluated the capacity of the plasticizer to break the bonds between polymer molecules by the dissolving temperature (DT) of a polymer in the plasticizer, and the interaction force between polymer and plasticizer molecules by the cloud point (CP) of the solution of the polymer in the plasticizer. Nakagami et al. [18] demonstrated clearly that the compatibility of plasticizers with polymers could be evaluated using the DT and CP values. The results with ethyl cellulose are shown in Table 7; a plasticizer (PEG 400, triethyl citrate, or diethyl *d*-tartrate) with a lower DT or CP value exhibits better compatibility, as can be expected. Diethyl *d*-tartrate is most compatible with ethyl cellulose among the six plasticizers shown in Table 7.

Examples of the effect of plasticizers on the softening temperature of commercially available acrylic copolymer latexes are shown in Fig. 4. The softening temperature of plasticized polymer must be above 50 to 60°C to avoid the agglomeration or adhesion not only during the coating process but also during storage and transportation. With Eudragit RS30D, the addition of plasticizers

Table 7   Dissolving Temperature (DT) and Cloud Point (CP) of 1% Ethyl Cellulose Solution in Plasticizer and Compatability Observed from Film Formed with Ethyl Cellulose and Plasticizer (1:0.5 w/w)

Plasticizer	DT (°C)	CP (°C)	Compatability[a]
Propylene glycol	174	—[b]	×
PEG 400	RT[c]	—	+
Triacetin	133	—	×
Polysorbate 80	139	—	×
Triethyl citrate	50	—	+
Diethyl d-tartrate	RT	<0	+ +

Source: Ref. 18.
[a]Compatibility: ×, no film formed; +, turbid film formed; + +, transparent film formed.
[b]Not determined.
[c]Room temperature.

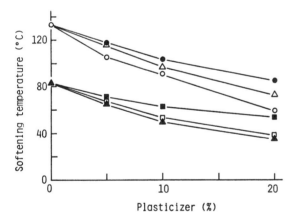

Fig. 4   Effect of plasticizers on the softening temperature of films cast from commercially available acrylic copolymer latexes. (From Ref. 23.)

Eudragit type	Symbol	Plasticizer
L30D	○	Triacetin
	●	Triethyl citrate
	△	PEG 6000
RS30D	▲	Triacetin
	□	Triethyl citrate
	▧	PEG 6000

had to be limited to 10%. On the other hand, Eudragit L30D had a softening temperature of 91°C upon 10% addition of triacetin (TA).

Coating was performed using the plasticized dispersions noted above and a spouted bed coater with a draft tube (Glatt GPCG-1, Wurster) [23]. The coating conditions are shown in Table 8. Lactose crystals with a mean particle size of 328 μm were coated at an inlet air temperature of 60°C. The release of lactose from microcapsules 60% coated with Eudragit L30D or RS30D plasticized by 10% TA and the effect of aging on the release are shown in Fig. 5. The release of lactose from RS30D microcapsules retarded by aging at 60°C suggests that film formation was not completed during the coating process despite its softening temperature being lower than the inlet air temperature. On the other hand, although it has a softening temperature of 91°C, L30D exhibited fairly good film formation even during the coating operation. This should be related to the low MFT value of L30D (i.e., 27°C) [6].

Plasticization is a potent method of facilitating film formation. Plasticization is, however, limited in practice to the level at which products are not sticky throughout coating, storage, and transportation. Consequently, there are some cases where the desired degree of film formation cannot always be achieved in the coating process through plasticization. In the Wurster process, the air distributor and its vicinity are heated to close to inlet air temperature. This gives the coat a chance to contact the chamber wall, which is heated to near the inlet air

Table 8 Operating Conditions in the Coating with Eudragit Copolymers by the Wurster Process (GPCG-1)

Core: Lactose (DMV 50 M) (g)	300	300	500	300
Membrane material	L30D	RS30D	E30D	RS30D:E30D (2:3)
Weight on a dry basis (g)	180	180	150	180
Plasticizer	TA	TA	—	—
Weight applied (g)	18	18	—	—
Diluent (talc) (g)	90	90	75	90
Surfactant (Tween 80) (g)	—	0.36	—	—
Volume of spray dispersion (mL)	1500	1500	900	1500
Inlet air temperature (°C)	60	60	27	60
Outlet air temperature (°C)	31	31	22	32
Inlet air rate (m³/min)	1.1	1.1	1.6	1.2
Spray rate (mL/min)	5.7	5.6	2.6	5.4
Spray pressure (atm)	2.2	2.2	2.1	2.1
Mean polymeric particle size (nm)	89	158	168	—
Softening temperature (°C)	91	49	18	53

Source: Ref. 23.

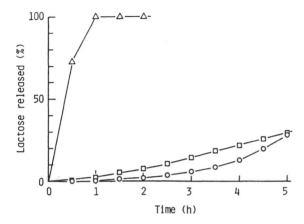

Fig. 5 Release of lactose from microcapsules 60% coated with Eudragit L30D or RS30D plasticized by 10% triacetin. (From Ref. 23.)

Eudragit type	Symbol	Aging	Dissolution
L30D	○	—	pH 1.2
RS30D	△	—	pH 6.8
	□	60°C, 12 h	pH 6.8

temperature, or to be exposed to the inlet air during fluidization. When the softening temperature of the coat is lower than the inlet air temperature, the particles tend to adhere to the bottom wall of chamber. In an extreme case, it becomes difficult to continue fluidization. This is a serious problem with fine cores, although it is only a minor problem with such large cores (328 μm) as those used in the example above.

D. Blend Polymer

Eudragit NE30D can be applied by blending it with RS30D, for example, to lower the softening temperature, since plain NE30D has a very low softening temperature, 18°C. The softening temperature of blend polymers of RS30D and NE30D is shown in Fig. 6. Below a 40% NE30D content, the softening temperature is changed only a little, but it is lowered rapidly above the 40% NE30D content. A blend polymer of RS30D/NE30D 2:3 with a softening temperature of 53°C was coated under the conditions shown in Table 8. The release profiles of aged and unaged products indicate very poor film formation during the coating operation (Fig. 7). On the other hand, in coating with NE30D alone, the particles were very sticky despite the low inlet air temperature of 27°C. Sustained release

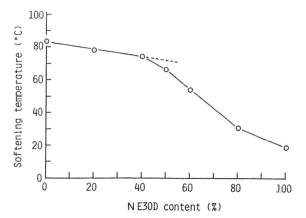

Fig. 6   Softening temperature of the films cast from blend polymer latexes of Eudragit RS30D and NE30D. (From Ref. 23.)

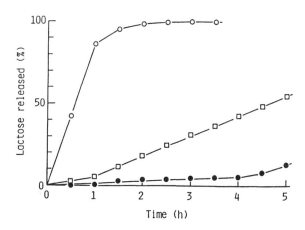

Fig. 7   Release of lactose from microcapsules 60% coated with blend polymer of Eudragit RS30D and NE30D (2:3) and 30% coated with plain NE30D in pH 6.8 aqueous medium. ○, RS30D:NE30D 2:3; ●, RS30D:NE30D 2:3 heated at 60° C for 12 h; □, NE30D. (From Ref. 23.)

from the particles produced suggests that film formation of NE30D is readily achieved during the coating operation (Fig. 7).

   Protzman and Brown [14] measured the MFT of copoly(EA-MMA) and blends of PMMA (MFG 92°C) and PEA (MFT −2°C) using apparatus of their own design (Fig. 8). The MFT of the copolymers increased continuously as the MMA content was increased. In contrast, the MFT of the blends increased

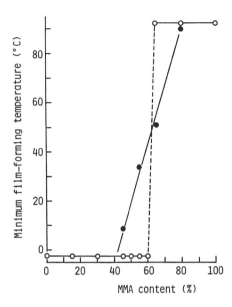

Fig. 8 Effect of composition of copoly(MMA-EA) and blend of PMMA and PEA on minimum film-forming temperature. ○, Blend; ●, copolymer. (From Ref. 14.)

sharply when the amount of PMMA in the blend reached 60%. They explained that above 60% PMMA content, film formation was not possible until temperatures that are high enough to deform the harder PMMA were reached. Although such a sharp change was not observed with RS30D and NE30D blends [14], the T_s value of the blends began to decrease rapidly after the amount of the harder RS30D content was reduced below 60% (Fig. 6). This means that a framework of hard polymeric particles may be embedded in the membrane above 60% RS30D content.

E. Hydration of Polymer

The excellent film-forming ability of Eudragit L30D may be due partially to the small size of polymeric particles (Table 8), but it seems to be insufficient for a full explanation. Eudragit RS30D and NE30D consist primarily of hydrophobic acrylic esters such as EA and MMA, although RS30D also contains a small amount of quarternary ammonium ester. On the other hand, Eudragit L30D contains the hydrophilic MA of 50% as the molar fraction of monomer [6]. As is well known, acetic acid is dissolved in hydrophobic carbon tetrachloride as the dimer formed with two hydrogen bondings [24]. This suggests that pendant carboxyl groups in the dried L30D membrane may be hydrogen bonded in the hydrophobic environment made by polymer backbones and pendant esters. This

supposition is supported by the high softening temperature (133°C) of L30D
(Fig. 4).

These hydrogen bondings can easily be broken by water molecules in
aqueous media. Hydration may cause the mechanical strength of particles to be
lowered and the polymeric particles to deform. In the film-forming process,
hydrogen bonding between carboxyl groups and water molecules can be a
driving force for easy film formation of L30D. By dehydration during the drying
process, the carboxyl groups should be hydrogen bonded to one another. This
suggests that molecular interactions can play an important role in the film-
forming process, in addition to physical factors. Weak molecular interactions
in aqueous media and hydrated polymeric phases such as ion-pair formation,
hydration of ions, hydrophilic hydration, hydrophobic hydration, and hydrophobic interaction may be involved in addition to hydrogen bonding. An understanding of the roles these interactions play may be important in discovering
additional advantages of aqueous dispersion systems.

Nakagami et al. [18] have shown that wet L30D films, whose water
content is 6.8 to 7.9%, have far lower glass transition temperatures than those of
dry L30D films, whose water content is 1.7 to 1.9% (Fig. 9). The glass transition
temperature of plain L30D film was changed from 91°C to 48°C by wetting. The
wet films were prepared by storing them at 30°C, 92% relative humidity. These
conditions would be realized at least locally in the coating chamber. Coating
under wet conditions yielded a more impermeable membrane [18]. Similar
wetting effects have been observed in experiments using a suspension of micro-

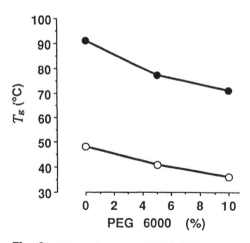

Fig. 9 Effect of water and PEG 6000 content on the T_g of Eudragit L30D cast films.
Water content (%): ●, 1.7–1.9; ○, 6.8–7.9. (From Ref. 18.)

nized HPMCAS powder [25]. Because the powder particle size is large, operation under wet conditions is critical for film formation.

Wet conditions are achieved by elevating the spray rate or lowering the inlet air temperature. However, coating continued for a long period under extremely wet conditions leads to agglomeration of core particles even if enhanced film formation is desired. Nakagami et al. proposed an intermittent spray process [18]. The change in granule water content during the operation is shown in Fig. 10. Spraying and drying were repeated intermittently to enhance film formation and avoid agglomeration.

As indicated earlier, the interaction of carboxyl groups of L30D with water molecules seems to contribute to excellent film-forming ability and the MFT far lower than the softening temperature. However, its application to pharmaceutical dosage forms is limited to enteric coating. Hence copoly(EA-MMA-HEMA) containing 2-hydroxyethyl methacrylate (HEMA) has been synthesized to develop a nonionic membrane material and thereby to elucidate the effect of monomer composition on film-forming ability [23,26]. HEMA has a hydroxyl group and exhibits high water solubility; hence cross-linked poly(HEMA) has been used as a gel-forming material [27–29]. The softening temperature of the cast films of copoly (EA-MMA-HEMA) is shown in Fig. 11. MMA elevated the softening temperature, which agrees with the results indicated in the literature [14,30,31]. The contribution of HEMA to increases in softening temperature is not as large. This suggests that the hydroxyl group of HEMA may not contribute as much to the hardening of dry films as does L30D.

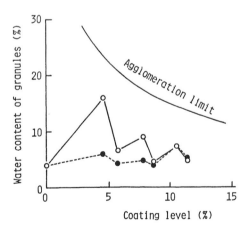

Fig. 10 Change of water content of granules during intermittent spray process in the fluidized-bed coating with Eudragit L30D. ●, Dry condition; ○, wet condition. (From Ref. 18.)

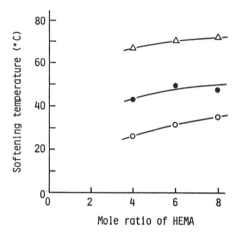

Fig. 11 Softening temperature of the films cast from copoly(EA-MMA-HEMA) latexes. Mole ratio (EA:MMA:HEMA): ○, 12:6:X; ●, 9:9:X; △, 6:12:X. (From Ref. 26.)

Coating was carried out under conditions similar to those listed in Table 8 using copoly(EA-MMA-HEMA) latexes that differed in HEMA content and had softening temperatures higher than the inlet air temperature [23]. The introduction of HEMA with a hydrophilic hydroxyl group led to enhanced water permeation through the membrane in cases of film-formable latex formulations or copolymers, as shown in Fig. 12 [26]. Film formation in the coating process

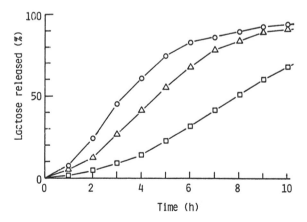

Fig. 12 Release of lactose from microcapsules 40% coated with copoly(EA-MMA-HEMA) in pH 6.8 aqueous medium. Mole ratio (EA:MMA:HEMA): ○, 9:9:8; △, 9:9:6, □, 9:9:4. (From Ref. 26.)

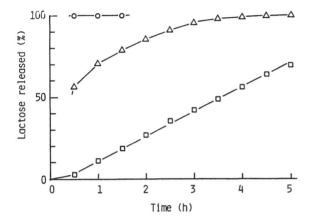

Fig. 13   Effect of HEMA on release of lactose from microcapsules 60% coated with copoly(EA-MMA-HEMA) in pH 6.8 aqueous medium. Mole ratio (EA:MMA:HEMA): ○, 6:12:4; △, 6:12:7; □, 6:12:8. (From Ref. 23.)

from 6:12:4 copoly(EA-MMA-HEMA) latex was poor, as shown in Fig. 13. However, as the content of HEMA was increased, the membrane became more impermeable despite the introduction of additional hydrophilic hydroxyl groups. This suggests that HEMA enhances film formation. The hydration and consequent softening of polymeric particles may be responsible for this facilitated film formation.

F.   Properties of the Core Surface

In practical coating of multiparticulates, porous granules are often used as cores. In addition, the coat of polymeric particles is more or less porous, due to incomplete film formation, and the polymer itself is sometimes water absorbable. Protzman and Brown [14] reported that the application of an aqueous dispersion of copoly(EA-MMA) to a porous surface into which water readily penetrated often resulted in appreciably higher MFT values. This indicates that the period during which capillary pressure is acting can be shortened by water penetration, resulting in poor film formation.

Most drugs are surface active. When they can migrate from the cores to film layers, the drug decreases the capillary pressure, which is a potent driving force in film formation. In addition, it is possible for the drug first dissolved and then precipitated by water evaporation to prevent the direct contacts of polymeric particles. Yang and Ghebre-Sellassie reported that these situations resulted in poor film formation [21], as shown in Table 6. As they pointed out, slow coating at initial stage is sometimes effective to avoid drug migration. Moreover, seal coating of cores is effective. Yang and Ghebre-Sellassie showed seal coating using the overcoating formulation in Table 5 prior to sustaining coating to be remarkably effective in preventing drug migration; the dissolution rate constant of 1.045 h^{-1} at 22°C was reduced to 0.536 h^{-1} by seal coating.

The materials used in subcoating (undercoating) or drug layering can also affect drug release. For example, when a fine phenacetin powder was layered on lactose crystals and subsequently coated with Eudragit L30D under the conditions given in Table 9, the release profiles of both phenacetin and lactose in the acidic medium were changed markedly by replacing the HPC used as a binder in the layering with PVP (Fig. 14). When microcapsules immersed in the dissolution medium were observed at 37°C using a photomicroscope, the microcapsules prepared using HPC were expanded by taking in water and finally burst. PVP interacted with L30D membrane and made the membrane very flexible or expandable. Consequently, the microcapsules took up more water and were enlarged even more. The release of water-soluble lactose which might move through water channels in the membrane was suppressed greatly by the increased or prolonged water influx. Inversely, the release of hydrophobic phenacetin which might migrate through the hydrophobic domains was enhanced by the decrease in membrane thickness due to increased film expansion and consequent increased surface area of each microcapsule.

The interaction between overcoated CAP or its hydrolytic breakdown products in Aquateric and underlying gelatin films during storage was reported

Table 9 Operating Conditions in Layering of Phenacetin and Subsequent Enteric Coating with Eudragit L30D by the Wurster Process (GPCG-1)

	Layering	Coating
Core: lactose (90 μm) (g)	216	
Layering powder: phenacetin (17 μm) (g)	250	
Binder: 3% HPC or PVP aquaeous solution (g)	1600	
Antiadherent: talc (inserted as powder) (g)	50	
Membrane material: Eidragit L30D (g)		1000
Plasticizer: triacetin (g)		30
Spacing agent: talc (g)		90
Water		added
Total (mL)		2400
Dry lacquer substance (g)		420
Inlet air temperature (°C)	80	60
Outlet air temperature (°C)	30–32	26–27
Spray rate (mL/min)	6.2–8.8	7.1–7.3
Spray pressure (atm)	2.5	2.5
Yield of product (%)		95–97

Source: Ref. 32.

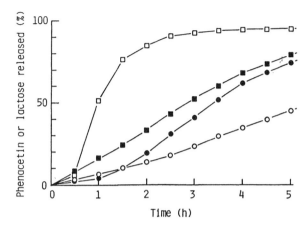

Fig. 14   Effect of binder used for drug fixing on release of phenacetin and lactose from microcapsules with lactose core, phenacetin layer, and Eudragit L30D coat (30% on a dry basis relative to cores) in aqueous acidic medium (pH 1.2). (From Ref. 32.)

Binder	Symbol	Released drug
HPC	○	Phenacetin
	□	Lactose
PVP	●	Phenacetin
	■	Lactose

[33]. This interaction led to increased drug release in 0.1 N hydrochloric acid and, inversely, slower dissolution in the phosphate buffer (pH 6.8). Anionic or amphoteric polymers such as sodium carboxymethylcellulose or acacia induce complex coacervation of gelatin [34]. It is possible that the interaction of CAP and gelatin in the solid state during storage has induced coacervation in the dissolution medium, although the details have not yet been studied.

G.   Additives

Solid additives that are insoluble in water are used as pigments, membrane diluents, antiadherents, or anticoagulants (separating agents) in coating with dispersions. Titanium dioxide, food dyes, and iron oxides can be used as pigments for coloring the coating. Diluents such as talc or magnesium stearate are used to thicken the membrane [35]. They are also effective as antiadherents. Eudragit L30D is, however, incompatible with magnesium stearate [36]. The anticoagulants are used to separate the interactive polymeric particles and consequently to avoid coagulation. The fumed silica in Surelease formulation and the oily acetylated monoglyceride (Myvacet 9-40) in the Aquateric formulation

(Table 1) act as the antiadherent and anticoagulant, respectively. A small amount of solid additives can suppress the membrane permeability, but an excessive amount should result in a discontinuous film structure [37,38].

In Table 10, recoveries of pigment and polymer are shown when latexes or polymeric solutions with suspended pigments were sprayed in the Wurster process [4]. In the case of HPC or HPMC aqueous solution, the recovery of pigment reached that of polymer (Table 10). On the other hand, while the polymer is recovered at high efficiency, one-third of the pigment particles in the aqueous latexes [Eudragit L30D and RS30D and 6:12:8 copoly(EA-MMA-HEMA)] are lost. When the softening temperature of copoly(EA-MMA-HEMA) was lowered (48°C) below the inlet air temperature (60°C) by changing the molar ratio of the monomer components from 6:12:8 to 9:9:8 [4], the core particle agglomeration was enhanced (from 0.5% to 9.6%), but the low level of pigment recovery was not improved. This suggests that the pigment particles may not be peeled away with films but may be spray dried. The same situation obtains when solid particles such as talc or magnesium stearate are mixed in aqueous dispersions as

Table 10 Operating Conditions in Coating With Typical Aqueous Coating Materials by the Wurster Process (GPCG-1) and Characteristics of Microcapsules Produced

Core: lactose 53–63 μm (g)		25				
	Solution		Latex			
	HPC	HPMC			EA-MMA-HEMA	
Membrane material	(L)	(TC-5R)	L30D	RS30D	(6:12:8)	(9:9:8)
Weight on a dry basis (g)	5	5	10	10	10	10
Pigment (2.9 μm)	0.2	0.2	0.2	0.2	0.2	0.2
Volume of spray dispersion (mL)	100	100	100	100	100	100
Inlet air temperature (°C)			60			
Outlet air temperature (°C)			27–33			
Inlet air rate (m³/min)			0.8–0.9			
Spray rate (mL/min)			1.2–2.1			
Spray pressure (atm)			1.5			
Yield (%)						
Product	94	93	93	94	91	94
Pigment	93	90	62	68	62	56
Polymer	94	98	92	89	89	97
Agglomerates (%)	68.6	80.8	1.1	2.8	0.5	9.6
T_g (°C)			133	88	72	48

Source: Ref. 4.

an antiadherent. As may be estimated from the fact that the surface roughness produced by talc incorpolated in EC solution has contributed little to suppression of core particle agglomeration [1], partially spray-dried solid particles may play the role of antiadherent.

Charged solutes in dispersion can cause polymeric particles to coagulate. Sodium lauryl sulfate and ammonium oleate are used as an anionic surfactant in Aquacoat and Surelease pseudolatex, respectively (Table 1). The addition to these ethyl cellulose pseudolatexes of salts of basic drugs such as propranolol HCl or chlorpheniramine maleate results in latex flocculation or coagulation. Bodmeier and Paeratakul [39] reported that replacing anionic surfactants with the nonionic surfactant Pluronic P103 overcame the coalescence of Aquacoat or Surelease particles that had been induced by interaction between the surfactant and the dissolved drug.

Unmodified films from commercially available latexes sometimes have exceedingly slow release rates, especially in large multiparticulates with small surface area. In such cases, pore-forming agents are used [40]. Such agents can also be used when a significant fraction of drug remains undissolved. Appel et al. [40] have demonstrated that urea can be used as a pore-forming agent with Aquacoat because it is a small, readily water-soluble, uncharged molecule. In Eudragit RL, RS, and NE30D, the release of active ingredient is increased or controlled by mixing nonionic, hydrophilic substances such as poly(ethylene glycol), methylcellulose, hydroxypropyl cellulose, poly(vinylpyrrolidone), lactose, sorbitol, and glycerine [36]. Eudragit NE30D provides coatings that are highly impermeable to drug when used alone [41]. The additives listed above are used to modify membrane permeability. They can cause NE30D membrane to distintegrate rapidly when a large amount is incorpolated [36,41].

The additives can affect the pH dependency of drug release from coated multiparticulates. Although the polymeric component of either Aquacoat or Surelease is ethyl cellulose, the release of propranolol HCl from beads coated with Aquacoat was higher at pH 7.5 than at pH 1.2 to 4.5 (Fig. 15), but the release from beads coated with Surelease was less affected by change in the pH [42]. Chang et al. reported similarly pH-dependent release of theophylline pellets coated with Aquacoat [3]. Goodhart et al. reported similar results with phenylpropanolamine HCl [35]. On the other hand, the release from beads coated with Eudragit RS30D or NE30D was affected little by changes in pH [3,35]. In the case of Aquacoat, pH-dependent release might be attributed to SDS, because the release of theophylline from pellets coated with Surelease exhibited pH dependency by adding SDS [3]. The mechanism of SDS-induced pH dependency is not clear at present. Appel et al. developed an osmotic device microporously coated with Aquacoat and pH-independently releasing diltiazem HCl [40]. This pH independency with Aquacoat came from making the drug solubility within the core constant through the use of buffering agents.

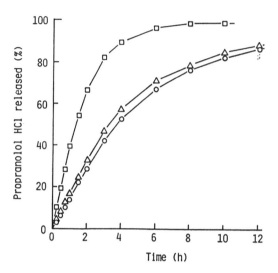

Fig. 15   Effect of dissolution fluid pH on the drug release from the granules prepared by fixing propranolol HC1 on nonpareil with 2% HPMC and by coating with 6% Aquacoat. pH of dissolution fluid: ○, 1.2; △, 4.5; □, 7.5. (From Ref. 42.)

H.   Coating of Fine Powder

In the preparation of multiparticulate dosage forms by the spray-coating process, particle agglomeration is a serious problem, especially with fine cores. Agglomeration takes place through the formation of interparticulate bridges of membrane materials dissolved or dispersed in spray droplets. Hence it must be related to the concentration of polymer in the spray dispersion, the size of sprayed droplets, and the strength of dried interparticulate bridges relative to the applied separation force [43]. It is very advantageous that the dispersion of a concentration of 20% or higher can be used with aqueous dispersion systems because of their low viscosity compared with polymeric solution systems. However, this high concentration can lead to easy agglomeration. The distribution of droplet size when a spraying system is used is shown in Fig. 16. The mass median diameter was 12 μm at a spray pressure of 2.3 atm and a spray rate of 4 mL/min. These spray conditions with a pneumatic nozzle can normally be used in a aqueous coating by the Wurster process (GPCG-1). Although a higher spray pressure will produce smaller droplets, it should not be expected that the mass median diameter became smaller 10 μm. When the interparticulate bridge formed with the polymer supplied by these droplets cannot be broken by separation force acting on the agglomerate, the agglomeration proceeds.

Figure 17 shows the schematic diagrams of typical types of commercially available coating machines [44,45]. In general, more intense separation force is

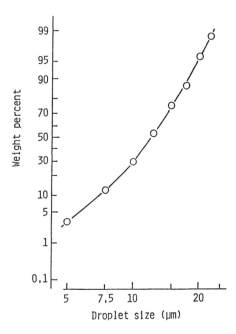

Fig. 16 Cumulative undersize distribution of the droplets produced by spraying water. (From Ref. 43.)

required for discretely coating smaller particles. The separation of particles by a conventional top-sprayed fluidized bed is usually too weak to discretely coat small particles. When the spouted bed was produced, the upward flow of particles in the central region become relatively intense (Fig. 17a). This type of spouted bed can be used in the coating of particles of size about 1 mm. Small particles of about 200 to 300 μm are usually coated by rotary fluidized-bed processors (Fig. 17b). The Wurster configuration (Fig. 17c) can produce further intense separation of particles due to the collision of particles against the partition. Brittle crystals such as phenacetin are easily fractured; even the harder crystals such as lactose become roundish [46]. Using the Wurster process, particles smaller than 100 μm may be coated discretely, but the smallest size of particles that the Wurster process can discretely coat will depend on the binding strength of membrane materials.

An interesting fact in Table 10 is that coating with polymeric dispersions exhibits the extremely low degree of agglomeration, whereas the coating efficiency of polymer is kept high. Film formation was not achieved during this coating process because the fine-core particles (53 to 63 μm) had to be coated under relatively dry conditions to avoid agglomeration, differing from the cases in Table 8 and Fig. 13, where large cores of 328 μm were used. This low degree

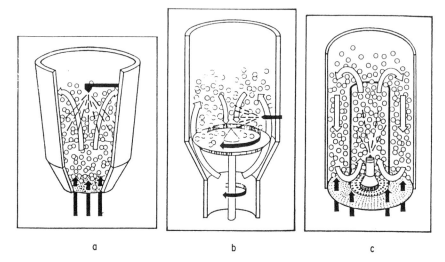

Fig. 17 Schematic diagrams of commercially available fluidized-bed machines applicable to coating of multiparticulates: (a) top-spray spouted bed; (b) rotary fluidized bed; (c) bottom-spray spouted bed with a draft tube (Wurster). (From Ref. 45.)

of film formation clearly resulted in a reduction in membrane binding strength. As the core particles become smaller, the agglomeration should be enhanced due to decrease in the inertia and momentum of particles. The property of polymeric dispersion demonstrated in Table 10 may be useful to avoid such a problem as agglomeration in the coating of fine-core particles [4].

The 6:12:8 copoly(EA-MMA-HEMA) seems to be a useful membrane material because its latex exhibits excellent film formation in coating under wet conditions (Fig. 13) and a low agglomeration tendency under dry conditions (Table 10). On the other hand, the dissolution of hydrophilic lactose from the fine microcapsules (6:12:8 copoly(EA-MMA-HEMA)) shown in Table 10 was too fast to achieve effective sustained release because the hydrophilic HEMA enhanced the water permeation. The release through copoly(EA-MMA-HEMA) film is suppressed by an increase in EA content [26]. However, the increase in EA (9:9:8) results in lowering the softening temperature to 48°C; consequently, agglomeration is enhanced (Table 10). To utilize both the low permeability of EA-rich copolymer and the low tendency of agglomeration and, if coated under wet conditions, the excellent film-forming ability of MMA- and HEMA-rich copolymer, a core-shell type of latex has been synthesized [4,47]. Each latex particle had a core of 12:6:4 copoly(EA-MMA-HEMA) and a shell of 6:12:8 copoly(EA-MMA-HEMA) (Fig. 18).

Core-Shell Latex (6:4)
    Core: EA:MMA:HEMA=12:6:4,
            Softening temperature: 26°C,
            Low water-permeability.
    Shell: EA:MMA:HEMA=6:12:8,
            Softening temperature: 78°C,
            High water-permeability,
            Noncohesive.

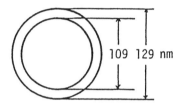

Fig. 18   Core-shell latex particle.

Regardless of its practical application, coating very fine powders will be of great interest. It may clarify the limit of application of air-suspension techniques to the coating process. Coating of micronized phenacetin powder by the Wurster process is described here. The coating conditions are shown in Table 11, where coating with the core-shell latex is compared to that with an organic solvent–based system of ethyl cellulose [2]. The micronized phenacetin powder had a mass median diameter of 11 μm. Its fluidization was very difficult, due to electrostatic charging. Hence, in coating with ethyl cellulose, the powder was slightly agglomerated at a high spray rate, up to 14 μm, and subsequently coated. Cholesterol in the membrane formulation was added to suppress agglomeration [1,2]. However, during the coating process, particles smaller than 20 μm were gradually agglomerated. Although the mass median diameter of product was 31 μm, its particle size distribution was very broad, due to the production of large agglomerates, as shown in Fig. 19. On the other hand, in coating with core-shell latex, the micronized powder was fluidized with the aid of very fine colloidal particles (Aerosil 200). The production of coarse agglomerates was reduced remarkably (Fig. 19) and the mass median diameter of product was 26 μm. In this case the powder could be fluidized by a very low inlet air rate (Table 11) and the powder seemed to be sucked into the partition and spouted by spray air. At a 20% coating level, the agglomerates were so easy to disintegrate that by screening with an air jet sieve their particle size distribution became identical to that of the original particles, whereas the corresponding product from

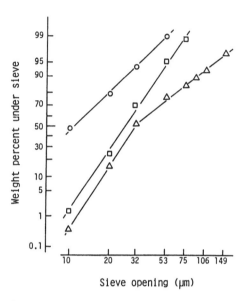

Fig. 19 Cumulative undersize distributions of the products by coating micronized phenacetin powder. ◯, Phenacetin powder; △, coated with organic system (run 1); □, coated with aqueous core-shell latex (run 2). (From Ref. 2.)

the organic solvent system of EC was tightly agglomerated and its mass median diameter was already 30 μm. This indicates a low binding strength for core-shell latex particles. An interesting fact is that the size of primary core particles would be almost identical to that of sprayed droplets, or the droplets would be somewhat larger. Several core particles could possibly be agglomerated by a droplet [43] and the agglomerates would be disintegrated by collision with other particles or with the partition. Using this core-shell latex, 12-μm corn starch particles could be coated with an agglomerate fraction of only a few percent and with a high polymer coating efficiency [4,47]. Composite latexes may have broad applications in the development of otherwise functional films [48–52].

V. SUMMARY

The mechanism of film formation from aqueous polymeric dispersions during the coating process has been discussed. The most potent driving force of film formation should be the capillary pressure generated by air–water surface tension. Film formation is a rate process that is dependent on the strength of surface tension, the mechanical properties of the polymeric phase, and the period during which capillary pressure can act. The behavior of water is a key to film formation. Water makes the capillary pressure generated and the polymer deform-

Table 11 Operating Conditions in Coating of Micronized Phenacetin Powder by the Wurster Process (GPCG-1)

| | Organic solvent | | |
	Agglomeration	Coating	Latex coating
Core: phenacetin (11 μm) (g)	250	—	250
Spray solution			
Ethyl cellulose (g)	3.75	75	
Cholesterol (g)	3.75	75	
Talc (g)	3.75		
Stearyltrimethyl ammonium chloride (g)		4.5	
Core-shell latex (g)			150
Ethanol (mL)	9.5	9.5	
Water (mL)			9.5
to make:	150	3000	750
Fluidizing agent			
Aerosil 200 (g)			19
Operating conditions			
Inlet air temperature (°C)	60	60	60
Outlet air temperature (°C)	28–29	26–29	30
Inlet air rate (m³/min)	0.2	0.5	0.06
Spray rate (mL/min)	14.2	11.3	2.9
Spray pressure (atm)	2.5	3.5	3.5
Product			
Yield (%)		90	89
Mass median diameter (μm)	14	31	26

able by hydration. Its evaporation rate is also decisively related to film formation.

In this chapter the traditional applications of commercially available aqueous dispersions have primarily been reviewed. Aqueous dispersion systems would seem to have possibly wider applications in the future. A few of these new trials have been discussed here.

REFERENCES

1. Y. Fukumori, H. Ichikawa, Y. Yamaoka, E. Akaho, Y. Takeuchi, T. Fukuda, R. Kanamori, and Y. Osako, *Chem. Pharm. Bull. 39*:164 (1991).
2. Y. Fukumori, H. Ichikawa, Y. Yamaoka, E. Akaho, Y. Takeuchi, T. Fukuda, R. Kanamori, and Y. Osako, *Chem. Pharm. Bull. 39*:1806 (1991).
3. R. K. Chang, C. H. Hsiao, and J. R. Robinson, *Pharm. Technol. 11*(3):56 (1987).

4. Y. Fukumori, Coating of fine powders with a core-shell type of aqueous latex by the Wurster process, *Proceedings of the Pre-World Congress on Particle Technology*, Gifu, Japan, 1990, p. 59.

5. H. Ichikawa, K. Jono, H. Tokumitsu, T. Fukuda, and Y. Fukumori, *Chem. Pharm. Bull. 41*:1132 (1993).

6. K. O. R. Lehmann, in *Aqueous Polymeric Coatings for Pharmaceutical Dosage Forms* (J. W. McGinity, ed.), Marcel Dekker, New York, 1989, p. 153.

7. C. R. Steuernagel, in *Aqueous Polymeric Coatings for Pharmaceutical Dosage Forms* (J. W. McGinity, ed.), Marcel Dekker, New York, 1989, p. 1.

8. K. L. Moore, in *Aqueous Polymeric Coatings for Pharmaceutical Dosage Forms* (J. W. McGinity, ed.), Marcel Dekker, New York, 1989, p. 303.

9. T. Nagai, F. Sekigawa, and N. Hoshi, in *Aqueous Polymeric Coatings for Pharmaceutical Dosage Forms* (J. W. McGinity, ed.), Marcel Dekker, New York, 1989, p. 81.

10. H. Sakamoto, H. Muto, and H. Kokubo, Granule coating with HPMCAS and EC in an aqueous system by using an agitating bed apparatus, *Proceedings of the Pre-World Congress on Particle Technology*, Gifu, Japan, 1990, p. 105.

11. R. E. Dillon, L. A. Matheson, and E. B. Bradford, *J. Colloid Sci. 6*:108 (1951).

12. W. A. Henson, D. A. Taber, and E. B. Bradford, *Ind. Eng. Chem. 45*:735 (1953).

13. G. L. Brown, *J. Polym. Sci. 22*:423 (1956).

14. T. F. Protzman and G. L. Brown, *J. Appl. Polym. Sci. 4*:81 (1960).

15. J. W. Vanderhoff, H. L. Tarkowski, M. C. Jenkins, and E. B. Bradford, *J. Macromol. Chem. 1*:361 (1966).

16. S. Muroi, *Chemistry of High Polymer Latices*, Kobunshi-Kankokai, Kyoto, Japan, 1970, p. 235.

17. I. Ghebre-Sellassie, U. Lyer, D. Kubert, and M. B. Fawzi, *Pharm. Technol. 12*(9):96 (1988).

18. H. Nakagami, T. Keshikawa, M. Matsumura, and H. Tsukamoto, *Chem. Pharm. Bull. 39*:1837 (1991).

19. A. O. Okhamafe and P. York, *J. Pharm. Sci. 77*:438 (1988).

20. G. I. Amer, J. N. Foster, and C. P. Iovine, European Patent Application 0292314 A2 (1988).

21. S. T. Yang and I. Ghebre-Sellassie, *Int. J. Pharm. 60*:109 (1990).

22. K. Toyoshima, in *Polyvinyl Alcohol* (C. A. Finch, ed.), Wiley, New York, 1973, p. 339.

23. Y. Fukumori, Y. Yamaoka, H. Ichikawa, Y. Takeuchi, T. Fukuda, and Y. Osako, ·*Chem. Pharm. Bull. 36*:4927 (1988).

24. E. E. Schrier, M. Pottle, and H. A. Scheraga, *J. Am. Chem. Soc. 86*:3444 (1964).

25. T. Fujii, Y. Doi, T. Takeda, and K. Inazu, Coating technique of granules with hydroxypropylmethylcellulose acetate succinate, *Proceedings of the Pre-World Congress on Particle Technology*, Gifu, Japan, 1990, p. 80.

26. Y. Fukumori, Y. Yamaoka, H. Ichikawa, Y. Takeuchi, T. Fukuda, and Y. Osako, *Chem. Pharm. Bull. 36*:3070 (1988).

27. K. V. R. Rao and K. P. Devi, *Int. J. Pharm. 48*:1 (1988).

28. C. G. Pitt, Y. T. Bao, A. L. Andrady, and P. N. K. Samuel, *Int. J. Pharm. 45*:1 (1988).

29. P. L. Lee, *J. Pharm. Sci. 73*:1344 (1984).

30. C. Bondy, *J. Oil Colour Chem. Assoc. 53*:555 (1970).

31. S. Muroi, *High Polymer Latex Adhesives*, Kobunshi-Kankokai, Kyoto, Japan, 1984, p. 83.

32. Y. Fukumori, Y. Yamaoka, H. Ichikawa, T. Fukuda, Y. Takeuchi, and Y. Osako, *Chem. Pharm. Bull. 36*:1491 (1988).

33. K. S. Murthy, N. A. Enders, M. Mahjour, and M. B. Fawzi, *Pharm. Technol. 10*(10):36 (1986).

34. G. L. Koh and I. G. Tucker, *J. Pharm. Pharmacol. 40*:309 (1988).

35. F. W. Goodhart, M. R. Harris, K. S. Murthy, and R. U. Nesbitt, *Pharm. Technol. 8*(4):64 (1984).

36. K. Lehmann and D. Dreher, *Int. J. Tech. Prod. Manuf. 2*(4):31 (1981).

37. J. M. Waldie, *Surface Coating*, Vol. 1, *Raw Materials and Their Usage*, Chapman & Hall, New York, 1981.

38. M. R. Harris and I. Ghebre-Sellassie, in *Aqueous Polymeric Coatings for Pharmaceutical Dosage Forms* (J. McGinity, ed.), Marcel Dekker, New York, 1989, p. 63.

39. R. Bodmeier and O. Paeratakul, *Pharm. Res. 6*:725 (1989).

40. L. E. Appel and G. M. Zentner, *Pharm. Res. 8*:600 (1991).

41. I. Ghebre-Sellassie and R. U. Nesbitt, in *Aqueous Polymeric Coatings for Pharmaceutical Dosage Forms* (J. W. McGinity, ed.), Marcel Dekker, New York, 1989, p. 247.

42. G. S. Rekhi, R. W. Mendes, S. C. Porter, and S. S. Jambhekar, *Pharm. Technol. 13*(3):112 (1989).

43. Y. Fukumori, H. Ichikawa, K. Jono, Y. Takeuchi, and T. Fukuda, *Chem. Pharm. Bull. 40*:2159 (1992).

44. D. M. Jones, *Pharm. Technol. 9*(4):50 (1985).

45. A. M. Mehta, M. J. Valazza, and S. N. Abele, *Pharm. Technol. 10*(4):46 (1986).

46. Y. Fukumori, T. Fukuda, Y. Hanyu, Y. Takeuchi, and Y. Osako, *Chem. Pharm. Bull. 35*:2949 (1987).

47. Y. Fukumori, T. Fukuda, and Y. Osako, Coating of 10 μm particles by the Wurster process, *Proceedings of the 7th Symposium on Particulate Preparations and Designs*, Shiga, Japan, 1990, p. 137.

48. M. Okubo and S. Yamaguchi, *Nippon Sechaku Kyokaishi 22*:276 (1986).

49. M. Okubo, *Shikizai 60*:281 (1987).

50. M. Okubo, A. Yamada, and T. Matsumoto, *J. Polym. Sci. 16*:3219 (1980).

51. S. Muroi, *High Polymer Latex Adhesives*, Kobunski-Kankokai, Kyoto, Japan, 1984, p. 63.

52. S. Yamazaki, *Kobunshi Ronbunshu 33*:663 (1976).

6

Coating of Multiparticulates Using Molten Materials
Formulation and Process Considerations

David M. Jones

Glatt Air Techniques, Inc.
Ramsey, New Jersey

Phillip J. Percel

Eurand America, Inc.
Vandalia, Ohio

I. INTRODUCTION

The vast majority of coating occurs using an application medium that must be evaporated. Available materials are applied as solutions or dispersions and typically contain only 30% solids or less. An alternative is to select a coating material that can be applied molten. Because there is nothing to be evaporated, processing times are comparatively short. In addition, many of the reasons for coating (improved stability, taste masking, sustained release) can be achieved with available materials, which have an added advantage of being relatively inexpensive.

II. COATING MATERIALS

A variety of coating materials are available for molten coating (Table 1). They are selected for their ability to protect a given substrate from a specific type of environment and to release that substrate, either gradually or instantly, in reaction to specific stimuli or via a deliberately altered environment. Substrate release may be accomplished as a function of heat, moisture, shear, pH, or

Table 1 Types of Coating Materials

Coating type	Coating trade name	Melting point (°C)	Color
Partially hydrogenated cottonseed/soybean oil	Van Den Bergh Foods K.L.X.	51–55	White
Partially hydrogenated palm oil	Van Den Bergh Foods 27 Stearine	58–63	White
Partially hydrogenated cottonseed oil	Van Den Bergh Foods 07 Stearine	61–65	Off-white to tan
Partially hydrogenated soybean oil	Van Den Bergh Foods 17 Stearine	67–71	White
Beeswax	—	62–65	Light tan
Paraffin wax[a]	Frank B. Ross Co. 130/135 AMP	55	White
Carnauba wax	Frank B. Ross Co. No. 1 Yelllow	84 (min.)	Yellow
Partially hydrogenated castor oil	Cas Chem Castor Wax	85–88	White
Poly(ethylene glycol)[a]	Union Carbide Carbowax 3350	54–58	White

[a]Various types and melting points available.

enzymes (e.g., digestion). It has been reported that release of substrates, barrier coated with beeswax, glyceryl monostearate, and acetylated monoglycerides, is probably due to diffusion and dialysis, pH-dependent dissolution, and possibly enzymatic breakdown [1,2].

The selection criteria for coating materials also include melting point, melting range, and viscosity (in a liquid state). Typically, coatings should have a melting point of less than 85°C. The liquid is maintained at a constant temperature during application, which is typically 40 to 60°C above its melting point. Therefore, the liquid temperature may be as high as 140 to 150°C, which presents some challenges to equipment users concerning liquid storage and delivery to the spray nozzle, as well as operator safety during processing.

Coatings with a broad melting-point range are usually more difficult to apply than a coating with a sharp melting point. A common problem is that they tend to be very tacky during spraying through a broad range of product temperatures. However, in some cases it is possible to blend two different materials to alter release properties (and melting point) without affecting the coating application process substantially.

III. PROCESSING

The fluidized-bed process is ideal for molten coating of multiparticulates. Particles as fine as 50 μm have been coated discretely. More typically, mean particle size ranges from 100 to 2000 μm. As is typical of all coating applications, a smooth spherical substrate is desirable, especially for sustained-release products. For the less stringent performance requirements, angular materials can be coated, but needlelike cores should be avoided due to friability concerns. A high-velocity airstream is used to separate the particles, which rush through a zone of finely atomized, molten liquid. Small (relative to the size of particles being coated) droplets impinge on the substrate surface, spread, and slowly congeal. Multiple passes through the coating zone eventually result in a continuous coating that is uniform in surface properties.

The quality of the coating applied is dependent primarily on the rate at which the coating droplets solidify on the particle surface. This is controlled by the liquid temperature during atomization and the temperature of the fluidizing product. The most continuous coating is achieved when the product temperature during spraying is maintained as close as possible to the congealing temperature of the coating material. This temperature varies depending on the fluidized-bed process chosen (conventional top spray, Wurster bottom spray, or rotary tangential spray).

IV. FLUIDIZED-BED TECHNIQUES

The top-spray fluidized-bed coater (Fig. 1) has evolved from the fluidized-bed dryer, which was commercialized more than 30 years ago. The substrate is placed in a product container that is an inverted truncated cone with a fine retention screen at its base. Heated air is drawn through the screen and product at a velocity sufficient to fluidize the substrate bed vigorously. The particles are accelerated upward past a nozzle that sprays the molten coating downward into the randomly fluidized substrate. Exiting the "spray zone," the product enters the expansion chamber, which is wider in diameter than the base of the product container. This results in a decreasing air velocity, allowing the cooling particles to decelerate and fall back into the product container to continue cycling throughout the duration of spraying.

The product container in the top-spray process is designed for no restrictions to product flow, an important consideration for the application of a molten coating. As mentioned previously, coat quality is highest when the product temperature during spraying is kept as close as possible to the congealing temperature of the coating material. In this condition, the substrate bed becomes

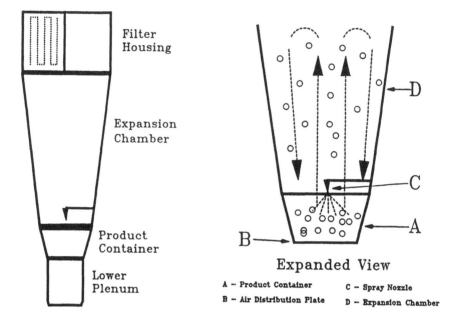

Fig. 1 Fluidized-bed top-spray coater.

very viscous and resistant to fluidization. Because the product container is a wide-open vessel, fluidization is not impeded.

The fluid temperature must be preserved until the liquid coating material is atomized into the bed. Figure 2 illustrates a nozzle assembly that accomplishes this task. The nozzle wand is triaxial. The center tube in the wand is the path for the molten liquid. It is surrounded by a small air space for delivery of a high-pressure, low-volume air signal to control a valve in the nozzle which opens when the spray pump is running. Both tubes are encircled by a larger air space through which the heated atomization air is supplied. The nozzle and wand extend into the fluidized bed from the wall of the expansion chamber. Vertically, it is positioned as close as possible to the substrate bed to minimize the distance the droplets must travel before impinging on the substrate. The bed is also very densely fluidized in this region. Because it is stainless steel and operating at a temperature far exceeding the melting point of the coating material, the nozzle wand must be insulated. This will protect any coated product from contacting the wand and remelting.

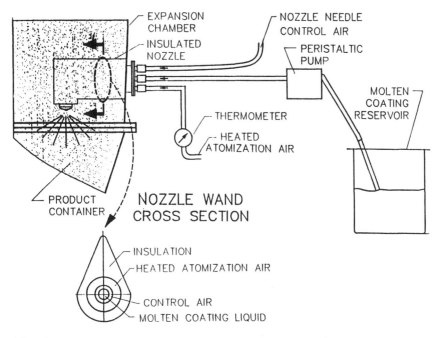

Fig. 2   Insulated nozzle and wand for top-spray hot-melt coating.

The top-spray fluidized bed is a system of choice for hot-melt coating, primarily because of its ability to operate with the product temperature closest to the congealing temperature of the melt (closer than with the other fluidized-bed techniques). However, there are limitations. These involve primarily the fluidization characteristics of the substrate bed. Figure 3 illustrates fluidization characteristics for particles of various sizes and densities [3]. The "C" or "cohesive" materials are generally smaller than 100 μm, with densities of 0.5 g/cm^3 and less. Many pharmaceutical materials fall into this category. Alone, they are difficult to fluidize fully and are very prone to adhering to machine wall surfaces due to static electricity. However, under the correct spraying conditions, the static charge is dispelled (visually) and even very fine substrates fluidize freely.

Substrates with particle sizes ranging from 100 μm to approximately 750 μm and with densities of 0.5 g/cm^3 and less will fluidize more easily than fine materials. As a result, they are not difficult to coat using the top-spray system. However, products with larger particle sizes and/or higher densities will begin to exhibit poor fluidization properties. These materials, classed as "sandlike" and "spoutable," may fluidize well in small machines due to shallow bed depth and wall effects as a result of the narrow diameter of the product container and

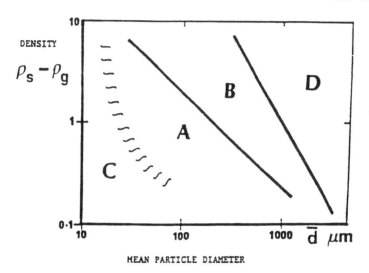

Fig. 3   Fluidization characteristics for particles of various sizes and densities.

expansion chamber. In lab machines it is possible to coat successfully particles of almost all sizes. Unfortunately, with this class of materials, scale-up may fail to duplicate results seen at the lab scale.

A fluidized bed of particles is created by bubbles of air being passed at some velocity through the bed. Particles are carried on the surface and in the drift and wake of the rising bubbles. The size of the bubbles has a strong influence on the quality of the fluidization pattern, and in general, the smaller, the better. Bubbles move through the bed of particles and coalesce, forming larger bubbles. The size to which they grow depends on their velocity and the resistance (from the bed) that they meet. Larger (and/or denser) substrates cause more rapid coalescence than do finer (and/or lighter) materials. Additionally, deeper beds, as would likely be encountered in larger machines, also lead to larger bubble sizes. Figure 4 shows the influence of bed depth and air velocity on bubble size [4]. In laboratory top-spray equipment (batch sizes up to approximately 10 kg), bed depth rarely exceeds 15 cm. With substantially different air velocities above that required for minimum fluidization, bubble diameter is still generally smaller than 10 cm. As a result, the fluidization pattern will be reasonable. With pilot and production equipment, bed depth is greater, in some cases exceeding 50 cm. Under these conditions, bubble diameter exceeds 20 cm and increases progressively to above 50 cm as air velocities well above the minimum fluidization velocity are used. The fluidization pattern that results can best be described as a "slugging" or "surging" bed. In an instant the nozzle, which is spraying smoothly and continuously, can be inundated with product, momentarily causing

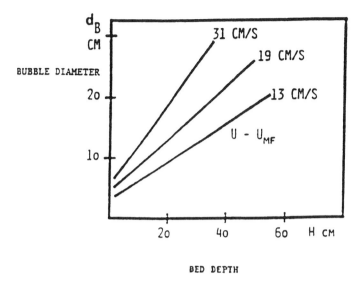

Fig. 4 Influence of substrate bed depth and air velocity on bubble size.

local overwetting and possibly agglomeration. Seconds later, the bed is at a substantial distance from the nozzle. Atomized droplets travel through the relatively cooler airstream, rapidly increasing in viscosity. When they contact the substrate, spreading may not occur, and the resulting coating will be porous. In the extreme case, some of the atomized liquid may spray congeal, or be carried upward, causing outlet air filter blinding and eventual collapse of the fluidized bed. Substrates that border the "sandlike" and "spoutable" classes are types that may be problematic in scale-up. It is suggested that a fluidization trial be conducted in production equipment early in the development phase of the project to avoid a possible catastrophe due to poor fluidization quality.

If particle size and/or density is in a potentially problematic range, the bottom-spray Wurster [5] coater (Fig. 5) should be considered. This process is used extensively for core material preparation by layering (solution or suspension) [6] and film coating using water [7–9] or solvent-based liquids. The product container consists of a cylindrical (or slightly conical) outer chamber fitted with an open cylindrical tube, or partition, generally one-half the diameter of the bottom of the container. The base of the product container is an orifice plate which allows the majority of the fluidization air to pass through the partition. As product is accelerated upward, it is sprayed by a nozzle mounted in the center of the plate, spraying concurrently with the flow of product. Particles exit the partition at high speed, entering an expansion area where they decelerate, eventually falling into the space outside the partition known as the downbed. The

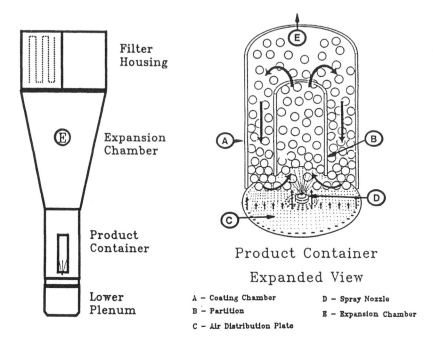

Fig. 5 Fluidized-bed Wurster bottom-spray coater. (From Glatt Air Techniques, Inc., 1987.)

orifice plate beneath this area has comparatively few holes, which are small and allow only enough air to keep the downbed aerated or in nearly weightless suspension. Horizontal transport of the product into the spray zone, which completes the cycle, occurs through a small gap at the base of the partition. Acting in concert, proper selection of orifice plate configuration and partition height should result in a rapid, smoothly flowing downbed. This will yield a continuous stream of product through the coating zone, producing high-quality applied film and rapid processing.

Use of the Wurster for molten coating of multiparticulates may require some slight changes in the configuration of the insert. Molten coatings are generally tacky during spraying, especially when attempting to congeal the coating slowly to achieve the highest quality. Consequently, it may be difficult to attain a rapid, smooth downbed without risking stalling in some region outside the partition. Substantially increasing the permeability of the downbed portion of the orifice plate is recommended. Also, horizontal transport of product at the base of the coating chamber is achieved primarily by the relatively small gap between the partition and orifice plate. The viscous nature of the bed inhibits movement through this small space, and as more coating is applied, less product will enter the coating zone at a given instant. Agglomeration may begin to occur as fewer particles are exposed to the consistent spray. A further consequence is that an increasing amount of coating could be sprayed through the sparsely fluidized upbed onto the outlet air filter, eventually affecting fluidization quality. Finally, product performance would be affected by the lower coating efficiency. Processing the batch with a partition height of approximately double that used for a nontacky coating of the same substrate will help. Initial fluidization will be quite vigorous and disorganized, but as coating is applied, the bed will become more viscous and fluidization will improve. The spray nozzle is surrounded by the fluidizing substrate. To keep the coating material molten during spraying, the atomization air is maintained at a temperature well above the melting point. Since the nozzle is stainless steel and very conductive, it must be insulated to prevent contact with coated product. Only a small insulation jacket should be used to avoid affecting the fluidization pattern inside the partition.

Scale-up of the Wurster process is not generally influenced by particle size and density, as may be the case using the top-spray technique described previously. When scaling up from laboratory- to pilot- or small-scale production equipment (18- and 24-in. Wursters), bed depth increases, as does fluidization height; hence attrition may become more of a problem (this is a concern for all types of coating substances, not only coating with molten materials). However, if the process is performed successfully at this scale, increasing to even larger machines (32- and 46-in. Wursters) should not be difficult. In all coating processes, the spray application rate is limited primarily by the performance of the coating zone. The spray is generally a solid cone of droplets contained in a relatively narrow diameter pattern. In Wurster processing, expanding the partition beyond 9 in. in diameter usually does not result in a higher application rate because the coating zone itself is limited in size. In fact, 12-in. and larger partitions will require a higher air velocity (and volume) to assure upward transport of product through the coating zone. The increased particle travel height may further exacerbate an attrition problem. Therefore, multiples of 9-in.-diameter partitions (and nozzles) are used in production equipment. The 32-in. Wurster uses three, and the 46-in. Wurster employs seven duplicate coating zones. The orifice plate configuration and partition height are selected such that air velocity and volume in each partition are similar to that used in pilot-scale equipment (18-in. Wurster). Accordingly, the product is not subjected to greater forces in the larger machine.

A relative newcomer to fluidized-bed processing is referred to as the tangential spray or rotary fluidized-bed processor (Fig. 6). Originally conceived for producing high-density granules or small pellets from powders, this technique excels at producing high-potency pellets by layering drug onto some type

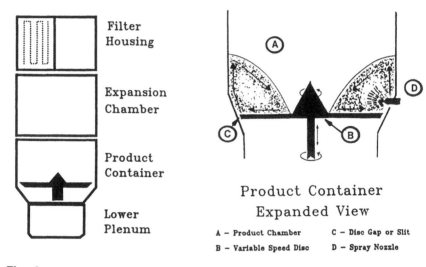

Filter
Housing

Expansion
Chamber

Product
Container

Lower
Plenum

Product Container
Expanded View

A – Product Chamber   C – Disc Gap or Slit
B – Variable Speed Disc   D – Spray Nozzle

Fig. 6 Tangential-spray or rotary fluidized-bed processor.

of core material via solution, suspension, or powder addition [6,10]. A controlled-release coating can subsequently be applied.

The product container consists of an unbaffled cylindrical chamber with a solid variable-speed disk at its base. The process air is drawn through a gap at the perimeter of the disk. The velocity is controlled by the disk height and fluidizing air volume, which are tailored to the particular process. Three forces combine to produce a pattern that is best described as a spiraling helix. Centrifugal force by the spinning disk causes the product to move forward and outward toward the chamber wall. The fluidization air provides acceleration upward, and gravity engenders the product to tumble in toward the disk surface once again. The nozzle is positioned tangentially, immersed in the bed, spraying concurrently with the rapidly tumbling product. Coatings can be applied using water, organic solvent, or via hot melt.

Process considerations with regard to molten coating in the tangential spray process involve the spray nozzle and the tacky nature of the bed during spraying. Again, because the atomization air is heated to keep the coating material molten, the nozzle must be insulated to avoid remelting product. Also, the insulation must be unobtrusive to avoid bed stalling in its surroundings. Second, the fluidization pattern is quite different from that encountered in both the top spray and Wurster processes. In these techniques, the product moves upward from the middle of the container and out to the expansion chamber walls. As it cascades downward, it will tend to clear the walls of any loosely attached particles. By contrast, the product flow is reversed in the tangential spray method. As the substrate temperature approaches the congealing temperature of the coating material, becoming tacky, particles may begin to adhere to the walls of the product container and lower segment of the expansion chamber. In laboratory experiments, this has led to sudden seizure of the substrate bed. To counter this problem, product temperature is maintained somewhat lower than that achievable using the top-spray technique.

The reverse product flow (outward to inward) may also influence the minimum particle size of the material to be coated. Small particles are more likely to adhere to machine surfaces as a result of static electricity, and they may remain uncoated. Additionally, the rapid rolling action may result in a tendency toward agglomeration or balling. For these reasons, tangential spray molten coating of particles smaller than 0.5 mm may be problematic.

Scale-up of molten coating using the tangential spray method depends on a few factors. The substrate should be robust, as the tangential spray process is probably the most mechanically stressful of all fluidized-bed processes. Of concern is that this also applies to the coating material itself. Some coatings become more brittle as the layer gets thicker or as the product is cooled prior to discharge. The severity of a coating cracking problem may well vary with machine size.

Further, a scale-up factor for disk speed must also be considered. Excessive radial velocity may worsen an attrition problem. Inadequate rotation could result in inefficient mixing, leading to poor coating distribution uniformity. One method for scaling up disk speed is to keep radial velocity constant regardless of the diameter of the disk. The larger-diameter disk will rotate fewer times per minute to achieve the velocity at its perimeter at the same value as that used in smaller equipment. A second method considers not only the speed of the disk (and/or particles) but also the radius of curvature. This term *radial acceleration* may more closely keep the forces to which particles are exposed as a constant in scale-up. Radial acceleration is expressed as follows:

$$a_n = \frac{V^2}{R}$$

where a_n is the normal acceleration (at constant velocity), V the velocity at the perimeter of the disk, and R the radius of the disk. As a result, radial velocity will increase somewhat as disk diameter widens.

V. PROCESS VARIABLES

Fluidized-bed coating, spraying molten materials, involves achieving a proper balance of process parameters that allow a rather simplified phenomenon to occur. Particles that are suspended and separated from each other by fluidization air enter a zone of finely atomized coating liquid. Coating occurs as the liquid droplets, which are substantially smaller in size than the substrate, impact the particles, spread, and solidify. This spray coating process continues, with layers building until the desired coating level has been achieved. The completion of spraying is followed by a final product stabilization or cooling step. Some general observations on the specific influences of process parameters follow.

A. Product Bed Temperature

The temperature within the coating zone is critical to the process. A product temperature that is too low may lead to overdry conditions, causing poor spreading of coating over the core material or, in the extreme, failure of the coating to adhere to the surface. The surface properties can easily be observed microscopically or by dissolution. Figures 7 to 9 show a progression from a poor to a good coating. Fine granular citric acid was encapsulated with partially hydrogenated soybean oil (Fig. 7). The coating was applied at a product temperature that was too cool for effective spreading. As the droplets were formed, they entered the cool fluidizing air, dropping quickly in temperature, and increasing in viscosity. When contacting the substrate surface, they retained their shape, resulting in a rough, porous texture, which is apparent when looking at the coated particles as 1000× magnification. Consequently, the release rate, as shown in Table 4, is faster than if the coating were applied at a product temperature that is closer to the congealing temperature of the coating material. Salt, which is cubic in shape, was also encapsulated with partially hydrogenated soybean oil (Fig. 8). The surface of the particle appears smoother than the citric acid but still exhibits a certain degree of unevenness. Potassium chloride, again with a hydrogenated soybean oil coating (Fig. 9), exhibits a smooth, uniform coating. Operating at a higher product temperature allows the coating material to spread on the substrate surface before congealing. In fact, the surface remains soft during spraying, and probably partially remelts as molten droplets are applied and the coating builds into a smooth, nearly continuous layer. Processing with a product temperature that is too high leads to overwet conditions and is typically manifested by excessive particle agglomeration or blinding of the product bowl screen (in top-spray or Wurster processing) or the outlet filter bags.

It is not practical to measure temperature within the coating zone; therefore, the process is controlled by regulating the fluidization air temperature to achieve the desired product temperature. Also, spray liquid and atomizing air temperatures may be adjusted in response to the observed spreading quality. Typically, inlet air temperatures for processing are 10 to 15°C less than the melting point of the coating, and atomizing air and spray liquid temperatures exceed the melting point by 40 to 60°C.

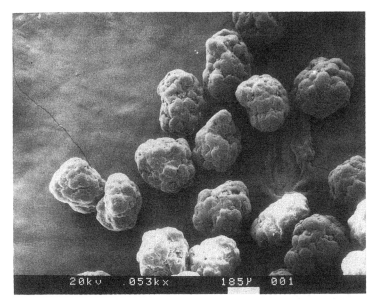

(a)

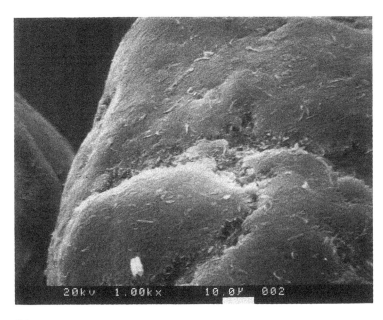

(b)

Fig. 7 (a) Scanning electron micrograph of wax-coated citric acid granule; particle at 53 × magnification. (b) Close-up SEM of surface morphology of wax-coated citric acid granule; particle at 1000 × magnification.

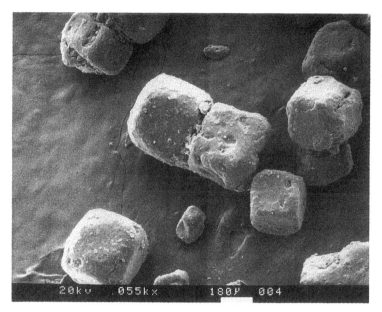

(a)

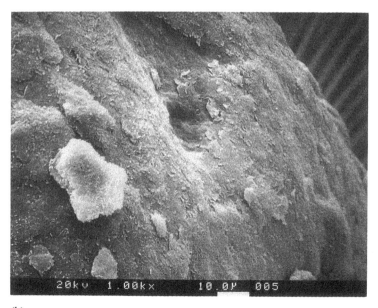

(b)

Fig. 8 (a) SEM of wax-coated salt granule; particle at 55 × magnification. (b) Close-up SEM of surface morphology of wax-coated salt granule; particle at 1000 × magnification.

126

(a)

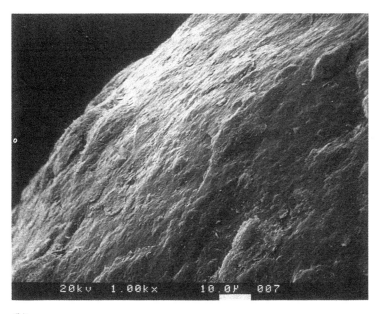

(b)

Fig. 9   (a) SEM of wax-coated potassium chloride granule; particle at 45 × magnification. (b) Close-up SEM of surface morphology of wax-coated potassium chloride granule; particle size at 1000 × magnification.

B. Droplet Size

The coating zone is small relative to the total area of particle activity within the fluidized bed and should be considered as a separate environment. For coating to occur, droplets must remain liquid within this zone. Droplet size and uniformity are critical. Size is influenced primarily by atomization air volume, which is measured indirectly by atomization air pressure. Small substrates require smaller droplet sizes, hence higher atomization air pressure, to minimize agglomeration [11]. Typically, atomization air pressure of 4 to 6 bar (1 bar = 14.7 psi) is used for small substrates (<200 μm), while 3 to 4.5 bar is typical for fine granular or granular materials (>200 but <2000 μm).

Two other factors that influence droplet size are liquid viscosity and spray rate. For small droplets, the viscosity of the molten material should be low. The spray rate is slow in comparison to applying a coating solution or dispersion; in fact, it is usually only 20 to 50% as fast. Of benefit is that the low liquid feed rate is well within the nozzle's ability to atomize to a very small droplet size. Additionally, since nothing is being evaporated, the solids application rate is high. The atomization air must be heated to a temperature that is the same as the temperature of the molten coating. To maintain this high liquid temperature during application, it is desirable for the liquid feed to be encircled completely by the atomization air.

Droplet size uniformity of the spray liquid is largely dependent on a properly functioning spray nozzle leading to a good spray pattern. Defective nozzle O-rings, partially blocked nozzle ports, and material buildup on the nozzle itself result in a poor spray pattern that affects spreading, agglomeration, and overall coating quality.

Nozzle port size selection is dependent on the type of coating, coating viscosity, and anticipated spray rate. At high atomization air pressures (3.0 bar and above) and the relatively low liquid spray rates, droplet size is not influenced by port size when using low-viscosity materials. However, larger port sizes may be employed when attempting to spray a somewhat thicker coating substance.

Spray rate is also a critical parameter, influencing coating quality (surface properties) as well as the degree of agglomeration. In general, slower spray rates result in less agglomeration and better coating distribution uniformity [11]. Processing time is not strongly affected because only solids are being applied and the solids addition rate remains high or on a par with coating processes using organic solvents or water. However, it is very difficult to keep the coating material molten and deliver it at rates lower than approximately 30 g/min. Fortunately, this rate is generally exceeded when using equipment for producing batches of approximately 5 kg and larger. In general, to assure adequate distribution of the coating substance, spraying time should be at least 20 min.

Upon completion of spraying, fluidization is reduced in vigor, and the product is cooled. It is desirable to decrease the product temperature, usually to

35 to 40°C (depending on the coating material and the initial inlet processing temperature). Generally, cooling time should be sufficiently short to avoid attrition of the coated product. However, if the fluidized bed unit uses outside air directly and the outside air temperature is less than − 10°C, cooling may be too rapid, resulting in cracks and fissures in the coating. In an extreme case such as this, the inlet air temperature should be adjusted to retain a slight amount of heat in the fluidization air (approximately 5 to 10°C).

C. Substrate Considerations

Particle size for substrates for molten coating can range from less than 100 μm to several millimeters. Small particles (up to 150 μm) may not be encapsulated as single particles but may agglomerate initially and then the aggregates are coated. Larger particles are typically encapsulated as single entities. Growth of the larger particles is dependent on coating quantity, but under ideal processing conditions, agglomeration can more easily be avoided. Table 2 shows the particle sizes of sodium chloride, citric acid, and potassium chloride. Each substrate was then coated with molten partially hydrogenated soybean oil in the following amounts: sodium chloride, 15% coating; citric acid, 50% coating; and potassium chloride, 30% coating (Table 3). Some fines may be generated during the coating process, as can be noted with the potassium chloride.

D. Capsule Release Rates

Assuming ideal processing conditions, release rates are influenced by initial particle size and shape, and obviously, coating quantity. Release rates of sodium chloride with a 15% coating, citric acid with a 50% coating, and potassium chloride with a 30% coating are presented in Table 4. The particle size of the initial substrate (Table 2) and the coated product (Table 3) are also presented. Particle shape varies considerably for these ingredients from irregularly shaped

Table 2   Particle Size: Uncoated

	Sodium chloride	Citric acid	Potassium chloride
Particle size: percent retained on			
U.S.S. 30 mesh (590 μm)	0.1	1.6	0.0
U.S.S. 40 mesh (420 μm)	34.9	39.5	34.7
U.S.S. 50 mesh (297 μm)	49.8	36.0	63.2
U.S.S. 60 mesh (250 μm)	9.5	11.8	1.9
U.S.S. 70 mesh (210 μm)	2.7	5.3	0.2
U.S.S. 80 mesh (177 μm)	0.2	4.1	0.0
Through U.S.S. 80 mesh	2.8	1.7	0.0

Table 3 Particle Size: Coated

	Sodium chloride	Citric acid	Potassium chloride
Coating level (quantity in finished product) (%)	15	50	30
Particle size: percent retained on			
U.S.S. 20 mesh (840 μm)	0.0	0.1	0.3
U.S.S. 30 mesh (590 μm)	11.4	9.4	4.9
U.S.S. 40 mesh (420 μm)	44.7	40.6	82.6
U.S.S. 50 mesh (297 μm)	41.3	46.3	8.8
U.S.S. 60 mesh (250 μm)	—	2.0	0.5
U.S.S. 70 mesh (210 μm)	1.5	0.6	0.7
Through U.S.S. 70 mesh	1.1	1.0	2.2

citric acid to almost spherical potassium chloride. The size of each ingredient is similar, with 75 to 95% of each ingredient between a 30- and 50-mesh particle size. The capsule release rate of potassium chloride is considerably slower than that for citric acid. These differences can be attributed to two factors. First, the irregular shape of the citric acid crystals results in a higher surface area and hence a thinner coating at an equivalent weight of substrate (compared to potassium chloride). An additional factor is the quality of the applied cooling. As stated previously, the smoother potassium chloride surface has lower porosity, leading to a slower release rate. Release rates were determined by placing a known quantity of encapsulate in a solvent for the inner substrate, agitating the material under controlled conditions, for a specific period, filtering the remaining cap-

Table 4 Capsule Release Rates of Lipid-Encapsulated Products

	Sodium chloride	Citric acid	Potassium chloride
Assay (%)	84.1	49.1	71.1
Capsule release rate: percent released in			
15 min	1.1	5.1	1.0
30 min	1.5	8.6	1.5
1 h	2.3	7.8	1.6
2 h	4.0	13.7	2.2
4 h	7.0	31.0	2.1
8 h	13.8	65.3	5.0

sules, and measuring the amount of substrate that has been dissolved by the solvent. The solvent used for the testing was a 1:1 mixture of propylene glycol and water.

VI. LIPID PROPERTIES

Characteristics of fats that are important in their identification are the iodine value, the acid value, and the saponification value. The iodine value is the number of grams of iodine that will react with 100 g of fat. The acid value is the number of milligrams of potassium hydroxide required to neutralize the free acids in 1 g of sample, while the saponification value is the number of milligrams of potassium hydroxide required to neutralize the free acids and to saponify the esters in 1 g of sample.

The major part of fat is the triglyceride. The triglyceride typically represents over 95% of the weight of most food fats. The minor components are mono- and diglycerides, free fatty acids, phosphatides, sterols, cerebrosides, fat-soluble vitamins, and other substances. The triglyceride is composed of glycerol and three fatty acids. Both the physical and chemical characteristics of fats are greatly influenced by the kinds and proportions of the component fatty acids and the way in which these are positioned on the glyceryl radical [12].

Process for Coating Drug-Loaded Spheres Using Molten Materials
in Top-Spray, Wurster Bottom-Spray, and Rotor Tangential-Spray
Fluidized-Bed Equipment

Examples of hot-melt coating using the various fluidized-bed processing techniques follow. In all experiments, a Glatt model GPCG 5 (Glatt Air Techniques, Inc., Ramsey, New Jersey) was utilized. The top-spray insert was 22 L in volume. A 9-in. Wurster insert was used, as was a 485-mm-diameter rotor insert. Chlorpheniramine maleate drug-loaded beads (solution layered in a Glatt model GPCG 60 using a 32-in. Wurster, 200 kg batch size) were coated using partially hydrogenated cottonseed oil (07 Stearine, Van Den Bergh Foods).

Dissolution for the chlopheniramine maleate–coated pellets was conducted by Eurand America, Inc. Vandalia, Ohio. The capsule release rates for sodium chloride, potassium chloride, and citric acid are from Van Den Bergh Foods, Joliet, Illinois. The method is described following the experimental data (Table 5).

The manufacturing procedure follows:

1. Assemble the Glatt model GPCG 5 using a 25-μm outlet air filter and the desired processing insert.
2. Insulate the nozzle using an appropriate nonheat conductive material to avoid remelting of coated product.

Table 5 Experimental Data

Insert	Top spray	Wurster	Rotor
Batch size (core) (kg)	8.0	8.0	8.0
Coating quantity (kg)	0.8	0.8	0.8
Atomization air pressure (bar)	3.0	3.0	3.0
Atomization air temperature (°C)	98–105	96–113	97–112
Spray rate (g/min)	38	36	37
Spray liquid temperature (°C)	104–111	106–113	101–112
Product temperature during			
spraying (°C)	46	48	50
Yield (kg)	8.65	8.77	8.76
Dissolution (%)			
6% coating			
15 min	9.3	8.8	5.0
30 min	21.4	21.2	13.8
1 h	47.2	42.8	33.2
2 h	76.2	73.4	62.4
4 h	94.5	93.2	85.0
6 h	99.1	98.7	91.1
8 h	100.3	102.0	93.2
10% coating			
15 min	2.5	1.2	1.3
30 min	6.2	3.6	4.8
1 h	19.0	9.3	20.2
2 h	48.2	24.4	56.4
4 h	82.5	49.9	89.4
6 h	93.2	65.1	97.6
8 h	98.1	73.8	100.0

3. Compress the machine tower and begin drawing fluidization air through the empty insert at 55°C inlet air temperature and the desired fluidization air volume.

4. Connect an atomization air heater in-line to the nozzle. Engage the heater with the temperature set at 110°C. Activate the pump to bring the atomization air pressure to 3.0 bar (to heat the air at the desired flow rate).

5. Melt at least double the required coating material to a temperature of 110°C. Maintain this temperature throughout spraying.

6. Using narrow-diameter peristaltic pump tubing (4 mm ID), recirculate the molten liquid to warm the tubing.

7. When the machine tower is warm (product temperature indicator shows at least 50°C), stop the machine and quickly load the substrate.

8. Begin fluidizing the substrate gently while continuing to recirculate the molten coating material.
9. When the substrate is within 2°C of the product temperature expected during spraying, increase the fluidization air volume to the desired rate, connect the liquid line to the nozzle, and begin spraying.
10. Adjust the inlet temperature up or down to achieve a viscous fluidized bed. The quality of the coating is best when congealing is slow.
11. When all of the coating has been applied, disconnect the liquid line, allow the remaining molten liquid to recirculate, decrease the fluidization air volume to allow a gentle tumble, and reduce both atomizing and fluidizing air temperatures to a minimum. Continue cooling until the product temperature falls below 40° C (not more than 10 min to avoid breakage of the coating).
12. Discharge and reconcile; submit for dissolution.

Dissolution conditions
1. 500 mL of deionized water containing 0.01% Tween 20
2. USP I (basket), 100 rpm, 37°C
3. 2-g sample, time points 15 min, 30 min, 60 min, 120 min, 240 min, 360 min, 480 min.
4. All collected fluids analyzed by ultraviolet spectrophotometry at 260.0 nm using a 1.0-mm quartz flow-through cell.

Also, assay for chlorpheniramine maleate by high-performance liquid chromatography was performed on all samples.

It was noted during processing that occlusion of the outlet air filter did not permit operation at a high product temperature when using the top-spray insert. As a result, the release rate of the top-spray-coated samples was faster than expected. Blinding of the outlet air filter was the limiting factor in achieving a high product temperature for the top-spray process, and bed stalling was the limit for the Wurster and rotor processes. As such, an additional top-spray experiment was conducted. In this case the outlet air filter was removed and spraying conducted at a substantially higher product temperature. Data for this experiment are given in Table 6 (with the previous top-spray results in parentheses).

Outlet air filters tend to become occluded more quickly when applying molten coatings. A lower processing temperature may minimize this problem, but the drug release is consequently faster, as seen in the first top-spray batch. When processing with coarse substrates (larger than 500 μm), it is recommended that rather than using an outlet filter in the machine, any fines be collected downstream of the processor.

The uncoated core pellet is illustrated in Fig. 10. Using each fluidized-bed technique, keeping as many of the processing conditions similar was a goal.

Table 6 Experimental Data

Insert		Top spray
Batch size (core) (kg)		8.0
Coating quantity (kg)		0.8
Atomization air pressure (bar)		3.0
Atomization air temperature (°C)		113–120
Spray rate (g/min)		40
Spray liquid temperature (°C)		113–120
Product temperature during		
spraying (°C)		53
Yield (kg)		8.64
Dissolution (%)		
6% coating		
15 min	2.8	(9.3)
30 min	7.2	(21.4)
1 h	20.9	(47.2)
2 h	51.0	(76.2)
4 h	82.7	(94.5)
6 h	92.2	(99.1)
8 h	96.3	(100.3)
10% coating		
15 min	1.1	(2.5)
30 min	1.6	(6.2)
1 h	2.8	(19.0)
2 h	6.1	(48.2)
4 h	11.3	(82.5)
6 h	14.2	(93.2)
8 h	16.2	(98.1)

However, it was not possible to keep product temperature during spraying the same, and the porous coatings are seen using the Wurster and rotor techniques in comparison to the top-spray product coated at 53°C product temperature (Figs. 11 to 16). Drug dissolution provides further evidence (Tables 5 and 6).

Summary

Lipids are being used in a variety of applications, including pharmaceutical products [13–17]. The fluidized-bed process is ideal for coating a broad range of particles using molten materials. Although each type of process (top spray, Wurster bottom spray, and rotor tangential spray) can be used, there are some limitations, due largely to core material properties such as particle size and density. Additionally, the quality of the coating (minimal porosity) is dependent

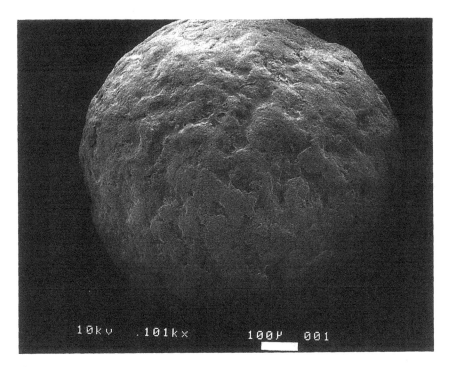

Fig. 10 Chlorpheniramine maleate drug–loaded pellets, 35 mg/g potency.

on product temperature during spraying. The unbaffled conventional top-spray-product container allows operation at the highest temperature, yielding the best quality. This is reflected in the dissolution results listed previously.

Additionally, it is very difficult to apply molten materials in small equipment (batch sizes less than 2 kg). The reason is that release properties are dependent on coating distribution uniformity. The coats should be applied for at least 15 to 20 min. If product release properties are achieved using a 10% coating, only 200 g would be applied to a 2.0-kg batch. Spraying for 20 min would require a spray rate of only 10g/min, and it is quite difficult to keep a material molten and at the desired temperature at such a low delivery rate. For this reason, it is suggested beginning with machinery capable of processing at least 5.0 kg per batch.

ACKNOWLEDGMENT

Special thanks to Mr. Dan Hensley of Eurand America for his help on dissolution.

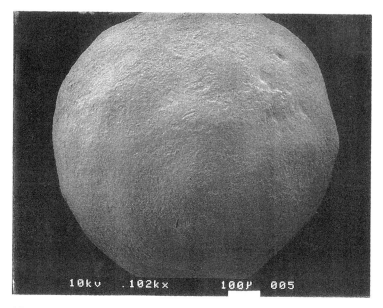

(a)

(b)

Fig. 11 Drug-loaded pellet coated to 6% using 07 Stearine, top-spray method; product temperature during spraying 53°C. (a) 102 × magnification; (b) 500 × magnification.

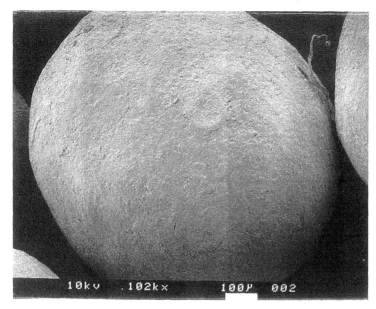

(a)

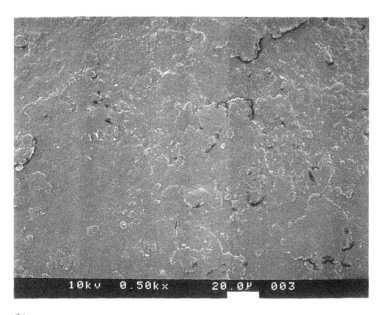

(b)

Fig. 12 Drug-loaded pellet coated to 10% using 07 Stearine, top-spray method; product temperature during spraying 53°C. (a) 102 × magnification; (b) 500 × magnification.

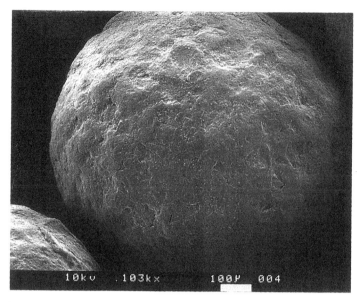

(a)

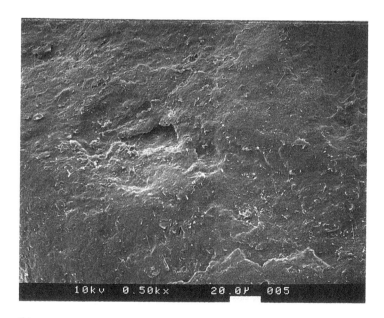

(b)

Fig. 13 Drug-loaded pellet coated to 6% using 07 Stearine, Wurster bottom-spray method; product temperature during spraying 48°C. (a) 103 × magnification; (b) 500 × magnification.

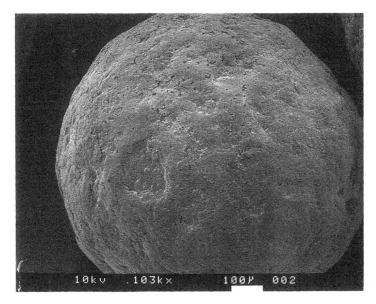

(a)

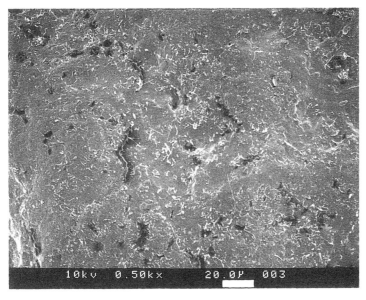

(b)

Fig. 14 Drug-loaded pellet coated to 10% using 07 Stearine, Wurster bottom-spray method; product temperature during spraying 48°C. (a) 103 × magnification; (b) 500 × magnification.

139

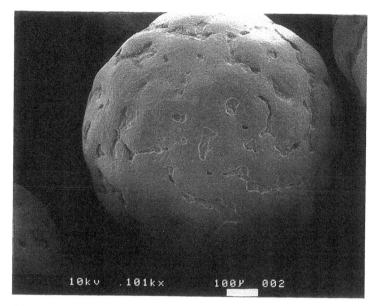

(a)

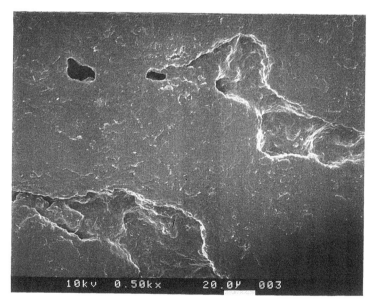

(b)

Fig. 15 Drug-loaded pellet coated to 6% using 07 Stearine, rotary fluidized-bed method; product temperature during spraying 50°C. (a) 101 × magnification; (b) 500 × magnification.

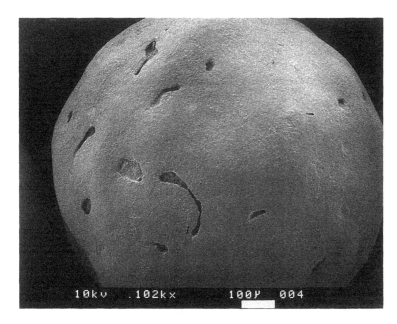

(a)

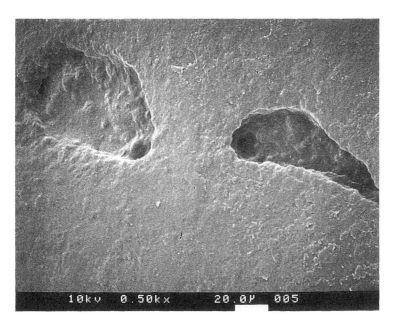

(b)

Fig. 16 Drug-loaded pellet coated to 10% using 07 Stearine, rotary fluidized-bed method; product temperature during spraying 50°C. (a) 102 × magnification; (b) 500 × magnification.

REFERENCES

1. U. Banaker and W. Speake, Fats and waxes in pharmaceuticals, *Manuf. Chem.*, September, p. 43 (1990).
2. C. L. Provost, H. Herbots, and R. Kinget, *Drug Dev. Ind. Pharm. 15*:25 (1989).
3. D. Geldart, *Powder Technol. 7*:285 (1973).
4. D. Geldart, *Powder Technol. 6*:201 (1972).
5. D. E. Wurster, U. S. Patent 2,648,609 (1953).
6. D. M. Jones, Solution and suspension layering, in *Pharmaceutical Pelletization Technology* (I. Ghebre-Sellassie, ed.), Marcel Dekker, New York, 1989, pp. 156–159.
7. A. M. Mehta et al., evaluation of fluid bed processes for enteric coating systems, *Pharm. Technol. 10*(4):46 (1986).
8. A. M. Mehta and D. M. Jones, Coated pellets under the microscope, *Pharm. Technol. 9*(6):52 (1985).
9. S. C. Porter, and L. F. D'Andrea, The effect of choice on process on drug release from non-pareils film coated with ethylcellulose, *12th International Symposium on Controlled Release of Bioactive Materials*, Geneva, July 1985.
10. F. W. Goodhart, and S. Jan, Dry powder layering, in *Pharmaceutical Pelletization Technology* (I. Ghebre-Sellassie, ed.), Marcel Dekker, New York, 1989, pp. 165–185.
11. M. J. Jozwiakowski, D. Jones, and R. M. Franz, Characterization of a hot-melt fluid bed coating process for fine granules, *Pharm. Res. 7*(11):1119 (1990).
12. Technical Committee of ISEO, *1974 Institute of Shortening and Edible Oils*, 4th ed., ISEO, 1974.
13. H. Hall and J. Wallace, Column film coating and processing, in *Controlled Release Systems: Fabrication Technology* (D. Hsieh, ed.), CRC Press, Boca Raton, Fla., 1988, pp. 63–67.
14. K. W. Leong, Synthetic biodegradable polymer drug delivery systems, in *Polymers for Controlled Drug Delivery* (P. Tarcha, ed.), CRC Press, Boca Raton, Fla., 1991, pp. 136,138.
15. H. Hall, Coating of granular bioproducts, in *Granulation Technology for Bioproducts* (K. L. Kadam, ed.), CRC Press, Boca Raton, Fla., 1991, pp. 283, 287–288.
16. S. Motycka, and J. G. Nairn, Influence of wax coatings on release rate of anions from ion-exchange resins, *J. Pharm. Sci. 67*(4) (1978).
17. I. Ghebre-Sellassie, Mechanism of pellet formation and growth, in *Pharmaceutical Pelletization Technology* (I. Ghebre-Sellassie, ed.), Marcel Dekker, New York, 1989, p. 129.

7

Suspensions and Dispersible Dosage Forms of Multiparticulates

Roland Bodmeier and Ornlaksana Paeratakul

College of Pharmacy
The University of Texas at Austin
Austin, Texas

I. INTRODUCTION

Considerable research effort has been spent on oral sustained- or controlled-release drug delivery systems, with the majority of these systems being solid dosage forms [1–3]. In recent years, multiparticulate dosage forms such as matrix or coated pellets or microparticles (microspheres or microcapsules) have gained in popularity for a variety of reasons [4–7]. When compared to single-unit preparations, such as tablets or capsules, multiparticulates distribute more uniformly in the gastrointestinal tract, thus resulting in a more uniform drug absorption and hence a reduced patient-to-patient variability. Multiparticulates minimize the risk of local irritation and possible intestinal retention of nondigestible polymeric materials upon chronic dosing. In addition, formulators have greater flexibility in achieving desired drug release patterns: for example, simply by mixing pellets with varying coating thicknesses and hence different release profiles.

Considering the final dosage form, multiparticulates could be filled into hard gelatin capsules or be compressed into tablets. The formulation of multiparticulates into these conventional oral dosage forms, however, may encounter several problems. The risk of tampering and the awareness of consumers of recent poisoning incidences will eliminate or significantly reduce the use of hard gelatin capsules in over-the-counter products. If multiparticulates are to be compressed successfully, good flow properties (a potential problem with small microparticles) are essential and the polymeric coating must be capable of resisting the severe mechanical stress during compression. Poor flow properties can cause problems in content uniformity. When compressed under high pressure, microcapsules may rupture and lose their protective or sustained-release action [8]. The poor compressibility often requires the addition of large amounts of easily compressible excipients or so-called cushioning agents. This dilution could result in a too-low drug content in the final dosage form. Fassihi studied the consolidation behavior of polymeric particles and reported both plastic deformation and particle fusion to be operative during compression [9]. The possible fusion of polymeric microparticles during compression could result in a nondisintegrating matrix with the loss of the character of a multiparticulate dosage form. Jalsenjak et al. reported that the tabletting of microcapsules resulted in a nondisintegrating matrix and in a reduction of drug release [10].

An alternative approach for the oral administration of multiparticulates is to suspend them into a liquid vehicle to form a suspension dosage form or into a dry powder system, which is to be reconstituted with water by the patient just prior to administration. These types of dosage forms overcome the above-mentioned problems and are often preferred among certain patient groups, such as infants, children, and the elderly, because of the ease in swallowing and the flexibility in the administration of dosages. In this chapter we discuss a variety of dosage forms consisting of suspended/dispersible multiparticulates or microparticles and various methods or approaches used, and the problems encountered in formulating these delivery systems.

II. LIQUID SUSPENSIONS OF MULTIPARTICULATES

The major challenges that have to be faced in the formulation of sustained-release multiparticulates into liquid suspensions are the diffusion or release of drug into the suspending vehicle during storage, and interactions between the suspending vehicle and the multiparticulates, resulting potentially in unwanted changes in the original properties of the multiparticulates during storage. Ideally, an inert suspending medium has to be found into which the drug does not diffuse during storage and which does not change important properties of the multiparticulates during storage. In particular, drug release from the microparticulates should remain unchanged over time.

Various approaches have been taken to minimize or eliminate drug diffusion into the external suspending vehicle during storage. The loss of drug into the storage vehicle will be determined by the solubility of the drug in this vehicle; a vehicle with minimal drug solubility could be selected. Multiparticulates containing water-insoluble drugs could be formulated into aqueous suspensions. Alternatively, edible oils or nonaqueous media could be selected as suspending vehicles for highly water-soluble drugs. Water-soluble drugs carrying ionizable functional groups have been bound to ion-exchange resins. These drug–resin complexes can be suspended into deionized water without drug leaching/loss from the complex over time. Various formulation approaches that have been used and reported in the literature are reviewed below.

Microspheres containing water-insoluble, nonsteroidal anti-inflammatory agents were suspended into aqueous vehicles without significant drug leaching during storage [11]. Polymeric microspheres containing indomethacin (ethyl cellulose/poly(ϵ-caprolactone), 7:3 w/w; 27% w/w indomethacin) were prepared by a solvent evaporation method and suspended in deionized water and stored at room temperature. Due to the low water solubility of the drug, less than 0.3% of the total dose was released or lost into the suspending medium after a 2-week period.

Slow-release theophylline–cellulose acetate butyrate microspheres were prepared by the same technique [12]. The drug-containing microspheres were formulated into various aqueous suspensions, which were then cycled between 30 and 40°C every 4 h to investigate the effect of various ingredients on stabilizing the suspension and inhibiting crystal growth. The stability of the microsphere suspension and the drug-crystal formation in the external phase were found to be dependent on the type of additives incorporated. Tragacanth was found to be most effective in stabilizing the suspension, while Veegum, methylcellulose, and silicone dioxide were ineffective.

The effect of storage temperature on possible changes in the release of ibuprofen from aqueous suspensions of ibuprofen–wax microspheres was investigated by the same authors. Temperature and type of wax affected the dissolution stability, with storage at 37°C resulting in an increase and storage at 45°C in a decrease in drug release [13]. Ibuprofen was chemically stable under these conditions.

Sustained-release ibuprofen–wax microspheres were prepared by a melt dispersion technique whereby the drug–wax melt was emulsified into a heated aqueous phase, followed by cooling of this oil–water emulsion to congeal the wax phase and to form solid, suspended wax particles [14]. The microparticles could either be separated by filtration, followed by drying to form a powder, or they could be kept in suspension. If left in suspension, drug crystals, which formed in the external aqueous phase during cooling as a result of a decrease in solubility, were separated by filtration or centrifugation.

A controlled-release suspension of ibuprofen–Eudragit RSPM microspheres was developed by Kawashima et al. [15]. The drug-containing microspheres were prepared by an emulsion–solvent diffusion method in which an ethanolic solution of drug and polymer was emulsified into an aqueous phase containing a sucrose ester surfactant [16]. The ibuprofen microspheres were then formulated into aqueous suspensions for oral use. Good dispersibility, as measured by the sedimentation volume, of the microparticles over a 6-month period was achieved by using a combination of a suspending agent, sodium carboxymethylcellulose (NaCMC), and a polyol, D-sorbitol. The viscosity of the dispersion was only 60 cP. An acidic pH was a prerequisite for the formation of a stable suspension. Caking was observed in media in which the pH was adjusted to more than 3. The low pH enhanced the adsorption of NaCMC onto and the electrostatic interactions with the positively charged microspheres as measured by zeta potential and NaCMC adsorption studies. The quaternary ammonium groups present in the acrylic polymer were responsible for the positive charge of the microspheres. The addition of D-sorbitol had an additional stabilizing effect. The amount of NaCMC adsorbed on the microsphere surface increased with increasing amounts of sorbitol due to the dehydration produced by the polyol. The formulated suspension possessed good fluidity, shear-thinning properties, and particle redispersibility during 6 months of storage. In addition to stabilizing the microparticle suspension, the acidity of the aqueous phase also prevented diffusion or loss of the acidic drug, ibuprofen (pK_a5.2), into the suspending vehicle. The storage in the acidic medium had no effect on the drug release.

Drug-loaded polymeric microparticles trademarked Pharmazome were developed by Elan Corporation [17]. The polymeric microparticles possessed a narrow particle size distribution, with a median diameter of 55 μm. A variety of drugs, such as theophylline, acetaminophen, and carbamazepine, have been incorporated in the micromatrix systems with loadings up to 55%. These controlled-release drug-containing microparticles were particularly useful for pediatric therapy and could be administered in a number of different dosage forms, such as liquid suspensions [18,19], edible foam dispersions, or sprinkle capsule [20]. The ready-made liquid suspension, suitable for dosing by spoon or other standard volumetric methods, provided the progressive liberation of the active ingredients as well as enhanced palatability due to the associated taste-masking characteristics of the micromatrix. An accurate dispensing of theophylline was achieved with an edible foam dispersion by using a metered dosing head of an aerosol system. This precise and effective dosing could be particularly useful in the case where a potent drug or a drug with a narrow therapeutic index has to be formulated. The sprinkle capsule form provided a wide flexibility in dosage titrations and could be particularly acceptable in patients 6 to 12 years of age. The controlled-release products showed good bioavailability compared with reference conventional dosage forms.

A prerequisite for the storage of multiparticulates in aqueous media is the chemical stability of the drug and polymer. With regard to the polymer selection, many pharmaceutically acceptable polymers (e.g., acrylic esters and ethyl cellulose) have been formulated into aqueous colloidal polymer dispersions for coating purposes; they are hydrolytically stable in the presence to water and are therefore good candidates as drug carriers for the formulation of aqueous suspensions of polymeric microparticles. With microparticles prepared from natural or semisynthetic waxes, hydrolysis during storage in aqueous vehicles or possible polymorphic transformations could be a concern.

Unless the drug is bound to an ion-exchange resin, multiparticulates of water-soluble drugs cannot be suspended into water. Microspheres containing pseudoephedrine HC1, a highly water-soluble drug, were prepared by modified solvent-evaporation methods and formulated into oral sustained-release suspension dosage forms [21]. Pseudoephedrine HC1 was entrapped within the polymeric microspheres by an oil-in-water (dispersion or cosolvent method) or an water-in-oil water (multiple emulsion method) emulsion–solvent evaporation method. The drug was dissolved, dispersed, or emulsified as a concentrated aqueous solution in a water-immiscible organic polymer solution, followed by emulsification of this phase into an external aqueous phase and solvent evaporation to form the microspheres. The microspheres were then formulated into oral suspension dosage forms using concentrated sucrose or sorbitol solutions, glycerin, propylene glycol, or Neobee M-5 oil (a mixture of triglycerides with C_8 to C_{10} fatty acids, accepted vehicle for ingestion) as suspending vehicles. The amount of drug leached into these media was followed over a 6 month period. Microspheres suspended in Neobee M-5 oil retained their original drug content after storage due to the minimal solubility of pseudoephedrine HC1 in the oil (<1 mg/mL). The drug solubilities in glycerin, propylene glycol, 85% sucrose, and 85% sorbitol solution were 218, 317, 416, and 384 mg/mL, respectively. The amount of drug lost into these suspending vehicles leveled off after 2 to 3 weeks and did not change significantly during further storage. Although the drug was highly soluble in these suspending media, more than 75% of the original drug loading was still present within the microspheres for all vehicles after 6 months. These vehicles had no affinity for the polymer and did not diffuse across the polymer. Drug loss into the suspending vehicle could therefore be eliminated by minimizing the solubility of the drug in the suspending vehicle or by choosing a vehicle that does not reach the drug during storage. The storage of the microspheres in different vehicles did not change the drug release profiles in 0.1 M pH 7.4 phosphate buffer, indicating minimal interactions of the polymer with the suspending vehicles. As opposed to aqueous-based media, oily suspending vehicles or nonaqueous solvents could potentially act as plasticizers for the polymer and could cause an unwanted change in its permeability characteristics.

Controlled-release liquid suspensions of dual-coated dosage forms were developed and patented by Benton and Gardner [22]. Drug-containing microparticles or matrix beads with a particle size of less than 1400 μm were coated with two layers prior to suspending them in a sugar-based acidic vehicle. The first coating consisted of an ingestible hydrophobic fat having a melting point of 101°F or lower. Low-melting fats were selected such that the materials would become softened or liquefied at body temperature, thus rendering the coatings permeable after ingestion. Various fatty materials and glycerides, such as theobroma, cocoa butter, or partially hydrogenated vegetable oils and their blends could be applied as the first coat. The second layer consisted of zein or the enteric polymer cellulose acetate phthalate. The overcoat materials were insoluble in the suspending medium but became readily permeable in the gastrointestinal tract. In addition, application of an overcoat helped to prevent agglomeration or clumping of the coated particles during manufacturing. High-fructose corn syrup was used as the liquid carrier, due to its limited water content (approximately 30%), natural acidity, high viscosity, and preservative activity. The formulated suspension displayed a shelf life up to at least 45 days while retaining their release profiles following intake. This dual coating technique has been used successfully in the formulation of a controlled-release suspension of vitamin C, a highly water-soluble and unstable compound, as well as antihistamines, analgesics, and geriatric drugs. However, long-term studies on the stability of the suspensions have not been included.

Zatz and Woodford patented a controlled-release liquid preparation based on the gel-forming ability of various macromolecular excipients upon contact with gastrointestinal fluids [23]. An aqueous or partially aqueous solution/suspension of water-soluble viscosity-enhancing agents, such as xanthan gum, sodium alginate, and the complex coacervate pairs such as gelatin and carragenan, could form semisolid or rigid gel matrices upon exposure to the stomach environment. After the gel formation, drug release from the preparation could be controlled by diffusion of the drug through the gelatinous matrix. Gas-producing salts such as calcium carbonate and other carbonates could be incorporated in the formulation to produce a "floating" gelatinous matrix, which could potentially prolong retention of the dosage form in the stomach, thus causing a delay in gastric emptying.

Another popular approach to the development of aqueous oral sustained-release suspensions is based on ion-exchange resins. The principle of ion exchange has been used for a long time in analytical and protein chemistry. The concept has been used in the preparation of pharmaceutical controlled-release water-soluble ionic drugs for more than 20 years. It is an attractive method for sustained drug delivery because, in theory, drug release characteristics depend only on the ionic environment and should therefore be less susceptible to environmental conditions, such as enzyme content and pH, at the site of absorption [2].

Resins are water-insoluble materials containing anionic or cationic groups in repeating positions on the resin chain. These resins are comprised of two principal parts: a structural portion consisting of a polymer matrix, usually styrene cross-linked with divinyl benzene, and a functional portion being the ion-active group to which the drug is bound. The functional group may be acidic (sulfonic or carboxylic) or basic (amino). The drugs of opposite charge will bind to the functional groups of the polymers. Upon administration, as the drug–resin complex reaches the gastrointestinal tract, the reverse reaction takes place through ion displacements and the drug is liberated. When in suspension form, the bound drug does not leach into the ion-free suspending medium during storage but is readily released upon ingestion. Typically, the resins have a particle size of approximately 150 μm. However, the particle size can vary from less than 100 μm up to 1 mm and, in addition, different porosity grades are available. The relatively small particle size of the resin materials enables the formulation scientists to formulate the drug–resin complexes into liquid suspension dosage forms with acceptable palatability and swallowability.

The drug–resin complex is prepared by mixing the resin with drug solution either by repeated exposure of the resin to the drug in a chromatographic column or by keeping the resin in contact with the drug solution for extended periods. The drug–resin is then separated and washed to remove unbound drug and dried to form the particles or beads. The release rate from the drug–resin complex can be controlled further by coating the drug–resin beads using a variety of microencapsulation or coating processes.

In the Pennkinetic system developed and patented by the Pennwalt Corporation [24], the drug-containing resin granules are first pretreated with an impregnating agent such as PEG 4000 to impart plasticity and to retard the rate of swelling of the granules in water. Pretreatment of the drug–resin complex was found to be essential for the particles to retain their geometry and coating during dissolution [25]. This impregnation step is followed by the application of a water-insoluble polymer such as ethyl cellulose via an air suspension technique to form an insoluble but permeable coating around the drug–resin complexes. The ethyl cellulose coating thus acts as a rate-controlling barrier for controlling the drug release. The coating thickness is inversely proportional to the drug release rate and thus could be varied to achieve a variety of dissolution profiles. In addition, a mixture or blend of uncoated and coated drug–resin complexes can be made to provide a combined drug release pattern; the coated material provides the sustaining profile and the uncoated material provides the rapidly released loading dose. The overall dissolution rate of the mixture will determine the rate of absorption and bioavailability of the dosage form. This blending technique has been used successfully with a sustained-release dextromethorphan suspension [26].

A sustained-release suspension preparation of codeine and chlorpheniramine was developed by the same inventors [27]. The sustaining coat was designed according to the biological half-life of each active ingredient. Codeine, having a short half-life, was incorporated as an ethyl cellulose–coated resin complex, and the drug with the longer half-life, chlorpheniramine, was incorporated as an uncoated complex. The formulation contained sweetening agents (corn syrup and sugar), suspending agents (xanthan gum and pregelatinized starch), a nonionic surfactant (polysorbate 80), and preservatives (parabens) in a flavored aqueous vehicle. This suspension was nonionic in nature and was found to be reproducible with regard to the dissolution rates and drug absorption in vivo without the potential for dose dumping. As with many other multiparticulate dosage forms, the main advantages of the system are claimed to be the possibility of tailoring the dosage regimens among individuals, especially with pediatric and geriatric patients, a reduced localized GI irritation, and the ability to mask unpleasant odors or tastes of the drugs. In addition, a recent study has shown that the release of pseudoephedrine from the Pennkinetic suspension and its absorption rate were independent of food intake or the contents in the stomach. This was due to the large concentration of ions in the gut relative to that required for ion exchange [28].

Besides the air suspension coating, various microencapsulation techniques have been used to prepare controlled-release drug–resin beads. Motycka and Nairn studied the influence of external wax coatings on the release from drug–resin complex [29]. The type of wax significantly affected the drug release rates, due to the polar character and solubility characteristics of the wax. The same authors employed several encapsulation procedures, such as complex coacervation, temperature change, and nonsolvent addition techniques, to prolong drug release from the resin beads [30].

Drug–resin complexes were microencapsulated with a water-insoluble polymer, cellulose acetate butyrate, by an emulsion solvent-evaporation method [31]. In this method, a dispersion of the complexes in solution of the polymer in acetone was emulsified into liquid paraffin containing the stabilizer magnesium stearate. Microcapsules containing chlorpheniramine, diphenhydramine, or pseudoephedrine were obtained after solvent evaporation and solidification of the polymer droplets. The complex-containing microparticles were formulated into aqueous suspensions with the aid of various suspending agents prior to storage at different temperatures. The accelerated stability studied showed that the microcapsules could be suspended up to 28 weeks without serious adverse effects on the sustained-release properties of the dosage form. The result indicated that the polymer membrane surrounding the core material remained intact throughout the storage period. The drug release from the microcapsules was a function of the type of drug, resin, and the rate-controlling polymer, the size of microcapsule, and the pH of the dissolution medium [32].

Other approaches to decreasing the loss of water-soluble drugs into aqueous suspending vehicles could include (1) increasing the osmotic pressure of the storage vehicle, thus inhibiting the diffusion of water into the multiparticulates during storage (water would be imbibed after dilution in the gastrointestinal tract), and (2) with ionizable drugs, adjusting the pH of the vehicle to a pH of minimal drug solubility. Extreme pH values, however, may not be tolerated by the patient and may cause degradation of polymers, thickening agents, and other ingredients.

General considerations to be addressed during the development of conventional drug suspensions also apply to the development of sustained-release suspensions. These considerations include the selection of proper suspending vehicles, suspending agents, surfactants, flavoring agents, preservatives, and other ingredients to formulate a stable, elegant, and pharmaceutically acceptable dosage form. The physicochemical properties of the internal and external phases, including those of the incorporated additives, are important factors that determine overall characteristics of the resultant suspensions. While the principle or theory of suspensions has been described extensively and can be found in numerous articles and reviews [33–36], this context provides a brief discussion on suspension theory with respect to the formulation approaches applicable for the development of oral sustained-release suspensions, with an emphasis on those containing drug-containing microspheres or microparticles as the dispersed phase.

   In the development of the formulations categorized as dispersed systems, a number of formulation factors must be taken into consideration. These include primarily the properties and nature of the ingredients incorporated as well as their possible interactions. A typical suspension contains a wetting agent, suspending agent, flocculating agent or protective colloid, sweetener, preservative, buffer system, flavoring agent, coloring agent, and formulation aids such as sequestering and antifoaming agents. Various formulation approaches have been used to reduce the sedimentation rate of the suspended particles. Modifications of the particle size and/or the viscosity of external liquid medium have been by far the most popular approaches employed by formulators to improve suspension stability. The particle size of the microparticles can be controlled primarily by the agitation rate with microspheres formed by emulsification processes (e.g., solvent evaporation, melt dispersion, and interfacial polymerization methods), by the atomization conditions with microparticles formed by spray drying or spray congealing, or by the size of the drug core with aqueous and nonaqueous phase separation methods. Changes in the size of the microparticles will, however, affect not only sedimentation behavior but also drug loading and release properties.

   With respect to the dispersion medium, an inverse relationship exists between the rate of particle settling and the viscosity of the external medium. Standard viscosity-enhancing agents could be added to the external aqueous or oil phase to eliminate sedimentation or floating of the microparticles. Agents possessing thixotropic properties are typically useful in the preparation of structured vehicles, in which the suspended particles can be entrapped such that particle sedimentation is prevented or severely retarded. Upon standing, a formation of the structured network results in an external phase of high viscosity. This semirigid structure undergoes breakdown temporarily under high shear stress such as shaking, but can be restored later upon further storage. The suspension viscosity can also be increased by increasing the volume fraction of the microparticle phase. Excessive viscosity may cause difficulties in transferring or redispersing the suspensions. In addition, minimizing differences between densities of the internal and external phases reduces the separation. The densities of the microparticles could be varied by preparing more porous particles or incorporating dense pigments.

Since most microspheres and microparticles are comprised of hydrophobic polymers as carrier materials, the wetting properties of these hydrophobic compounds can potentially affect the redispersibility and suspendability of microparticles. Wettability of the drug particles or of the polymeric materials can be assessed and evaluated by contact angle measurements. A wetting agent is often incorporated in the formulation to improve particle wettability and/or spreadability of the vehicle. Low concentrations of surfactants can be used as the dispersion aids; however, high surfactant concentrations may result in undesirable product features such as unpleasant tastes or excessive foaming.

The addition of protective colloids, such as proteins or gums, into suspensions can result in particle flocculation. Flocculation is defined as the formation of loose particle aggregates, called flocs, which can easily be redispersed under mild agitation. As a result of particle settlement, the flocculated product typically appears as a system containing a loose sediment and a clear supernatant. The technique of controlled flocculation refers to the use of electrolytes to reduce electrical charges or zeta potential between the suspended particles. The reduction of particle charge density, via the addition of a flocculating electrolyte, allows the dispersed particles to approach each other and form flocs. An effective flocculating agent is usually a multivalent electrolyte possessing a charge opposite to that of the particle surface. This approach may be particularly useful in the formulation of a suspension containing microparticles having charged polymer as the carrier material.

III. DISPERSIBLE MULTIPARTICULATES

A dosage form based on controlled-release multiparticulates to be dispersed in water just prior to administration has recently been introduced under the trade name FluiDose (European patent application No. 88902818.9). One advantage claimed for this dosage form is the possibility of administering large amounts of drugs in a convenient way, this being especially suitable for the elderly. Placebo pellets, produced by extrusion and spheronization, were coated in a fluidized bed with three layers of polymeric materials [37]. The first coat consisted of water-permeable insoluble acrylic polymer (Eudragit NE 30D, rate-controlling membrane), the second layer being the acrylic polymer admixed with calcium sulfate dihydrate, and the third coat a layer of water-soluble hydroxypropyl methylcellulose (HPMC). The coated pellets were then mixed with a powder containing the anionic polysaccharide sodium alginate. For reconstitution, it was recommended the dry composition be poured into water followed by stirring for ½ min, leaving it for 1 min, and stirring again immediately before intake. HPMC hydrated and formed a gel layer around the pellets. The hydration of HPMC provided a delay in the release of calcium ions from the pellets. Calcium then reacted with the dissolved sodium alginate, forming a gelled network around the cores and in the medium. The gelled network reduced the tendency of the particles to settle and resulted in more complete emptying of the particles from the glass. In addition, the gelled layer around the particles improved the swallowability and reduced the tendency of the pellets to be retained in the mouth. This dosage form has proven acceptable in patient tests, comparing favorably with capsules and tablets.

IV. DRUG-CONTAINING NANOPARTICLES

Nanoparticles are colloidal particles consisting of polymeric materials in which drugs, enzymes, antigens, or other active components are dissolved, entrapped, encapsulated, and/or adsorbed [38–41]. These nanoparticles are frequently formulated into aqueous colloidal suspensions. A variety of natural and synthetic polymers, both biodegradable and nonbiodegradable, have been used as carrier materials in the preparation of the drug-containing nanoparticles. The preparation techniques include emulsion polymerization, micelle polymerization, desolvation of macromolecules, and emulsion–solvent evaporation methods. The particle size of the resultant polymeric nanoparticles is generally in the nanometer range (10 to 1000 nm) and is dependent primarily on the method of preparation employed. The development of nanoparticles or nanocapsules has been investigated initially and focused for use in parenteral applications and drug targeting. However, in recent years, their potential applications have been extended to other routes of administration, such as oral, topical, or ophthalmic drug delivery. The possibility of delivering drugs in the form of nanosuspensions or in dry (redispersible) form via these routes of administration has been investigated by a number of researchers.

Emulsion polymerization is by far the most widely used technique in the preparation of nanoparticles or latex carriers. In this process, the monomers are emulsified in water with stirring and addition of emulsifiers that help stabilize the monomer droplets. The polymerization is then started with addition of the initiator and the process continues to its completion. The loading of drugs into or onto the nanoparticles could be achieved either by incorporation of the drug during the polymerization stage or by adsorption of the drug onto the surface of already polymerized particles. The major problems in using this type of colloidal carriers are residual catalysts or monomers and/or the toxic metabolites formed after degradation of the carriers.

A colloidal delivery system based on biodegradable natural macromolecules was described by Marty et al. [42]. The production technique was derived from a coacervation method and included the solubilization of the macromolecules, such as gelatin, in an aqueous medium prior to desolvating and cross-linking steps. The nanoparticles could be purified by gel filtration or could be further freeze-dried to obtain the dry powder form. Lipophilic drugs could be entrapped into the systems more successfully then water-soluble drugs, which would be incorporated more efficiently by using a nonaqueous system.

Colloidal dispersions of preformed polymers can be formed by emulsification of polymer melt or polymer solution in water followed by high shear homogenization. The polymers include polyurethanes, epoxy resins, polyesters, polypropylene, and various cellulose ethers and esters, such as cellulose acetate and ethyl cellulose. Polymeric nanoparticles or nanosuspensions containing indomethacin, intended for oral use, were prepared by a microfluidization–solvent evaporation method [43]. In this technique a water-insoluble polymer and the drug were codissolved in a volatile, water-immiscible organic solvent. After emulsification of this organic phase into a surfactant-containing aqueous phase and subsequent evaporation of the solvent, a drug-loaded polymeric dispersion was obtained. Prior to the evaporation of organic solvent, the particle size of the primary emulsion could be reduced into a colloidal size range by a homogenization step called microfluidization. The drug-containing nanosuspensions were evaluated with respect to total drug content, drug content in the polymer and aqueous phase, particle size, drug crystallization in the aqueous phase, in vitro drug release, and their stability against flocculation in different media. Nanosuspensions with a total drug content of 35 mg indomethacin per milliliter of nanosuspension could be prepared without drug crystallization with more than 98% of the drug being present within the polymer phase. Unwanted drug crystallization in the aqueous phase was found to be dependent on the drug loading, the drug–polymer compatibility, the organic solvent, and the type and amount of surfactant used.

Nanoparticles containing ibuprofen, indomethacin, or propranolol could be formed spontaneously after the addition of solutions of the drugs and acrylic polymers (Eudragit RS100 or RL100) in the water-miscible solvents acetone or ethanol to water without sonication or microfluidization steps [44]. The colloidal dispersions were stabilized by quaternary ammonium groups and did not require the addition of surfactants or polymeric stabilizers. With drug-containing nanoparticles, conversion of the nanosuspension into a redispersible powder could be done by freeze- or spray-drying techniques to improve the shelf life of the product.

Suspensions of nanoparticles are pharmaceutically elegant and have milk-like consistency and appearance. The drug is incorporated into a polymeric matrix; potential uses include taste masking of poorly tasting drugs and sustained release. In addition, since primarily lipophilic drugs are incorporated, increased rate of dissolution and absorption may be seen because of the significant increases in the surface area. Upon oral administration, the nanoparticles may remain in the GI tract or may actually be absorbed. In the case of absorption, biodegradable polymers have to be used.

V. CONCLUSIONS

Suspension of multiparticulates or microparticles and their dispersible preparations offer a variety of advantages over the conventional controlled-release solid dosage forms. The preference of liquid preparations among pediatric and geriatric patients could substantially increase patient compliance as well as the efficacy and accuracy of dose titrations and administration dosages. Although this category of dosage form is still considered to be in its infancy, the trend toward multiparticulate preparations and the necessity of liquid preparations for specific patient groups will ensure increasing growth in the development and applications of this type of dosage form.

REFERENCES

1. R. W. Baker, *Controlled Release of Biologically Active Agents*, Wiley, New York, 1987.
2. H.-W. Hui, J. R. Robinson, and V. H. L. Lee, in *Controlled Drug Delivery: Fundamentals and Applications* (J. R. Robinson and V. H. L. Lee, eds.), Marcel Dekker, New York, 1987, pp. 412–414.
3. F. Theeuwas, P. S. L. Wong, and S. I. Yum, in *Encyclopedia of Pharmaceutical Technology*, Vol. 4 (J. Swarbrick and J. C. Boylan, eds.), Marcel Dekker, New York, 1991, pp. 303–348.
4. H. Bechgaard and G. H. Nielsen, *Drug Dev. Ind. Pharm. 4*:53 (1978).
5. C. Eskilson, *Manuf. Chem.*, March, pp. 33, 36, 39 (1985).
6. J. Sjögren, in *Rate Controlled in Drug Therapy* (L. F. Presscott and W. S. Nimmoe, eds.), Churchill Livingstone, Edinburgh, 1985, pp. 38–47.
7. B. C. Lippold, in *Oral Controlled Release Products: Therapeutic and Biopharmaceutic Assessment* (U. Gundert-Remy and H. Möller, eds.), Wissenschaftliche Verlagsgesellschaft, Stuttgart, 1990, pp. 39–57.
8. S. Y. Lin, *J. Pharm. Sci. 77*:229 (1988).
9. A. R. Fassihi, *Int. J. Pharm. 44*:249 (1988).
10. I. Jalsenjak, C. F. Nicolaidou, and J. R. Nixon, *J. Pharm. Pharmacol. 29*:169 (1977).
11. R. Bodmeier and H. Chen, *J. Controlled Release 10*:167 (1989).
12. A. R. Nanda and J. C. Price, *Pharm. Res. 7*:S-96 (1990).
13. C. M. Adeyeye and J. C. Price, *Pharm. Res. 6*:S-71 (1989).
14. R. Bodmeier, J. Wang, and H. Bhagwatwar, J. Microencapsulation, in press.
15. Y. Kawashima, T. Iwamoto, T. Niwa, H. Takeuchi, and Y. Itoh, *Int. J. Pharm. 75*:25 (1991).
16. T. Niwa, Y. Kawashima, H. Takeuchi, and T. Hino, Designs of novel drug delivery devices (microsponge and microballoon) by emulsion-solvent diffusion method, *Proceedings of the 2nd World Congress on Particle Technology*, Kyoto, Japan, 1990, pp. 506–513.
17. R. T. Sparks and E. J. Geoghegan (to Elan Corporation), U. S. Patent 4,952,402 (1990).

18. A. L. Boner, G. Vallone, E. Valletta, D. Bernocchi, and M. Plebani, *Int. J. Clin. Pharmacol. Res. 7*:345 (1987).

19. J. Devane and S. Mulligan, In-vitro and in-vivo characteristics of controlled-release liquid formulations of acetaminophen, theophylline and carbamazepine, *Proceedings of the 16th International Symposium on Controlled Release of Bioactive Materials*, Chicago, 1989, pp. 372–373.

20. J. Devane and S. Mulligan, Characteristics of different pediatric presentation forms of a single theophylline microparticle formulation, *Proceedings of the 16th International Symposium on Controlled Release of Bioactive Materials*, Chicago, 1989, pp. 370–371.

21. R. Bodmeier, H. Chen, P. Tyle, and P. Jarosz, *J. Controlled Release 15*:65 (1991).

22. B. F. Benton and D. L. Gardner (to Battelle Development Corporation), U.S. Patent 4,876,094 (1989).

23. J. L. Zatz and D. W. Woodford (to Research Corporation), U.S. Patent 4,717,713 (1988).

24. Y. Raghunathan (to Pennwalt Corporation), U.S. Patent 4,221,778 (1980).

25. Y. Raghunathan, L. Amsel, O. Hinsvark, and W. Bryant, *J. Pharm. Sci. 70*:379 (1981).

26. *Delsym Data Sheet*, Fisons Corporation, Pennwalt Pharmaceutical Division, Rochester, N.Y., 1982.

27. L. P. Amsel, O. N. Hinsvark, K. Rotenberg, and J. L. Sheumaker, *Pharm. Technol. 8*:28 (1984).

28. D. A. Graves, M. T. Wecker, M. C. Meyer, A. B. Straughn, L. P. Amsel, O. N. Hinsvark, A. K. Bhargava, and K. S. Rotenberg, *Biopharm. Drug Dispos. 9*:267 (1988).

29. S. Motycka and J. G. Nairn, *J. Pharm. Sci. 67*:500 (1978).

30. S. Motycka and J. G. Nairn, *J. Pharm. Sci. 68*:211 (1979).

31. O. L. Sprockel and J. C. Price, *Drug Dev. Ind. Pharm. 15*:1275 (1989).

32. O. L. Sprockel, J. C. Price, R. Jennings, R. L. Tackett, S. Hemingway, B. Clark, and R. E. Laskey, *Drug Dev. Ind. Pharm. 15*:1393 (1989).

33. L. Kennon and G. K. Storz, in *The Theory and Practice of Industrial Pharmacy* (L. Lachman, H. A. Lieberman, M. M. Rieger, and J. L. Kanig, eds.), Lea & Febiger, Philadelphia, 1976, pp. 162–183.

34. M. J. Falkiewicz, in *Pharmaceutical Dosage Forms: Disperse Systems*, Vol. 1 (H. A. Lieberman, M. M. Rieger, and G. S. Banker, eds.), Marcel Dekker, New York, 1988, pp. 13–48.

35. R. A. Nash, in *Pharmaceutical Dosage Forms: Disperse Systems*, Vol. 1 (H. A. Lieberman, M. M. Rieger, and G. S. Banker, eds.), Marcel Dekker, New York, 1988, pp. 151–198.

36. C. M. Ofner III, R. L. Schnarre, and J. B. Schwartz, in *Pharmaceutical Dosage Forms: Disperse Systems*, Vol. 2 (H. A. Lieberman, M. M. Rieger, and G. S. Banker, eds.), Marcel Dekker, New York, 1989, pp. 231–264, 317–334.

37. F. Sørensen, A. V. Børsting, A. Skinhjøj, and F. N. Christensen, A novel fluid dosage form for administration of large doses of controlled-release particles, *Proceedings of the 18th International Symposium on Controlled Release of Bioactive Materials*, Amsterdam, 1991, pp. 353–354.

38. J. Kreuter, *Pharm. Acta. Helv.* *53*:33 (1978).
39. R. C. Oppenheim, in *Drug Delivery Systems* (R. L. Juliano, ed.), Oxford University Press, New York, 1980, pp. 177–188.
40. R. C. Oppenheim, *Int. J. Pharm.* *8*:217 (1981).
41. D. J. Burgess, in *Encyclopedia of Pharmaceutical Technology* (J. Swarbrick and J. C. Boylan, eds.), Marcel Dekker, New York, 1990, pp. 31–63.
42. J. J. Marty, R. C. Oppenheim, and P. Speiser, *Pharm. Acta Helv.* *53*:17 (1978).
43. R. Bodmeier and H. Chen, *J. Controlled Release* *12*:223 (1990).
44. R. Bodmeier and H. Chen, *J. Microencapsulation* *12*:223 (1990).

8

Multiparticulate Encapsulation Equipment and Process

Donald K. Lightfoot

SmithKline Beecham Pharmaceuticals
King of Prussia, Pennsylvania

I. OVERVIEW OF CAPSULE FILLING

A. Historical Perspective

The development of machinery to fill gelatin capsules can be traced to Arthur Colton, who in 1913 patented a machine capable of filling powders automatically into two-piece hard gelatin capsules. Colton was also responsible for developing machinery for the mass production of empty capsules from 1895 to 1920. Today's high-speed automatic capsule-filling machinery was not developed until the late 1950s. Before reviewing the actual machinery, it is important to understand the basic principles employed by all modern capsule-filling machines.

B. Basic Operations Common to All Filling Machines

All capsule-filling machines perform the following operations in encapsulating product (Fig. 1):

1. *Rectification.* Capsules fall by gravity into feeding tubes or chutes. They are aligned by mechanically gauging the diameter differences between the cap and body, and the aligned capsules are transferred into a two-section housing or bushing.

2. *Separation.* The overall diameter of the upper bushing or housing is slightly larger than the diameter of the capsule cap. The base diameter of this housing is smaller than the cap diameter and slightly larger than the capsule body diameter, which also matches the diameter of the lower capsule body bushing or housing. Vacuum pulls the capsule body into the lower housing, while the cap is

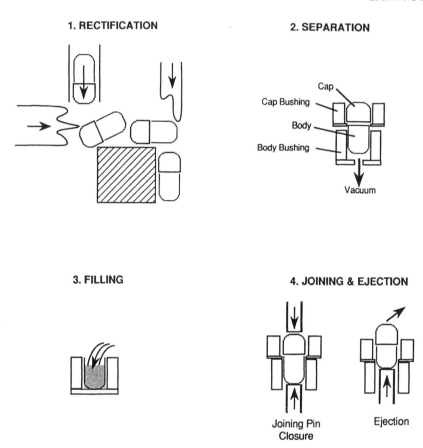

Fig. 1 Basic operations common to all filling machines.

retained in the upper bushing or housing. Once the capsule is opened, the upper and lower housing or bushing is separated to position the capsule body for filling.

3. *Filling*. The open body is then dosed with the medicament. The various types of dosing mechanisms are described later in this chapter.

4. *Joining and Ejection*. The cap and body housing or bushing are realigned for joining. An upper plate or pin will hold the cap stationary in the housing while a lower closing pin will push the body upward into the caps until the capsules are completely joined, engaging the locking mechanism of the capsule shell. A lower ejection pin will ascend, ejecting the capsule from the housing or bushing into an exit chute. Compressed air nozzles are frequently used to assist the ejection of the capsules from the machine, particularly with fully automatic filling machines.

C. Methods of Dosing

Various methods are employed in filling or dosing of hard-shell gelatin capsules. Multiparticulate dosing methods can be categorized into the filling of powders or fine granules (usually in the particle size range smaller than 250 μm) and the filling of pellets or beads (larger than 250 μm). Capsules can also be dosed with tablets or caplets and liquids or pastes. In addition, it is possible to multiple-dose capsules with various combinations of powders, pellets, tablets, or capsules.

Filling of Powders and Fine Granules

Auger Fill Method.   This method is employed by the Model 8 (Fig. 2) and Elanco-Fill filling machines and is illustrated in Fig. 3. A rotating auger blade at the discharge of the drug hopper forces the powder into the open capsule bodies, housed in a rotating ring beneath the drug hopper. The factors controlling the fill are the rotation speed of the body ring and the amount of powder in the drug hopper. The auger design and auger speed can also influence fill weights.

Dosator Method.   The dosator principle is employed by numerous inter- mittent-motion and continuous-motion capsule-filling machines and is illustrated in Fig. 4. The dosator consists of a hollow tube and a piston that is volumetrically adjusted for the proper dose. The dosator descends to the bottom of the powder bed, being maintained at a constant level, precompressing the powder in the chamber formed by the tube and piston. The spring-loaded piston then descends, compressing the powder into a slug. The dosator moves out of the powder bed over the capsule body. The piston descends to the bottom of the tube, discharg- ing the slug into the capsule body.

Powder formulations for a dosator machine should have good flow charac- teristics to maintain a uniform powder bed volume and prevent cavitation of the powder bed by the dosators. The powder should also have cohesive properties to allow for slug formation and prevent loss of product from the open end of the dosator during the transfer to the capsule body. Lubrication of dosator powder formulations is also necessary to allow for a smooth discharge of the compressed slug and prevent binding of the spring-loaded piston. The density of the powder is also an important factor. Very fine powder formulations that tend to entrain air will cause inconsistent dosing. One of the dosator machines (Zanasi-Matic) has overcome this problem by using vacuum ports at the base of the powder bed to precompact the powder, creating a more uniform density. Another dosator ma- chine (MG2) holds the piston at a constant point while the tube descends into the powder bed (Fig. 5). This creates a syringe effect, enabling this dosator design to pick up very fine or bulky particles without compaction and cleanly transfer this loose powder to the capsule body.

Tamping and Dosing Disk Method.   This principle is currently utilized by the Hoefliger & Karg GKF series, the SeJong, and the Elanco-Fill F-80 machin-

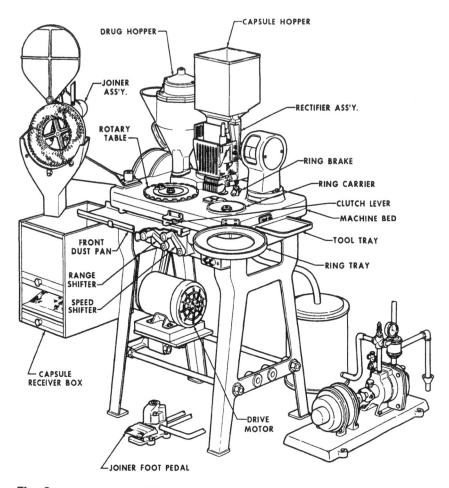

Fig. 2 Model 8 capsule-filling machine.

ery. This is illustrated in Fig. 6. The dosing disk is a rotating steel plate with precisely bored holes which form the dosing chamber. The disk is situated at the base of a powder container or dosing bowl. The power is auger fed into the bowl from a powder supply hopper, maintaining a constant powder bed level directly above the dosing disk. The dosing disk indexes the holes beneath five sets of spring-loaded tamping punches that compact the powder into the holes in the disk forming the powder slug. A stationary sealing or base plate beneath the disk retains the powder in the holes during compaction. Following the five tamps, a deflector removes the powder from the disk surface and the holes index directly over the capsule bodies. A set of transfer pins penetrate through the holes in the

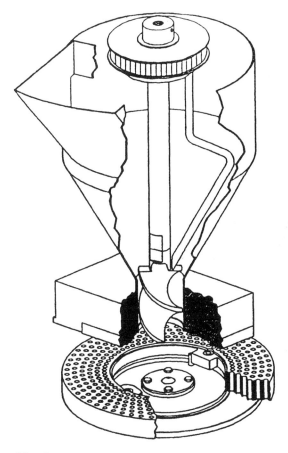

Fig. 3 Auger fill (Model 8 machine).

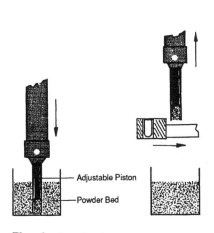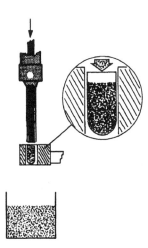

Adjustable Piston
Powder Bed

Fig. 4 Powder dosator.

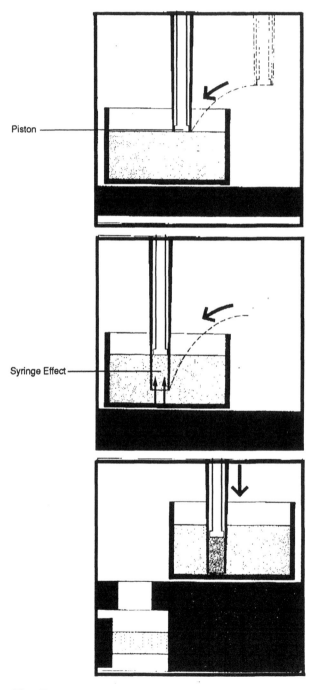

Fig. 5 MG2 dosator.

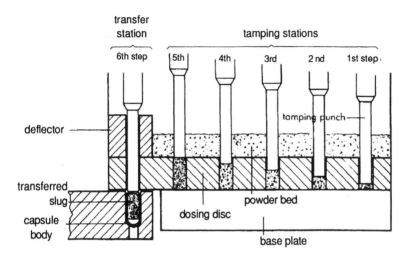

Fig. 6 Tamping and dosing disk method.

disk, discharging the powder slug into the capsule body. The main factor controlling the dosage weight is the thickness of the dosing disk. Further weight adjustment is achieved by the penetration depth of the five series of tamping pins, which control the tamping pressure required to form the slug.

Formulations for the tamping and dosing disk machines should have good flow characteristics. The diameter of the dosing bowl on the GKF 1500 is 340 mm, and poor flowing powders will not form a uniform level over the holes in the dosing disk, which are located around the perimeter of the bowl. Formulations should also be adequately lubricated to prevent powder from adhering to the dosing disk, tamping pins, and sealing plate. Heat-sensitive products or ingredients with a low melting point may be problematic on this type of machine, due to the frictional heat and pressure generated by the scrapers and sealing plate. Sticky powders and products with low melting points may fuse to these surfaces, causing machine binding. Since the proper functioning of this machine requires compacting a powder slug, the formulation should also possess cohesive properties. This allows for a clean transfer of the slug into the capsule body and prevents excess loss of powder between the bottom of the dosing disk and the sealing plate. The machine can be used to fill loose or noncompressible powders, using the dosing disk bores to volumetrically measure the dose; however, this will usually result in increased weight variation and product loss.

Vibrating Feed Frame Method.   This method is employed by the Osaka filling machine and resembles the tablet press methodology. The feed frame has a perforated resin plate that is activated by a vibrator. The powder flows into the vibrating feed frame, which facilitates the flow of the powder into the capsule body passing underneath. The dosing of the capsule is controlled by the vibration amplitude of the feed frame, as well as the weight cam, which positions the bodies in the holding ring as they pass under the feed frame. Machine speed controlling dwell time under the feed frame is also a critical factor. As illustrated in Fig. 7, the level of powder may initially extend above the filled capsule body. The capsule body pins push the capsule up against a scraper plate at the surface of the capsule ring, compacting the powder into the capsule body. Excess powder flow from the feed frame is removed at the end section of the feed frame and recycled. Free-flowing powder formulations are essential for this type of dosing mechanism and play a significant factor in the machine speed.

Filling of Pellets or Beads

Feed-Frame Method.   This method is utilized by the Osaka R180 and the Elanco Rotofil (Fig. 8). The pellets flow through the feed frame into the open capsule body passing underneath. The positioning of the body and the speed of the machine control the dosage weight. Most pellet formulations for this method are designed with a bulk density to fill completely the capsule body that is used to measure the dose. Partial dosing of the capsule is usually not possible with this type of mechanism.

Dosing Chamber Method.   This method in various configurations is widely utilized by many machine manufacturers (i.e., MG2, Hoefliger & Karg, Macofor, and the Elanco-Fill F-80). The pellets flow into an adjustable volumetric chamber (Fig. 9). The pellets are then discharged out of the chamber, through a discharge tube or plate, into the capsule body. Many machine manufacturers provide for the attachment of a second dosing chamber for dosing capsules with two types of pellets or beads. Pellet formulations for this type of mechanism should be free flowing and free from agglomerations or electrostatic charge, which will interfere with the discharging of the pellet dose out of the chamber into the capsule body. Uniform particle size is also essential for accurate dosing.

Vacuum Dosator Method.   This method is unique to the Zanasi capsule-filling machines (Fig. 10). The dosator is the same basic design as the powder

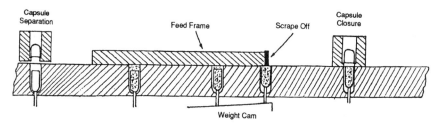

Fig. 7 Vibrating feed frame.

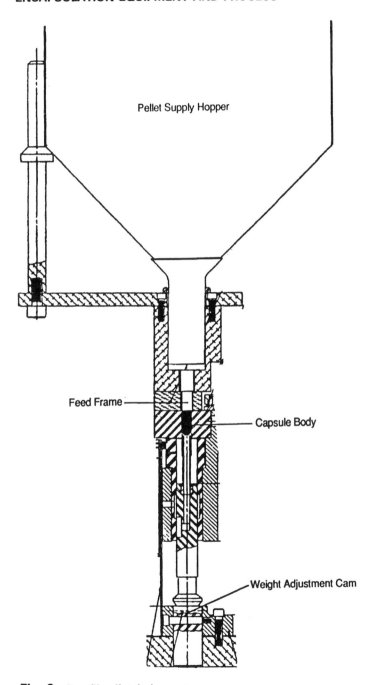

Fig. 8   Rotofil pellet dosing system.

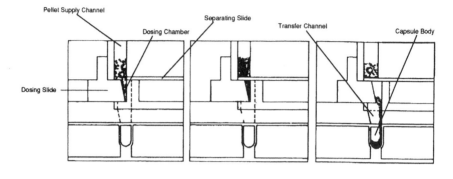

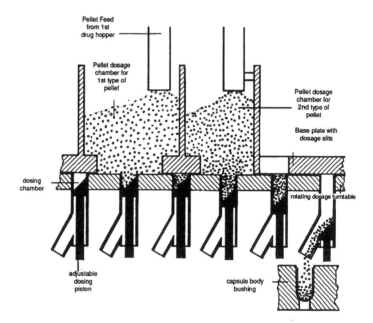

Fig. 9   Pellet dosing chambers.

dosator described above. As the dosator descends midway into the pellet bed, vacuum is applied to the inside of the tube suctioning the pellets into the dosator. As the dosator moves out of the pellet bed, surplus product is removed from the end of the tube with either a wiper arm or jet of air. The dosator moves over the capsule body, and the piston descends to the bottom of the tube, discharging the pellet dose. Pellet formulations for this type of mechanism should be dust free to prevent blocking of the vacuum system.

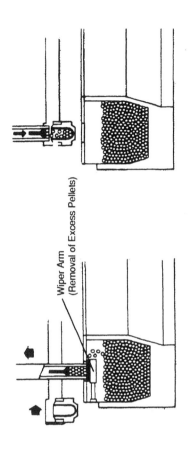

Wiper Arm
(Removal of Excess Pellets)

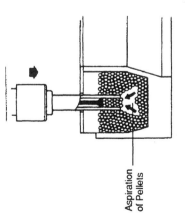

Aspiration
of Pellets

Fig. 10 Vacuum dosator.

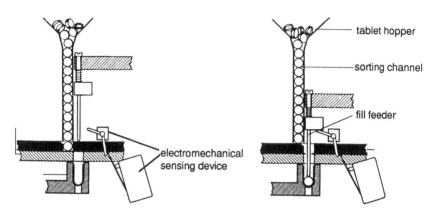

Fig. 11   Tablet dosing mechanism.

Filling of Tablets

Filling of solid dosage forms (i.e., tablets into hard gelatin capsules) has been a recent trend, offering some distinct formulation advantages. The basic feed tube/ bushing transfer mechanism employed is illustrated in Fig. 11. Usually, an electromechanical device is incorporated to verify the dosing and is coupled with a reject mechanism for capsules with missing tablets. To assure proper tablet feed, it is essential that the tablet dimensions and hardness specifications are maintained within strict tolerances. The capsule-filling machine manufacturer should be consulted to verify the compatibility of tablet shape with design of tablet feed mechanism. Spherical tablets are the optimal design.

Filling of Liquids or Pastes

Capsule-filling machine manufacturers have developed positive-displacement liquid pumps for accurate dosing of capsules with oils, pastes, and thermomelt- ing and thixotropic substances. The capsules should be sealed after filling to prevent product leakage.

Multiple Dosing of Capsules

Many capsule-filling machines can be equipped with more than one type of dosing unit to provide for filling different dosage combinations into a capsule as illustrated in Fig. 12. Overencapsulation of tablets or capsules has become an effective technique for blinding product to be used in clinical trials. The most novel machine recently developed for this purpose is the MG Futura, which has four dosing stations (Fig. 13).

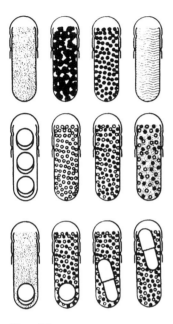

Fig. 12 Combinations of fill material possible in hard gelatin capsules.

II. CAPSULE MACHINERY

A. Classification of Machinery

Capsule-filling machines can be classified by level of automation and type of machine motion into the following categories:

1. *Totally manual.* With this type of machine, both the empty capsule feed and the product feed are controlled manually by the operator.

2. *Semiautomatic.* With this type of machine, the feeding of the empty capsules into the feed ring, dosing of the product into the capsule bodies, and the joining of the capsule are controlled by the machine. The operator must initiate these steps and must manually transfer the rings, then separate and rejoin them during the filling process. Output of this machine is very operator dependent.

3. *Automatic—intermittent-motion.* All operations are machine controlled. Machine motion is regulated by a geneva cam, which allows the machine to index continuously in a stop/start mode. The transfer and separation of the empty capsules, compacting of the powder slug, dosing of the capsule, joining, and ejection all occur during the stationary mode. This provides for greater dwell time for these critical functions.

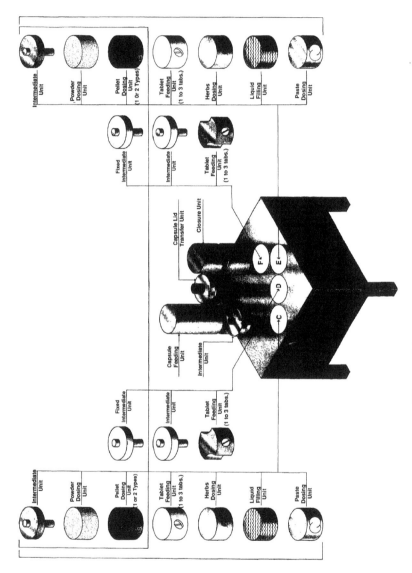

Fig. 13 MG Futura capsule-filling machine.

4. *Automatic—continuous motion*. These machines, in principle, operate the same as the intermittent-motion machines. The capsule feeding, dosing, and joining elements operate independently at a speed proportionate to the machine cycle. Since there is no idle dwell time, it is usually possible to achieve higher output speeds with this type of machinery.

B. Listing of Capsule-Filling Machinery

The commercially available capsule-filling machines throughout the world are summarized in Table 1.

C. Automation

The major focus for automation of capsule-filling machinery has been in the process control area of statistically sampling and adjusting the filled capsule weights. By interfacing the balances with computer systems, a feedback loop is established to activate servomotors, which adjust the dosing elements to maintain the weights within a programmed range. If the machine cannot be adjusted properly, the machine will shut down, and the malfunction will be indicated by the computer. These systems also provide hardcopy documentation of the in-process weights and an end-of-batch statistical summary.

Other types of automation employed include systems to sense and clear jams in the empty capsule feed system, electronic inspection of empty capsules for defects, and in-process collection and identification of samples. The ultimate in automation is an unattended capsule-filling machine employing all of the foregoing features. These systems have been reported to operate completely unattended for up to 8 h. Automated handling systems to feed the empty capsules, to provide product mix, and to remove the filled capsules have also been developed for these stand-alone machines (Fig. 14).

III. ENCAPSULATION PROCESS

A. Capsule-Fill-Weight Determination

The capsule-fill-weight capacity is governed primarily by the capsule size and calculated values listed in Table 2. The tapped bulk density of the powder to be encapsulated needs to be determined. For the filling of pellets or beads, the untapped bulk density values should be used.

B. In-Process Controls

Filled-capsule-weight measurements of the average and range of individual weights is the traditional in-process control method employed in monitoring capsule-filling-machine dosing accuracy. Acceptable limits should be developed

Table 1 Capsule-Filling Machinery

Manufacturer	Type of machine	Model	Hourly output	Method of dosing		Other dosing possibilities
				Powders	Pellets	
Bonapace, Milan, Italy	Manual	Minicap	2,000	Hand-filled with vibrator and pressing device	Hand-filled with vibrator assistance	None
	Semiautomatic	B/B-6/S	14,000 (2 oper.)			None
	Automatic intermittent motion	RC 530	34,000	Dosator	Dosing chamber	Tablets, liquids, or pastes
Robert Bosch GmbH, Waiblingen, Germany	Automatic intermittent motion	GKF-130	7,800	Tamping and dosing disk	Dosing chamber (2 units)	Tablets, liquids, or pastes
		GKF-400	24,000			Tablets, liquids, or pastes
		GKF-800	48,000			Tablets, liquids, or pastes
		GKF-1500	90,000			Tablets, liquids, or pastes
Capsugel, Greenwood, South Carolina	Semiautomatic intermittent motion	Type-8	15,000	Auger fill	Feed shoe (similar to feed frame)	Tablets (TFR-8) and Caplets (ACF-8)
		Ultra-8	30,000			Tablets (TFR-8) and Caplets (ACF-8)
Shionogi Qualicaps, Indianapolis, Indiana	Semiautomatic intermittent motion	Model 8S	15,000	Auger fill	Feed shoe (similar to feed frame)	Tablets, caplets, oil/pastes (Rotop 8)
		Elanco-Fill	30,000			Tablets, caplets, oil/pastes (Rotop 8)
	Automatic continuous motion	Rotofil	60,000	Not designed for powders	Feed frame	None
	Automatic intermittent motion	Elanco-Fill F-80	80,000	Tamping and dosing disk	Dosing chamber	Liquids or pastes

Manufacturer	Operation	Model	Output	Dosing mechanism	Metering	Product
Harrow Hofliger, Allmersbach, Germany	Automatic intermittent motion	KFM III-1	4,500	Tamping and dosing disk	Dosing chamber (2 units)	Tablets, liquids, or pastes
		KFM III	13,200			Tablets, liquids, or pastes
Macofar, Bologna, Italy	Automatic intermittent motion	MT 5	6,000	Dosator	Dosing chamber (2 units)	Tablets
		MT 20	20,000			Tablets
		MT 40	40,000			Tablets
		MT 80	80,000			Tablets
MG2, Bologna, Italy	Automatic continuous motion	G36	9–36,000	Dosator (syringe effect design)	Dosing chamber	Tablets, liquids, and pastes
		G38N/G60	60,000			Tablets
		G37N/G100	100,000			Tablets
		G120	120,000			Tablets
		Compact	6–48,000	Dosator	Dosing chamber (4 units)	Tablets, liquids, or pastes
		Futura	6–48,000	Dosator (2 dosing stations)	Dosing chamber (4 units)	Tablets (4 units) / Liquids and pastes (2 units)
Osaka, Osaka, Japan	Automatic continuous motion	S-40	40,000	Vibratory feed frame	Feed frame	None
		R-50	50,000			None
		NR-100	100,000			None
		NR-180	180,000			None
Se Jong, Seoul, Korea	Automatic intermittent motion	SF-35	18–43,200	Tamping and dosing disk	Dosing chamber (2 units)	Tablets
		SF-70	36–86,400			Tablets
Nuova Zanasi, Bologna, Italy	Automatic intermittent motion	Zanasi 6	6,000	Dosator (suction dosage tank to precompact powder)	Vacuum dosator	Tablets
		Zanasi 12	12,000			Tablets
		Zanasi 25	25,000			Tablets
		Zanasi 40	40,000			Tablets, liquids, or pastes
		Zanasi 70	70,000			Tablets
	Automatic continuous motion	Matic 60	60,000	Dosator (suction dosage tank to precompact powder)	Vacuum dosator	None
		Matic 90	90,000			None
		Matic 120	120,000			

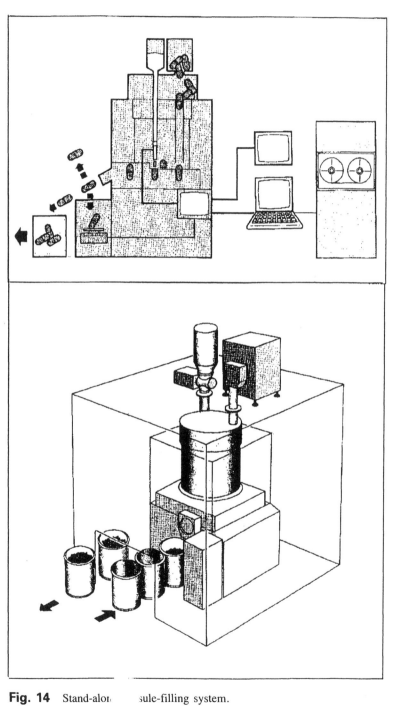

Fig. 14 Stand-alone capsule-filling system.

Table 2 Calculated Capsule-Fill-Weight Capacity

Capsule size	Approximate body volume (mL)	Weight capacity (mg) for tapped bulk density (g/cm^3) of powder[a]						
		0.6	0.7	0.8	0.9	1.0	1.1	1.2
000	1.37	822	959	1096	1233	1370	1507	1644
00	0.95	570	665	760	855	950	1045	1140
0 el	0.76	456	532	608	684	760	836	912
0	0.68	408	476	544	612	680	748	816
1	0.49	294	343	392	441	490	539	588
2	0.37	222	259	296	333	370	407	444
3	0.29	174	203	232	261	290	319	348
4	0.21	126	147	168	189	210	231	252
5	0.13	78	91	104	117	130	143	156
Supro A	0.68	408	476	544	612	680	748	816
Supro B	0.50	300	350	400	450	500	550	600
Supro C	0.37	222	259	296	333	370	407	444
Supro D	0.30	180	210	240	270	300	330	360
Supro E	0.21	126	147	168	189	210	231	252

[a]For pellet and bead products, use untapped bulk density.

based on the statistical process capability of the product–machine combination. Other in-process factors that should be routinely monitored are filled-capsule joining lengths and overall physical quality.

IV. FURTHER PROCESSING OF FILLED CAPSULES

A. Cleaning/Polishing

Filling capsules with very fine powders may cause a dust residue on the capsule surface. There are three basic commercially available methods that can be employed to clean the capsules:

1. Rolling capsules in a coating pan with salt
2. Passing capsules through a rotating polishing belt that employs lamb's wool or other suitable material to wipe dust off the capsule
3. Passing the capsules through a brush-type cleaning machine immediately following the encapsulation process

B. Weight Sorting

Capsules can be 100% weight sorted at filling machine speeds to remove any unacceptable weight capsules. High-speed machines available for this purpose include:

1. *Vericap 2110* (Mocon, U.S.): 2100 capsules/min
2. *Bosch KKE 1500* (Robert Bosch GmbH, Germany): 1500 capsules/min
3. *Anritsu K525D Capsule Autochecker* (Japan): 1250 capsules/min

C. Banding/Sealing

There are two basic purposes for banding or sealing hard-shell capsules:

1. To create a leakproof closure for the filling of oils, pastes, and liquids
2. To create a tamper-resistant dosage form that cannot be opened without visible signs of damage to the capsule

Applying a gelatin band to the seam of the cap and body has been the most common technique to achieve this. Commercial machinery such as the Qualiseal (Shionogi) is available for this purpose.

D. Recovery of Product from Capsules

The capsule-fill material can be recovered by the following techniques:

1. Reopen capsules on the filling machine by employing high vacuum at the separation station and ejecting the capsules unjoined. This is not very effective with grooved locking-style capsules.
2. Cut capsules open using a mill or commuting machine. This will usually result in an excessive amount of gelatin fines and may cause damage to the product.
3. Pass the capsules through the Urschel Slitter (Urschel Laboratories, Inc.). This unit has been designed specifically for this purpose. The capsules are fed through channels where cutting knives slit the surface of the capsule along its longitudinal axis (Fig. 15). This process will cause little damage to the internal product and has proven effective in recovering controlled-release pellets from capsules.

IV. CONCLUSION

The hard shell gelatin capsule offers significant advantages as a dosage form for multiparticulates. An assortment of capsule filling machines have been developed, with various dosing mechanisms to encapsulate multiparticulates, as well as powders, tablets, pastes and liquid products. The multiple dosing mechanisms available on many capsule filling machines also provides for some unique formulation possibilities.

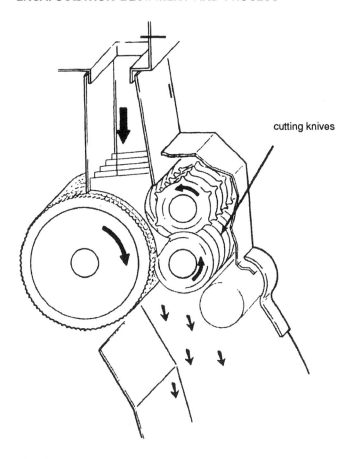

cutting knives

Fig. 15 Urschel capsule slitter.

BIBLIOGRAPHY

1. Anritsu Brochure, *Capsule Autochecker Series, No. K3008-1*.
2. L. L. Augsburger, in *Sprowls' American Pharmacy* (L. W. Dittert, ed.), J. B. Lippincott, Philadelphia, 1974, pp. 323–326.
3. Bosch Brochure, *Capsule Checking System*, KKE 1500.
4. Bosch Brochure, *The Economic Machinery Program for the Filling and Closing of Hard Gelatin Capsules*, GKF130-400-800-1500-3000.
5. Capsugel Bulletin, *General Specifications Capsugel Hard Gelatin Capsules*, February 1991.
6. Capsugel Publication, *A Guide to the Hard Gelatin Capsule*, 2MK83.
7. Capsugel Publication, *All About the Hard Gelatin Capsule*, CAP 126E, 1991.

8. Capsugel Publication, *Installation, Operation and Maintenance of the Type 8 Capsule Filling Machine*, Warner Lambert Co., 1989.
9. Dott. Bonapace & Co. Brochures, *Minicap Capsule Filling Setup*; *Capsule Filling Set-up*; B/B-6/S, A/B-1/s.
10. Elanco Qualicaps Brochure, *From the Leader in Capsule Machine Technology* (*Elanco-Fill, F-80, Rotofil, Model 8S, Qualiseal*).
11. Elanco Qualicaps, *Capsule Technical Information Manual*, B5.450.
12. V. B. Hostetler and J. Q. Beilard, Capsules, in *Theory and Practice of Industrial Pharmacy* (L. Lachman, H. A. Lieberman, and J. L. Kanig, eds.), Lea & Febiger, Philadelphia, 1970, pp. 346–359.
13. Macofar Cem Brochures, *Automatic Capsule Filling and Closing Machine*, MT-80, MT-40.
14. MG2 Brochures, *G37N*, MG Futura, Bologna, Italy.
15. Nuova Zanasi Brochure, *6-12-25-40-70, Intermittent Motion Capsule Filling Machines*.
16. Nuova Zanasi Brochures, *Matic 60-90-120, Production Systems for Continuous Motion Capsule Filling Machines, Z5000R*.
17. Osaka Brochure, *R-180*, Osaka Automatic Machine Mfg. Co., Osaka, Japan.
18. K. Ridgway, *Hard Capsules: Development and Technology*, Pharmaceutical Press, London, 1987.
19. Se Jong Machinery Co. Brochure, *Se Jong Auto Capsule Filling Machines*.
20. Urschel Laboratories Brochure, *Urschel Model CD-A Capsule Slitter*, 1322, March 1988.

9

Compaction of Multiparticulate Oral Dosage Forms

Metin Çelik

College of Pharmacy
Rutgers—The State University of New Jersey
Piscataway, New Jersey

I. INTRODUCTION AND SCOPE

The design and development of multiparticulate dosage formulations in the form of compressed tablets rather than hard gelatin capsules are becoming increasingly important. Hard gelatin capsules are very elegant dosage forms but have higher production cost, lower production rate, and tampering potential (Tylenol and Sudafed-12hour) than those of compressed tablets. When a multiparticulate dosage product is developed in the form of a tablet, it is often desirable to produce compacts that disintegrate into many subunits soon after ingestion, to attain more uniform concentrations of active substances in the body. It is not necessary to emphasize the fact that the coated subunits in the formulation must withstand the process of compaction without being damaged, since, for example, the existence of a crack in the coating may have undesirable effects on the drug release properties of that subunit. The type and amount of coating agent, the size of the subunits, selection of the external additives, and the rate and magnitude of the pressure applied must be considered carefully to maintain the desired drug release properties of that subunit. To avoid such a problem, formulation scientists must have comprehensive knowledge of how that formulation will behave during tableting, as well as of how the other material and/or process-related parameters will affect the performance of that formulation as a drug delivery system.

The goal of this chapter is to deal with the phenomena and mechanisms involved during compaction of a particulate system. The techniques that have been utilized to assess the compactional behavior of pharmaceutical materials are also presented. Particular emphasis is given to the compaction properties of the coated and uncoated pellets that are the main components of multiparticulate dosage forms. In the literature (and in this chapter), the terms *pellet*, *microsphere*, and *bead* are used interchangeably.

II. DEFINITION AND STAGES OF COMPACTION

Compaction can be defined as "the compression and consolidation of a two-phase (particulate solid–gas) system due to an applied force." *Compression* is a reduction in the bulk volume of the material as a result of the gaseous phase. *Consolidation* is an increase in the mechanical strength of the material resulting from particle–particle interactions [1]. The phenomena and mechanisms involved during compaction of powders have been examined by several workers in the fields of pharmaceutics and metallurgy [2–6]. It can be deduced from these works that the principal physicomechanical process involved in the compaction of particulate matter can be assumed to occur as shown in Fig. 1.

A. Particle Rearrangement

Prior to penetration of the upper punch into the die, the particles flow with respect to each other, with the finer particles entering the voids between the larger ones, resulting in a closer packing arrangement. At this stage, the energy involved in the process is due to interparticulate friction. The packing characteristics of a solid material, which define the number of contact points (i.e., potential bonding areas) between the particles, are governed primarily by the density, size, size distribution, shape, and surface properties of the individual particles as well as by the process variables, such as the rate of flow, and the relationship between the die-cavity diameter and particle diameter. The packing characteristics of spherical, cubical, needlelike, or filamentous particles will be dramatically different. For a monosized spherical model powder, the theoretical powder voidage of a closely packed system is approximately 40%. If a secondary powder component with a diameter equal to or smaller than that of the first powder component is added, the voidage can be reduced to approximately 16%. However, in practice, this spherical model packing arrangement will be limited by the size and shape of the die cavity [7]. For round tooling, the ratio of die-wall diameter to the particle diameter affects the repacking stage, and wide variations in packing fraction can be observed if this ratio is less than 10. Powders with poor flow characteristics protect relatively large voids by bridging that may cause a considerable rearrangement at low compaction pressure during the next stage of compaction.

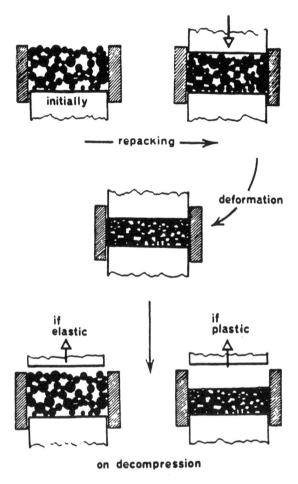

Fig. 1 Diagrammatic representation of the stages of compaction.

B. Deformation of Particles

As the upper punch penetrates the die containing the powder bed, initially, there are essentially only points of contact between the particles. Utilizing the punches, application of an external force to the bed results in forces being transmitted through these interparticulate points of contact, where the stress (i.e., the result of force) is developed and where local deformation of the material occurs. If there is any energy loss, it would be due to interparticulate and die-wall friction as well as deformation of particles. The deformation will feature either one or a combination of the following: elastic, plastic, and/or fragmentation. The type of deformation depends on the rate and magnitude of the applied force as well as the duration of the locally induced stress and physical properties of the material.

If the applied load is released before the stress reaches a yield value (Fig. 2), the particles deform elastically, that is, regain their original form, and return to the closely packed rearrangement state. The yield value is referred as the elastic limit above which deformation of the particles becomes irreversible. The slope of the linear portion of the plot given in Fig. 2 is termed the modulus of elasticity (i.e., Young's modulus). As the volume of the powder bed is reduced progressively with further load application, either plastic deformation or fragmentation becomes the dominant mechanism of compaction. Plastic deformation usually occurs with powders in which the shear strength is less than the tensile strength, whereas fragmentation becomes dominant with hard, brittle materials in which the shear strength is greater than the tensile strength. If the latter deformation occurs, the surface area of the powder and the potential bonding area between the particles increase with the introduction of new, clean surfaces due to fragmented particles. This furthers densification with the infiltration of these small fragments into the voids between the large particles.

There is no pharmaceutical powder that exhibits only one of the deformation mechanisms noted above, although the spectrum ranges from highly elastically deforming to highly plastically deforming or highly brittle materials. Even

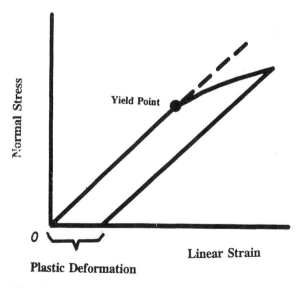

Fig. 2 Diagrammatic representation of the stress-strain relationship.

for materials that are known to be brittle, smaller particles of these materials may deform plastically. Although, so far, the physics of compaction were explained as sequential events, in practice, different regions of a powder bed can undergo different stages of compaction. For example, while particle rearrangement is occurring in certain regions of the powder bed, elastic deformation can take place in another region.

C. Bonding

Attraction between particles is inversely proportional to the distance between them, and when the particles are in sufficiently close proximity they can become permanently attached to each other. This is called bonding, and there are several mechanisms that will contribute to the strength of the bond, although they never act independently. The simplest type of bonding, mechanical interlocking, is essentially entanglement of particles in such a manner as to promote adhering. This mechanism is facilitated by irregular particle shape and surface roughness.

Particles can also bond as a result of a phase transition (i.e., melting) at the points of contact where the magnitude of the pressure is tremendously high as a result of the force applied on such a small area, and the temperature at those points may reach the melting points of the materials to liquefy the solid particles. This results in a significant increase in the contact area and a decrease in the pressure (and therefore in the temperature) at those points and subsequent solidification of the particles. This mechanism is called fusion or cold welding because the overall temperature of the tablet is still cold since typical pharmaceutical materials are poor heat conductors.

The third mechanism, intermolecular forces, encompass three known types of molecular bonding: van der Waals force, hydrogen bonding, and ionic bonding. Van der Waals forces are essentially typified by gravity and are dependent on the mass and distance of separation of the particles, whereas the other mechanisms actually involve the electrons.

D. Further Deformation of the Compact Followed by Its Expansion

When a compact has been formed in the die, a further increase in the applied pressure, or application of a constant stress during dwell time, may cause further deformation of a time-dependent material as the compact consolidates by viscoelastic and plastic flow. When the pressure is removed during decompression, which is the final stage of a compaction event, the compact undergoes a sudden elastic expansion followed by a much slower viscoelastic recovery as the compact is ejected. It is well known that it is the elastic expansion that is the main cause of some tableting problems, such as capping and lamination. Such problems can be minimized by altering the formulation and/or process conditions since the magnitudes of both elastic and viscoelastic recoveries depend on both the time-dependent properties of that material and the machine characteristics, such as the speed of compaction and the geometry of the punch travel guide.

III. EQUIPMENT AND MONITORED PARAMETERS

Instrumented single-station and multi-station presses have been used widely in compaction studies [8–13]. During the last two decades, isolated punch-die assemblies and integrated compaction research systems (ICRS), so-called compaction simulators, also became available [13]. The latter systems are designed specifically to be capable of mimicking the exact cycle of any tableting process (as well as following a customized profile) in real time and to record all important parameters during the cycle. As can be seen from Table 1, which compares the equipment on which tableting studies can be performed, integrated compaction research systems have many advantages over the others. In addition, the ICRS can be used in pharmaceutical research, development, and production for many purposes, such as studying basic compaction mechanisms, process variables, scale-up parameters, establishing compaction data banks, and finger-printing of pharmaceutical powders. The parameters monitored during compaction vary widely from one study to the other. Data obtained from the measurements of forces on and displacements of the upper and lower punches [11–15], axial to radial load transmission [16–21], die-wall friction [15], ejection force [10], temperature changes [22–24], and other miscellaneous parameters [10, 25,26] have been used to assess the compaction behavior of a variety of pharmaceutical powders and formulations.

Table 1 Comparison of Equipment for Tableting Studies

Feature	Single-station press	Multi-station press	Isolated punch and die set	ICRS
Mimic production conditions	No	Yes	Maybe	Yes
Mimic cycles of many presses	No	No	Maybe	Yes
Require small amount of material	Yes	No	Yes	Yes
Easy to instrument	Yes	No	Yes	Yes
Easy to set up	Yes	No	Maybe	Maybe
Data base in literature	Yes	Yes	Some	Some
Used for stress-strain studies	No	No	Yes	Yes
Equipment inexpensive	Yes	No	Maybe	No

Source: Ref. 13.

IV. DATA ANALYSIS TECHNIQUE

It has been pointed out that the number of compaction equations proposed to characterize the compressional process approximates the number of workers in this field [13]. However, many of the equations have been shown to have applicability over only a limited range of applied force and for only a few types of materials. Certainly, no universal relationship has yet emerged and is unlikely to do so, since a comprehensive analysis of the mechanisms involved is difficult due to the complexity of the systems being compacted. More than one data evaluation technique may have to be applied in order to increase the validity of the conclusions drawn from the results of a compaction study. Table 2 summarizes the information required concerning compaction of a solid system, and the

Table 2 Recommended Methods of Studying Important Compaction Parameters

Information required	Preferred method
Stages of compaction	Heckel equation
	Kawakita equation
Compressibility	Percent porosity versus pressure
	Leuenberger equation
	Kawakita equation (at low pressures)
	Heckel equations (at high pressures)
Consolidation	Heckel equation
	Leuenberger equation
	BI (bonding index)
Elastic deformation/elastic expansion	E (percentage elastic recovery)
	Work done on the lower punch in a second compression
Plastic deformation/brittle fracture	Heckel equation
	Compression cycles
Yield strength/yield pressure	Heckel equation
	Compression cycles
Work of elastic deformation	F-D curves
Work of friction	
Net work of compaction	
Lubricity	R value
	Compression cycles
	Unckel/Shaxby-Evans equations
Capping tendency/lamination tendency	ERI (elastic recovery index)
	PC/ER (plastoelasticity index)
	BFI (brittle fracture index)

Source: Ref. 27.

method(s) recommended to obtain that information. It would be convenient at this stage to classify the compaction data analysis techniques into three groups in order to evaluate these methods as those that utilize the data obtained from the measurements of (1) the applied pressure and punch displacements, (2) pressure transmitted, and (3) miscellaneous parameters.

A. Utilization of Applied Pressure and Punch Displacement Data

A number of mathematical equations, many of which are empirical in nature, have been developed from the relationship between applied pressure and the volume of the powder bed to assess the densification and consolidation mechanisms involved during the compaction of solid materials. These equations have been evaluated and compared by several authors [27–30].

Percentage Porosity

One of the methods used to determine the compressibility of a powder bed, that is, the degree of volume reduction due to an applied pressure, is measurement of the porosity changes during compaction. The following equation can be used for percentage porosity (ϵ) determination:

$$\epsilon(\%) = 100\left(1 - \frac{V_t}{V_c}\right) \tag{1}$$

where V_t is the specific solid volume (i.e., true volume) of the material and V_c is the compact volume at a given pressure. Assuming that radial die-wall expansion during compaction is negligible, this equation can be written in the form

$$\epsilon(\%) = 100\left(1 - \frac{H_t}{H_c}\right) \tag{2}$$

where H_t is the theoretical true thickness of the compact at zero porosity and H_c is the compact thickness at a given pressure (corrected for punch deformation).

The percentage porosity method, among others, was utilized in a comprehensive study by Maganti and Çelik of the compactional behavior of the uncoated and Surelease (aqueous ethyl cellulose dispersion)-coated microspheres [31,32]. In part I of the study [31], they compared the percentage porosity profiles of the compacts made from the powder and the uncoated microspherical forms of microcrystalline cellulose. As shown in Fig. 3a, microcrystalline cellulose powder exhibited a higher initial porosity change at low pressures than that of its uncoated microspheres. This was attributed to the higher initial porosity of the powder (geometric mean diameter $D_g = 50$ μm), which required low pressures to undergo repacking, whereas microspheres ($D_g = 650$ μm) exhibited a shorter particle rearrangement stage and proceeded to the subsequent steps of compaction. Upon compaction, the microspheres required lower pressures than the powder did to obtain the same porosities. The difference between the magnitude of the applied pressure to achieve the corresponding porosities increased as the minimum porosity was approached. However, the powder produced stronger compacts than the pellets, suggesting that the ability of a material to be reduced in volume when compressed does not ensure the formation of a strong compact.

In part II of the study [32], Maganti and Çelik compared the percentage porosity profiles of the uncoated pellets consisting of microcrystalline cellulose (80%), propranolol HCl (10%), and dicalcium phosphate or lactose (10%) with the Surelease-coated form of these pellets. They observed that the uncoated pellets required higher applied pressures to produce compacts of the same in-die porosities as the coated pellets (Fig. 3b). The maximum applied pressure required to compact the coated pellets to the predetermined in-die porosity decreased as the amount of coating on the pellets increased, indicating that pellets coated with higher amounts of coating were more easily compressed.

Walker and Bal'shin Equations

Most of the earliest attempts to study the degree of consolidation of powders were made in the field of powder metallurgy [3,33,34]. As early as 1923, using materials such as lead shots, ammonium chloride, sodium chloride, and potassium nitrate, Walker [33] observed a logarithmic relationship between the applied pressure (P_a) and the relative volume (V_r) of the compact as

$$V_r = C_1 - K_1 \log P_a \qquad (3)$$

where C_1 and K_1 are constants. In this work, K_1 values were found to be greater for plastically deforming materials than for brittle fracturing materials. Walker related the ratio of C_1/K_1 to the compactional behavior of powders and reported that relatively high ratios were generally found for materials producing weak tablets. In a recent work [13], the C_1/K_1 ratios were found to be higher for elastically deforming materials than for plastically deforming or brittle fracturing substances.

Kawakita Equation

Another equation that has received considerable attention in the field of powder compaction was developed by Kawakita [35] and is expressed as

$$C = \frac{V_i - V_p}{V_i} = \frac{abP_a}{1 + bP_a} \qquad (4)$$

where C is the degree of volume reduction, V_i the initial apparent volume, V_p the powder volume under applied pressure P_a, and both a and b are constants that can be calculated from the P_a/C versus P_a plots. The constant a does not correlate to any properties of the material being compacted, while the constant b, the

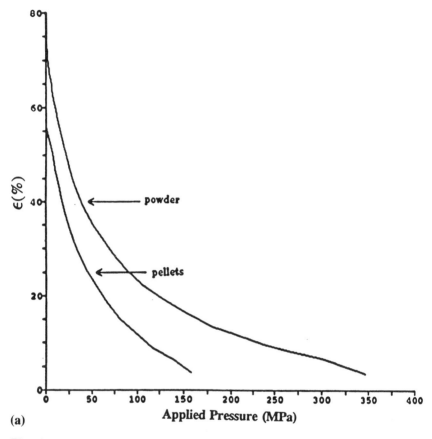

(a)

Fig. 3   Porosity versus pressure plots for compacts of (a) microcrystalline cellulose powder and pellets (after Ref. 31), and (b) uncoated and Surelease-coated microspheres of F-I formulation consisting of microcrystalline cellulose (80%), lactose (10%), and propranolol HCl (10%) (after Ref. 32).

coefficient of compression, is related to the plasticity of the material [36]. Another limitation of the Kawakita equation is that using this method, the compaction process can be described only up to a certain pressure, above which the equation is no longer linear [37].

Heckel Equation

The most commonly used equation in pharmaceutical compaction studies was developed by Heckel [38,39], who considered that the reduction in voidage obeys a first-order kinetics type of reaction with applied pressure. This relation-

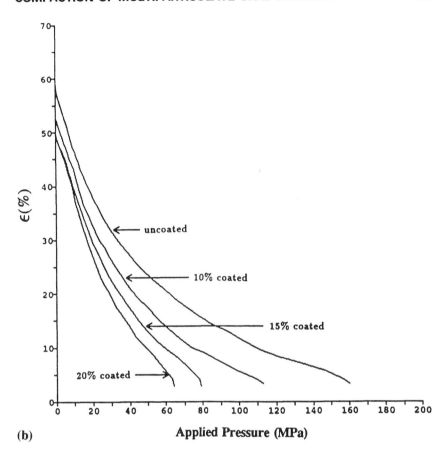

(b)   **Applied Pressure (MPa)**

ship between porosity and applied pressure is similar to those proposed previously, by some other workers [40–42]. Heckel's equation is expressed as

$$\ln \frac{1}{1 - \rho_r} = KP_a + A \tag{5}$$

where ρ_r is the relative density of the compact, and K and A are constants that can be determined from the slope and intercept of the extrapolated linear region of the plot, respectively, as shown diagrammatically in Fig. 4a. Hersey and Rees [43] related the constant K to the mean yield pressure P_y as

$$K = \frac{1}{P_y} \tag{6}$$

Hence K is related inversely to the ability of a material to deform plastically under pressure. The constant A is a function of the initial compact volume and

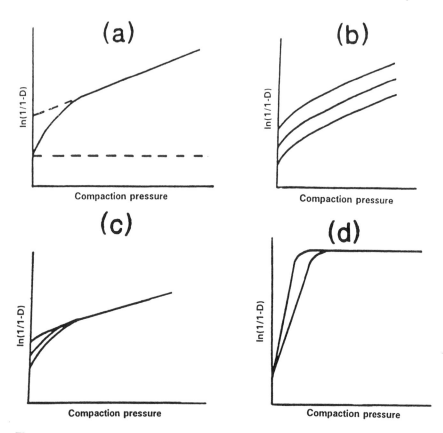

Fig. 4   Heckel plots: (a) diagrammatic representation; (b) type 1 (after Ref. 43); (c) type 2 (after Ref. 43); (d) type 3 (after Ref. 46).

can be related to the densification during die filling and particle rearrangement prior to the bonding. The Heckel equation was found to be similar to several other compaction equations [29,44]. These equations differ only in the nature of the constant A, which was considered by Heckel [39] to be the sum of two terms rather than a true constant.

As can be seen in Fig. 4a, a typical Heckel plot is linear only at high pressures. Heckel [39] suggested that the initial curved region of the plots can be attributed to the particle movement and rearrangement in the absence of inter-particulate bonding. Therefore, he proposed that this stage corresponds to the packing process discussed by Seelig and Wulff [3]. The linear portion of a $\ln[1/(1 - \rho_r)]$ versus P_a plot was attributed by Heckel to both plastic deformation and cold welding.

The application of Heckel plots to the compaction of pharmaceutical powders permits an interpretation of the consolidation mechanisms and a measure of the yield pressure of the powder under examination. Heckel plots have been classified into three types [43,45,46]. The type 1 plots are exhibited by different particle-size fractions of a given material that consolidates by plastic flow (Fig. 4b). Variations in initial powder bed density result in different final bed densities under any particular applied pressure. In the type 2 plots, exhibited by materials that consolidate by particle fragmentation, a single relationship occurs above a certain pressure irrespective of the initial bed density (Fig. 4c). This feature is also independent of particle size and is thought to be due to the progressive destruction of the particles by fragmentation and their subsequent compaction by plastic deformation. The type 3 plots are attributed to the absence of a particle rearrangement stage coupled with plastic deformation and the possible melting of asperities (Fig. 4d).

Using Heckel plots, Fell and Newton [47] pointed out that density measurements made under load include an elastic component that will increase the value of $\ln[1/(1 - \rho_r)]$ for a given compaction force and therefore will result in a false (low) yield pressure value that is inversely related to the ability of a material to deform plastically under pressure. Roberts and Rowe [48] investigated the effect of punch velocity on the compaction behavior of a number of pharmaceutical materials by determining the change in their yield pressures. They observed that plastically deforming materials exhibited an increase in the yield pressure with increasing punch velocity. This was attributed to a change from ductile to brittle behavior or a reduction in the amount of plastic deformation due to the time-dependent nature of plastic flow.

Çelik and Marshall [13] suggested that compaction data collected under static conditions, where porosity values are calculated from out-of-die measurements on tablets, would be expected to produce results different from those obtained under dynamic conditions, where porosity values are determined from in-die measurements, since presumably, the elastic recovery of a tablet contributes to its porosity. It must be kept in mind that the fidelity of in-die measurements depends on whether the compaction data are corrected accurately for deformation of the punches and associated machine components. The number of data points used to obtain a Heckel profile is another factor that affects the accuracy of this method. Rue and Rees [49] suggested that the Heckel equation should be used with caution since the Heckel profiles obtained for a given material vary depending on the experimental conditions. Therefore, they proposed use of the area under the Heckel plots obtained at varying contact times as a quantitative method to compare the degree of plastic deformation for a range of materials. Düberg and Nyström [50] applied the Heckel equation to both compression and decompression phases in an attempt to distinguish plastic and elastic deformation characteristics of the materials.

Several workers [37,43,44,51] treated their compaction data using both Kawakita and Heckel equations, to compare the applicability of these methods. Some of them [43,44] suggested that the Kawakita equation does not yield as much information as does the simpler Heckel equation. Others [37,51] proposed the use of both methods together, to describe the mechanisms of compaction more accurately, since the linearity is observed at low pressures with the Kawakita equation and at high pressures with the Heckel equation. On the other hand, some authors [52] noted that the Heckel equation applies only to pressures so high that they actually are not in the range of pressures applied to pharmaceutical tablets.

Maganti and Çelik [31,32] applied the Heckel equation to data obtained for the compacts of microcrystalline cellulose powder and pellets as well as uncoated and coated pellet formulations. The slopes of the linear portion of the Heckel plots differed for the powder and pellet forms (Fig. 5a), suggesting that changing the shape, size, and surface properties of microcrystalline cellulose particles may have affected the compaction properties (such as degree of bonding) of this material. When the Heckel plots were compared for coated and uncoated pellet formulations, these workers observed an increase in the slopes of the linear portion with increasing amounts of coating (Fig. 5b). However, the authors pointed out that using this method, it was difficult to distinguish between elastic and plastic deformation. This was in agreement with the opinion of Düberg and Nyström [50], who suggested that the density values at pressure contain both elastic and plastic components; therefore, the yield pressure reflects the total ability of the particles to deform.

Cooper and Eaton Equation

Cooper and Eaton [53] developed an equation relating the fractional consolidation, defined as the decrease in voidage as a proportion of the original voidage, with pressure for the compaction of certain powders:

$$\frac{V_i - V_p}{V_i - V_t} = C_2 \exp\left(\frac{-K_2}{P_a} \right) + C_3 \exp\left(\frac{-K_3}{P_a} \right) \tag{7}$$

where C_2, C_3, K_2, and K_3 are constants. The two terms on the right-hand side of the equation are related to the slippage of particles at early stages of compaction and to the subsequent elastic deformation, respectively. Their findings supported the view that the yield strength values of metal powder compacts are related to the linear portion of the Heckel plots. However, the difficulty in practical use of Eq. (7) is the assignment of some physical significance to the constant parameters of this equation. Another drawback of this method is that this equation applies only to a single-component system [52].

Leuenberger Equation

A relatively new method used in the field of powder compaction was proposed by Leuenberger [54], who related the two important indices of powder compression: compactability (the ability of a material to yield a compact with adequate strength) and compressibility (ability of the material to undergo volume reduction under pressure). This relationship can be expressed as

$$P_{dh} = P_{max}[1 - \exp(-\Upsilon P_a \, \rho_r)] \tag{8}$$

where P_{dh} is the deformation (Brinell) hardness, P_{max} denotes the theoretical maximum deformation hardness that would be attained as P_a approaches infinity and relative density approaches 1, and Υ is the compression susceptibility. Leuenberger and others [55–57] observed a good correlation when they applied Eq. (8) and its modified versions to the single-component powders and their binary mixtures. It was noted by these workers that a low P_{max} value shows a relatively poor compactability, and this limiting value cannot be exceeded even at very high compaction pressures. A high value of Υ indicates that the theoretical limit of hardness and a sharp decrease in compact porosity may be attained with relatively low compaction pressures.

F-D Curves; Energy and Power Involved in Compaction

A common method for assessment of the compaction behavior of materials is the use of compression force versus punch displacement profiles (F-D curves), from which the work involved during tablet compaction can be calculated (Fig. 6a). Krycer and others [58] suggested that since powders with different packing characteristics and different elastic–plastic deformation properties will absorb varying amounts of energy, it might be more useful to measure the work of compaction rather than other characteristics. Fell and Newton [59] formed the opinion that the work done on compaction of a powder mass is utilized for both volume reduction and particle bonding but that only the latter contributes to the strength of the tablet. The total work done on compaction is, therefore, not necessarily a criterion of tablet strength. However, this conclusion was not supported by the work of Çelik and Marshall [13], who observed that the rank order of total energy involved during compaction of the powders and the tensile strength of their compacts were similar.

In the literature, the method of obtaining F-D curves and the definition and utilization of work phenomenon vary considerably. In Fig. 6a, the area *OAB* represents the work exerted by the top punch on the material compressed. As the compact expands during decompression, some of this work will be transferred back to the upper punch. This work of expansion during decompression is represented by the area *ABC*. Hence the area *OAC* may be considered to correspond to the total work involved in compaction. However, as Jones [60]

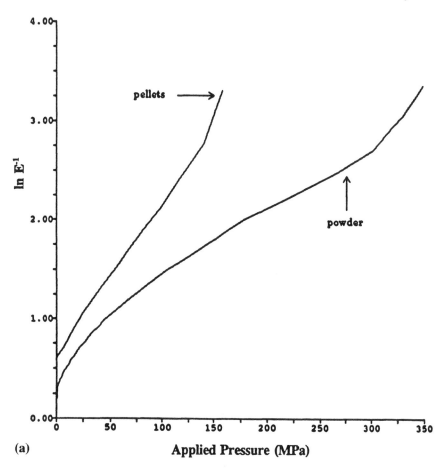

Fig. 5   Heckel plots for compacts of (a) microcrystalline cellulose powder and pellets (after Ref. 31), and (b) uncoated and Surelease coated microspheres of F-I formulation [consisting of microcrystalline cellulose (80%), lactose (10%), and propranolol HCl (10%)] and F-II formulation [consisting of microcrystalline cellulose (80%), dicalcium phosphate dihydrate (10%), and propranolol HCl (10%)] (after Ref. 32).

pointed out, compact expansion due to elastic recovery may continue *after* the top punch has lost contact with the compact, and thus the *measured* work of compaction may not truly represent the *complete* work of compaction. In addition, the work required to overcome friction at the die wall can be calculated from the difference between the area under the curves obtained for the upper and lower punches (Fig. 6b). If this work is also deducted from the total work, the overall net work of compaction can be calculated [61]. Lammens and others [62]

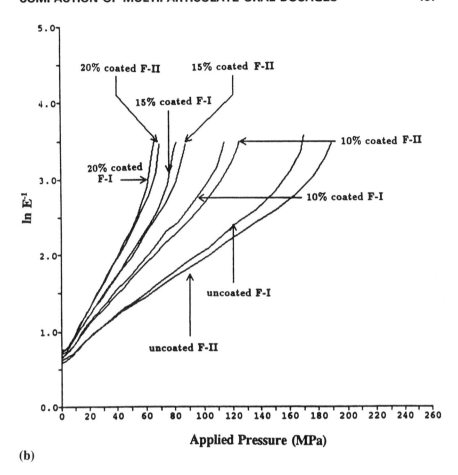

(b)

suggested that although the force exerted by the upper punch is generally used for obtaining F-D curves, measurement of the force transmitted to the lower punch may be of more value.

Çelik and Marshall [13] obtained F-D curves (for both single- and double-ended compaction studies) by plotting the arithmetic mean of the upper and lower punch forces versus changes in the height of the powder bed during compaction. They observed hysteresis in such curves when the F-D curves were obtained by plotting the upper punch force versus its displacement. They then calculated the total work of compaction (TWC) using the following equation, in which the contributions of the upper and lower punches to the total work were estimated separately:

$$\text{TWC} = \int_{X=0}^{X_{\text{max(up)}}} F_{\text{up}} \, dX_{\text{up}} + \int_{X=0}^{X_{\text{max(lp)}}} F_{\text{lp}} \, dX_{\text{lp}} \qquad (9)$$

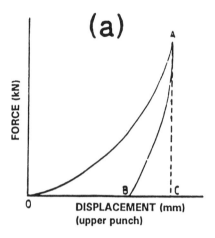

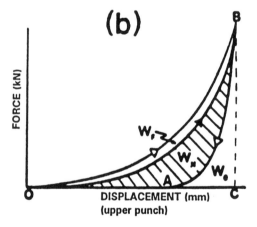

Fig. 6 Force versus displacement profiles (F-D curves): (a) diagrammatic representation; (b) W_f, work done in overcoming friction; W_e, elastic deformation energy; W_n, net mechanical energy.

where F_{up} and F_{lp} are the forces on the upper and lower punches, respectively; X_{up} and X_{lp} are the contributions of the upper and lower punches, respectively, to the decrease in the distance between them; $X = 0$ is the point where the porosity corresponds to the initial porosity and maximum applied load is attained at $X_{max(up)}$ and $X_{max(lp)}$.

Assuming that the table has completed its plastic deformation and fully recovered elastically prior to a second compression (recompression), de Blaey and his co-workers [11,63,64] reported that the work done on the lower punch during the recompression phase can be used as a measure of the elastic energy recovered upon removal of the upper punch in the first compaction. However, Krycer and others [58] suggested that the assumptions above may not be strictly applicable in practice.

In their compaction study of the powder and pellets of microcrystalline cellulose, Maganti and Çelik [31] reported that the TWC values and the strength of the compacts of microcrystalline cellulose pellets were both significantly less than those of the powder form, suggesting that the degree of bonding of this material has been affected considerably by the changes in its shape, size, and also possibly by the reduction in the number of potential bonding sites that occur due to the pelletization process. They proposed that the pellets, which are large and spherical in shape as compared to the small, irregular powder particles, have a low surface/volume ratio, and this might have resulted in a decreased area of contact between the particles as they consolidated. These workers also compared the TWC values of the uncoated and Surelease-coated pellet formulations and reported that increasing amounts of coating exhibited relatively lower TWC values at corresponding pressures [32]. A similar correlation was observed in the strength of their ejected compacts.

An alternative way of presenting work of compaction data is the use of power profiles (i.e., energy per unit time), on which very little information exists in the literature [13,65–67]. Çelik and Marshall [13] employed two different methods to analyze their power data. The first technique involved the calculation of instant power (φ_n) by using the following equation, in which the force and displacement data of the total tableting cycle are analyzed in small fractions:

$$\varphi_n = \frac{\Delta d_n}{t_n} \left(\frac{\Delta F_n}{2} + F_n \right) \tag{10}$$

where Δd_n is the distance the punch moves, ΔF_n the change in the applied force during the interval t_n, and F_n the force at the beginning of t_n.

It was observed that the peak of the instant power versus pressure plots corresponded to the point of inflection on the force versus time profiles. The second technique was termed the average power of consumption (APC), which was calculated by dividing the cumulative total work of compaction by the corresponding contact time using the equation

$$\text{APC} = \int_{X=0}^{X_{max}} \frac{F \, dX}{t} \tag{11}$$

where F is the applied force; X the decrease in the distance between the upper and lower punches; $X = 0$ and X_{max} are the points where ϵ (the porosity) = ϵ_0 and ϵ_{max}, respectively; and t is the time for which the powder has been compressed. In their work, the slopes of the average power consumption versus pressure plots could clearly be divided into two groups, those where strong or

weak tablets were produced. However, this was later contradicted by the findings
of Maganti and Çelik [31], who reported that the compacts of the pellets of
microcrystalline cellulose exhibited higher APC values at corresponding pressures than those of the powder form of this material. This was attributed to the
shorter contact time (the time during which the upper punch remained in contact
with the material) due to the absence of a particle rearrangement stage, which
was observed only in the case of the powder. This suggests that the method used
to determine the time interval variable and to present the power data can change
both the magnitude and rank order of power consumption significantly. Applying
three different power equations to the same data, Armstrong and Palfrey [66]
also obtained different total power expenditures. Therefore, power data must be
used with caution if power consumption is to be correlated with the compaction
characteristics of a powder.

B. Utilization of Transmitted Pressure Data

Upper-to-Lower Punch Stress Transmission
During single-ended compaction of a powder mass, a rather complicated force
transmission pattern, with major components of F_a, the axial force causing
compression of the powder, and F_r, the radial force exerted on the die wall,
occurs [68], as shown in Fig. 7. The force sensed by the lower punch, F_b, is less
than F_a because of the existence of a frictional force, F_d, between the die wall and
the solid being compacted.

 If μ is the coefficient of friction between the particulate solid and the die
wall, then

$$F_d = \mu F_r \tag{12}$$

Shaxby and Evans [69] found a logarithmic relationship between applied pressure, P_a, and transmitted pressure to the lower punch, P_b, as

$$P_a = P_b \exp\left(\frac{4H_c K_4}{D} \right) \tag{13}$$

where H_c is the thickness of the compact, D the diameter, and K_4 a material
constant. Unckel [70] and Toor and Eagleton [71] modified this equation to
include the ratio of radial to axial stress, Ω (which is equal to the Poisson ratio,
v), and the coefficient of friction, μ, as

$$P_a = P_b \exp\left(\frac{4\mu\Omega H_c}{D} \right) \tag{14}$$

This equation can also be expressed as

$$\ln \frac{P_a}{P_b} = \frac{4\mu\Omega H_c}{D} \tag{15}$$

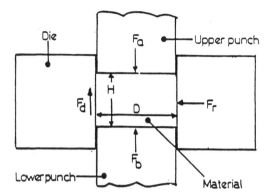

Fig. 7 Diagrammatic representation of the forces operating in a die. (After Ref. 68.)

In practical terms, the equations above are useful to show the importance of a low H_c/D ratio for the tablet being compacted and the influence of the coefficient of friction between the powder and the die wall. When the H_c/D ratio is high, there will be a considerable decay of the applied pressure down the compact length, resulting in a poorly compacted region adjacent to the stationary punch. In the case of a rotary machine, the effect is reduced by employing two moving punches to apply pressure to the compact.

Since Nelson and others [72] elucidated the potential usefulness of the ratio

$$R = \frac{F_b}{F_a} \tag{16}$$

where R, which is termed the R value, is the coefficient of lubricant efficiency. A number of authors [73,74] have used this ratio to study the efficiency of lubricants in tablet formulations. However, studies by de Blaey and Polderman [11] indicated that since force transmission is not constant during the complete compaction cycle, a large variation occurs in the amount of work required to overcome the friction with the die wall, especially during the initial phase of compaction. Consequently, it was concluded that R values measured at the end of the cycle may not truly represent the values prevailing during compaction. Guyot and others [75] have suggested that the work input calculated from the area under the curves of the upper punch displacement and the mean value of upper and lower punch forces can be used as a lubrication index instead of the R value.

Axial-to-Radial Stress Transmission (Compression Cycles)

Examination of the axial-to-radial stresses during a complete compression cycle (i.e., application, relaxation, and dissipation phases of the applied pressure) can yield data to obtain quantitative values for a number of compaction properties of powders. Using load cells and strain gauges, Nelson [16] studied the transmission of forces to the die wall during the application of axial pressure. However, this work was not extended to cover the relaxation and dissipation stages of the applied pressure.

Long [17] employed a split die to measure the radial pressure over a complete compression cycle. Assuming that the die was perfectly rigid, and there was no die wall friction, he described two possible compression profiles for ideal isotropic systems: a body with a constant yield stress in shear (Fig. 8a) and a Mohr's body (Fig. 8b). Initially, as the upper punch descends into a filled die, the powder particles are rearranged to form closer packing and then begin to undergo elastic deformation. During this phase, the axial pressure (P_a) transmitted through the powder mass generates a radial pressure (P_r) that can be determined from the following equation:

$$P_r = \nu P_r \qquad (17)$$

where ν is Poisson's ratio.

With further increase in P_a, the elastic limit (i.e., the yield point, which is denoted by A and A' in Fig. 8a and b, respectively) of the material is exceeded, beyond which either brittle fracture or plastic deformation takes place and bonds form between the particles. At this stage, the radial force is no longer determined by Poisson's ratio but either exhibits constant yield in stress where the relationship between P_a and P_r is unity (AB), or behaves as a Mohr's body in which the yield stress is a function of the normal stress across the shear plane and the slope of $A'B'$ may vary in magnitude.

In the first case, after the maximum force has been applied (B and B' in Fig. 8a and b, respectively), the change in P_r is once again the Poisson's ratio times the rate of change of P_a. Therefore, the slope BC is equal to the slope of OA. In the second case (the Mohr's body), on release of P_a, the radial pressure diminishes following the line $B'C'$ and its slope is equal to slope of the initial curve OA'. Then the yield takes place (C and C' in Fig. 8a and b, respectively) and it continues until P_a has returned to zero. At this point (D and D' in Fig. 8a and b, respectively), there is a residual radial stress still exerted on the compact and in the case of a Mohr's body, this residual die wall stress is somewhat larger.

Leigh and others [18] suggested a third possible compression profile presenting the behavior of a perfectly elastic body or a body compressed below its elastic limit (Fig. 8c). On unloading, the radial force returns to zero. There will be no residual radial force exerted on the die wall and the body will be free to move out from the die. These authors also suggested that the concepts above, developed for solid isotropic bodies, can be explained in particulate systems in terms of the following stages of compact deformation: OA (or OA') = elastic deformation; AB (or $A'B'$) = plastic flow/crushing.

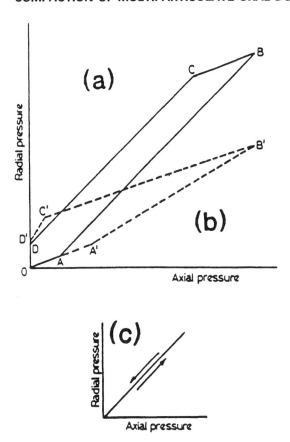

Fig. 8 Diagrammatic representation of radial versus axial pressure cycles: (a) body with a constant yield stress in shear (after Ref. 17); (b) Mohr body (after Ref. 17); (c) ideal elastic body (after Ref. 18).

Leigh and others [18] also reported that the analysis of the compression cycles of materials may indicate the formation of satisfactory or unsatisfactory compressed tablets. These workers observed that acetaminophen granulations behaving as a Mohr's body tended to cap and laminate and found that the change from a compact which readily caps and laminates to an intact one was associated with the transformation from a Mohr's body type of compression cycle to one with a constant yield stress is shear. Contrary to these findings, Shotton and Obiorah [76] observed no tendency to cap or laminate, although their materials behaved like a Mohr's body as well. Instead, they suggested that materials possessing low residual die wall pressure would yield weak tablets that might cap on ejection.

C. Miscellaneous Techniques

Measurements of Elastic Expansion of the Compacts

It has been pointed out that [77–79] on release of compression force, the compact can expand within the die and may continue to do so after ejection. This expansion may generate stresses that are the main cause of some tableting problems, such as capping and lamination. Employing the acoustic emission technique, Rue and others [79] observed that both capping and lamination occurred completely within the die during decompression and not during ejection. Numerous techniques have been employed to determine the magnitude and the disruptive effects of the elastic recovery of formed compacts. Some of these were evaluated by Krycer and others [80]. Several workers [80–82] used the percentage recovery (*E*), defined as

$$E(\%) = 100\left(\frac{H_e - H_c}{H_c} \right) \tag{18}$$

where H_c and H_e are the heights of the compacts under pressure and after ejection, respectively. However, the expansion of compacts has been measured at varying times following their ejection by different authors [80,83,84], and therefore, the values may include the slower viscoelastic recovery as well as the rapid elastic expansion. Another drawback of this equation is that the radial expansion of the compact is not considered. Recently, Maganti and Çelik [31] modified this equation to take into account the radial expansion, which was later termed volumetric strain recovery (VSR) by Çelik and Okutgen [85] as follows:

$$VSR(\%) = 100 \left(\frac{D_e^2 H_e - D_c^2 H_c}{D_c^2 H_c} \right) \tag{19}$$

where D_c and D_e are the diameters of the compacts at maximum load in the die and after ejection, respectively.

Using Eq. (19), Maganti and Çelik [31] compared the VSR values of the compacts of powder and pellet forms of their formulations, and observed that the compacts of microspheres had higher VSR values than those made from the corresponding powder formulations. This was attributed to the greater expansion of the compacts of pellets during the decompression and ejection phases of the compaction event. A large decrease in the mechanical strength of the compacts of the pellets suggested that many of the bonds formed during compaction did not survive the unloading and ejection phases. Later, Maganti and Çelik [32] reported that both the VSR values and the mechanical strength of the compacts of pellets decreased when the pellets were coated with 10% of Surelease. This was attributed to an increase in the number of bonds between these phases due to introduction of stronger binder–substrate bonds by Surelease. However, further increases in the amount of coating (15% and 20%) on the pellets diminished the mechanical strength of the resulting compacts. Although their mechanical strength was relatively higher, the VSR values of these compacts were found to be higher than those of the uncoated pellets. The "apparent" contrast in the rank order of VSR values and mechanical strength values was attributed to the differences in the elastic properties of the uncoated pellets and the film coating.

Another technique of studying elastic expansion is the use of the work done on the lower punch in a second compaction [65,86]. As cited earlier, this method assumes that the tablet fully recovers elastically prior to recompression and that only elastic deformation occurs on the recompression stage. However, it is questionable whether the foregoing assumptions apply in practice. Summers and others [86] employed an elastic modulus (EM) defined as

$$EM = \frac{P_a H_c}{\Delta H_c} \tag{20}$$

where ΔH_c is the decrease in the height H_c.

An alternative technique to measure the disruptive effects of elastic expansion was suggested by Çelik [21] and Çelik and Travers [87], who proposed the elastic recovery index (ERI), defined as

$$ERI = \frac{ER}{SM} \tag{21}$$

where ER is the magnitude of elastic recovery undergone by the compact during decompression (corrected for punch deformation) and SM is the strain movement (i.e., the magnitude of further plastic and viscoelastic deformation under a constant stress). It was suggested that if a material has a high ERI value, it forms weak, possibly capped and/or laminated tablets. Using a similar ratio, ER/PC, where PC is the plastic compression determined under constant strain conditions, some others workers [88,89] observed high ratio values for the compacts of materials with capping tendencies.

Percolation Theory and Fractal Geometry

A recent development in the field of pharmaceutical compaction is the application of percolation theory and fractal geometry by Leuenberger and co-workers [90–92]. Percolation theory deals with the number and properties of clusters of randomly occupied sites in a geometric lattice [93]. It describes a phenomenon whereby a property of a system as a function of a continuously varying parameter which diverges, vanishes, or just begins to be manifested at one sharply defined point. This point, referred to as the percolation threshold, is where a component of the system just begins to percolate or to form continuous structure throughout the length, width, and height of the system [94].

Three types of percolation have been described [94]:

1. *Site percolation*, where a cluster is a group of neighboring occupied sites.
2. *Bond percolation*, where all sites are occupied and the cluster is a group of neighbors connected by bonds.
3. *Site–bond percolation*, which is a combination of the above two.

Depending on dosage form design, important changes in mechanical and bio-pharmaceutical properties of the tablet occur at the percolation threshold. The percolation threshold of a powder system varied with the material characteristics of different substances depending on the geometric arrangement of the particles. Therefore, the particle size, size distribution, and shape may also influence the value of the percolation threshold [95]. It has been suggested that the data obtained from the compaction of binary powder systems can be explained satisfactorily with the help of the percolation theory [95]. However, more work is needed to verify the applicability of this theory to the pharmaceutical compaction.

Measurements of Tablet Strength

The need to quantify the strength of tablets has long been recognized as a quality control parameter both in compaction research and in industrial practice. The mechanical strength of the tablets has been described by several means, including the crushing strength [96,97], the axial [98–100] or radial tensile strength [101–103], hardness [104–107], and the work required to cause tablet failure [108]. A common method of assessing the strength of tablets involves measurement of the force required to break a tablet in a diametrical compression test. Using a load transducer in place of a tablet, Brook and Marshall [96] compared several crushing strength testers and observed variations in crushing strength values due partially to inaccuracies in the instrument scale values and varying methods of applying load.

The radial tensile strength (T_s) of tablets can be calculated from the following equation:

$$T_s = \frac{2F_c}{\pi D H_e} \qquad (22)$$

where F_c is the force needed to break the tablet, and D and H_e are the diameter and the thickness of the ejected tablet, respectively. Several precautions must be taken when using this equation. Various factors (e.g., test conditions, deformation properties of the material, adhesion conditions between the compact and its support, tablet shape) may influence the measurements of the tensile strength [98,102]. A modified form of Eq. (22) can be used for measurement of the tensile strength of convex-faced tablets [103].

Millili and Schwartz [109] reported that the strength and physical properties of microspheres containing microcrystalline cellulose were affected by the granulating solvent. In their work, water granulated microcrystalline cellulose pellets were found to be strong, hard, and uniform in shape, whereas the 95:5 ethanol/water granulating solvent resulted in microspheres with lower strength and less uniform shape. On the other hand, the former pellets exhibited poorer compressibility than the latter ones. This was attributed to the weak bonding of 95% ethanol-granulated pellets, which ruptured upon compaction, exposing more smooth surface-to-surface contacts for bonding. The water-granulated pellets resisted rupturing due to their strong bond strength and allowed fewer surface-to-surface contacts for bonding to occur.

Maganti and Çelik [32] observed that the radial tensile strength values of the compacts of uncoated microspheres were less than those of the Surelease-coated pellets. However, further increases in the amount of coating caused a reduction in the strength of the compacts. This was attributed to the adhesive binding properties of Surelease. The increase in the radial tensile strength of the compacts with the addition of a small amount (10%) of Surelease as a coating agent to the microspheres was presumed to cause the development of some binder–binder and binder–substrate bonds, in addition to the substrate–substrate bonds between the fragmented neighboring microspheres. On the other hand, further increases in the amount of coating caused an increase in the overall binder concentration and in the relative ratio of binder–binder bonds to substrate–binder bonds, thereby producing compacts with lower radial tensile strength values due to the lack of cohesive properties of Surelease.

Some authors [90–100] suggested the determination of axial tensile strength because of the sensitivity of radial tensile strength measurements to crack propagation variations. The axial tensile strength (T_{sx}) can be calculated from the relationship

$$T_{sx} = \frac{4F_c}{\pi D^2} \tag{23}$$

The measurements of tablet strength in axial direction were found to be valuable for the assessment of capping and/or lamination tendencies of the materials [110].

Crushing strength is often imprecisely termed "hardness," which is, in fact, a surface property measured by the resistance of a solid to local permanent deformation. Hardness can be determined by either static (e.g., the Brinell, Vickers, and Rockwell hardness tests) or dynamic methods [56]. The static indentation methods involve the formation of a permanent indentation on the surface of the material being tested and the hardness is determined by means of the load applied and the size of the indentation formed [21,104,106]. In the dynamic indentation tests, either a pendulum is allowed to strike from a known distance or an indenter is allowed to fall under gravity onto the surface of the test material. The hardness is then determined from the rebound height of the pendulum or the volume of the resulting indentation. Using an apparatus consisting of a steel sphere pendulum acting as an indenter, Hiestand and others [105] estimated the hardness (i.e., the mean deformation pressure) of compacted materials by dividing the energy consumed during impact by the volume of the indentation.

Hiestand and Smith [111] used the indentation hardness and the tensile strength measurements to formulate the following three dimensionless indices to characterize the relative tabletability of single components and mixtures:

1. *Bonding index* (BI) is defined as the ratio of the tensile strength to the dynamic indentation hardness and is claimed to be the indicator of the survival success of areas of contact.
2. *Brittle fracture index* (BFI) is a measure of brittleness, which is the principal cause of capping and lamination.
3. *Strain index* (SI) is obtained during the determination of dynamic indentation hardness and is indirectly related to the proximity of the surfaces that remain in contact after decompression.

V. OTHER COMPACTION STUDIES ON THE MULTIPARTICULATE DOSAGE FORMS

There is only a small amount of literature available on the compaction characteristics of microspheres or microcapsules, some of which were referred to earlier in this chapter. As can be seen from the following, most of these studies report the effects of compaction on the drug release profiles. Juslin and others [112] studied the feasibility of achieving controlled release of a drug from compacts of coated spheres. In their study, phenazone spheres were coated with acrylate plastic, mixed with different additives and compacted into tablets. They observed that the drug release rate increased with an initial increase in the applied pressure. This was attributed to the cracks in the coat formed during compaction. However, the authors claimed that further increases in the pressure again retarded the release profile possibly due to the closer interparticulate contacts within the tablet, which partly compensated for the leaks of the pellet coats. Badwan and others [113] reported an increase in the drug release profile of sulfamethoxazole beads when compact into a tablet.

In a compaction study of microcapsules (125 to 250 μm) containing phenylpropanolamine–resin complexes [114], it was found that the time required for 50% drug release (T50%) from the compacts of microcapsules with various external diluents was affected by the magnitude of the pressure, the amount of microcapsules, and the type of external additive. The drug release rate from the microcapsule compacts increased with an increase in the magnitude of the pressure, suggesting that the diffusion-retarding membrane was ruptured during compaction. An increased percentage of microcapsules also caused a decrease in the T50%. None of the external additives used, including microcrystalline cellulose, spray-dried lactose, and dextrates, was able to retain the retarded drug release from the microcapsules. However, formulations containing micro-crystalline cellulose as the diluent were found to be least affected by the pressure effects while capable of accepting larger quantities of microcapsules.

Bodmeier and Chen [115] compacted the biodegradable spheres prepared from polylactides [low-molecular-weight poly(DL-lactide) or a relatively high-molecular-weight poly(L-lactide)]. They observed that the energy imparted during compaction caused fusion of the low-molecular-weight polylactide particles resulting in transparent pellets without visible particle boundaries. Béchard and Leroux investigated [116] the effect of compaction on the drug release from the compacts of varying mesh cuts of coated microspheres containing chlorpheniramine maleate. In the study, microspheres were coated with Aquacoat (aqueous ethyl cellulose pseudolatex dispersion), and their mesh cuts of 20/30 (590 to 840 μm), 30/40 (420 to 590 μm), or 40/60 (250 to 420 μm) were compacted with external additives of microcrystalline cellulose, dicalcium phosphate anhydrous, or compressible sugar. The workers reported that massive film fracture occurred at high pressures regardless of the microsphere particle size and the external additives used. Total loss of the controlled release was observed. They pointed out that smaller particles appeared to be more fragile than larger ones. This was attributed to the differences in film thickness, which was found to be 15 μm for the 40/60-mesh microspheres, as opposed to 20 to 25 μm for the 20/30- and 30/40-mesh pellets. However, the formulations containing the smallest spheres were found to be more compatible, in terms of particle size, with direct compression excipients.

Maganti and Çelik [31] observed significant changes between the compaction properties of the powder and pellet forms of the same formulations. In this study, the powder formulations deformed plastically and produced strong compacts, whereas their pellet forms exhibited elastic deformation and brittle fragmentation, which resulted in compacts of lower tensile strength. The compacts made from powder formulations exhibited long disintegration times. These authors reported that the compacts of the powder formulations had released 95% of the drug (propranolol HCl) at 60 min, whereas this release occurred in 5 min in the case of the compacts of pellets, suggesting that the pellets lost their sustained release properties during compaction. In the same study, the effect of external additives and punch velocity on the compaction properties of their pellets was also studied. It was observed that the mechanical strength of the compacts increased with the inclusion of microcrystalline cellulose as an external diluent and decreased when pregelatinized starch or soy polysaccharide was used. The internal lubrication of the formulation with 0.5% magnesium stearate also caused a reduction in the strength of the compacts of pellets. The drug release profiles of the compacts with different external additives were found to be similar to that of compacts made from pellets alone.

In part II of their study, Maganti and Çelik [32] also reported that the addition of Surelease as a coating material altered the deformation characteristics of uncoated pellets by introducing plastoelastic properties into their previously brittle and elastic nature. An increase in the amount of coating applied caused a

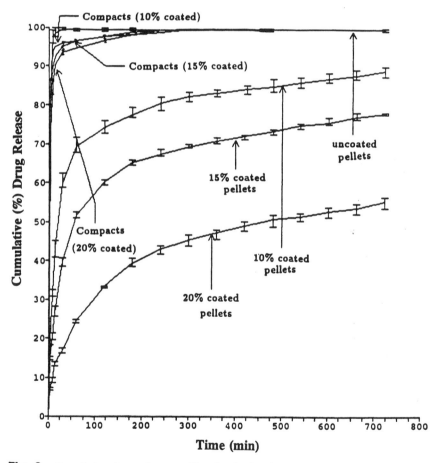

Fig. 9 Cumulative drug release profiles for the Surelease-coated microspheres of F-I formulation, consisting of microcrystalline cellulose (80%), lactose (10%), and propranolol HCl (10%), and their compacts. (After Ref. 32.)

reduction in the mechanical strength of the resulting compacts. A decrease was observed in the consolidation of Surelease-coated pellets as the rate of load application increased. The optimization method used in their study showed that the dependency on the rate of load application was a function of the amount of coating added to the pellets. Pellets with increased amounts of coating exhibited more time dependence. These workers reported that regardless of the amount of coating applied, the Surelease-coated pellets lost their sustained-release properties when compacted into tablets (Fig. 9). This was attributed primarily to the formation of cracks within the coating during compaction and to the fragmentary/ elastic nature of the pellets.

VI. CONCLUDING REMARKS

When a multiparticulate dosage form is developed in the form of tablet, the following factors must be considered to produce tablets with the desired compaction and postcompaction properties. The selection of external additives, such as diluents, is of importance in the design of multiparticulate tablets since these additives are expected to prevent the occurrence of film cracking in the coated subunits. Their compatibility with the subunits in terms of particle size is also very critical since nonuniform size distribution can cause segregation, resulting in many tableting problems, such as weight variation and poor content uniformity. To minimize the occurrence of such problems, placebo microspheres, with good compaction and cushioning properties, can be used as diluents if the size of the active microspheres is much larger than that of the external powder additives. Another alternative is to produce subunits of smaller sizes. At present, the technology is available to manufacture coated sub-100-μm microspheres. Small active subunits also improve the content uniformity of low-dose drugs. However, the surface area to be coated will increase as the size of the microspheres decreases. Any problems or advantages associated with the increased amount of total coating substrate must also be considered carefully in the multiparticulate dosage development.

The size and shape, as well as surface properties, of microspherical particles may differ from those of their powder form. This may also cause a change in their deformation mechanisms. For example, microcrystalline cellulose, which deforms primarily by plastic deformation, can produce pellets that exhibit elastic and/or brittle fragmentation. Any undesirable change in the mechanism of compaction may cause a reduction in the mechanical strength of the ejected compacts and/or can alter the drug release characteristics. Since the surface area of the spherical particles will be a minimum as compared to the other shapes, microspherical multiparticulate formulations may require only a very small amount of lubricant. Therefore, the amount of, and mixing time with, the lubricant must also be carefully considered.

It is important that coated subunits in the formulation be able to withstand the process of compaction without being damaged. The type and amount of coating agent, size of the subunits, selection of the external additives, and rate and magnitude of the pressure applied are the most critical factors to be considered to maintain the desired drug release properties of that subunit.

REFERENCES

1. K. Marshall, in *Theory and Practice of Industrial Pharmacy*, 3rd ed. (L. Lachman, H. Lieberman, and J. Kanig, eds.), Lea & Febiger, Philadelphia, 1986, p. 66.
2. W. D. Jones, *Principles of Powder Metallurgy*, Edward Arnold, London, 1937.
3. R. P. Seelig and J. Wulff, *Trans. Am. Inst. Min. Metall. Eng. 185*:561 (1946).
4. D. Train, *J. Pharm. Pharmacol. 8*:745 (1956).
5. D. Train, *Trans. Inst. Chem. Eng. 33*:258 (1957).
6. C. L. Huffine and C. S. Bonilla, *J. Am. Inst. Chem. Eng. 8*:490 (1962).
7. R. K. McGeary, *J. Am. Ceram. Soc. 44*:513 (1961).
8. T. Higuchi, E. Nelson, and L. W. Busse, *J. Am. Pharm. Assoc. Sci. Ed. 43*:344 (1954).
9. E. Shotton, J. J. Deer, and D. Ganderton, *J. Pharm. Pharmacol. 15*(suppl.):106T (1963).
10. F. W. Goodhart, G. Mayorga, M. N. Mills, and F. C. Ninger, *J. Pharm. Sci. 57*:1970 (1968).
11. C. J. de Blaey and J. Polderman, *Pharm. Weekbl. 106*:57 (1971).
12. J. T. Walter and L. Augsburger, *Pharm. Technol. 10*(2):26 (1986).
13. M. Çelik and K. Marshall, *Drug Dev. Ind. Pharm. 15*:759 (1989).
14. R. F. Lammens, T. B. Liem, J. Polderman, and C. J. de Blaey, *Powder Technol. 26*:169 (1980).
15. G. Ragnarsson and J. Sjogren, *J. Pharm. Pharmacol. 37*:145 (1985).
16. E. Nelson, *J. Am. Pharm. Assoc. Sci. Ed. 44*:494 (1955).
17. W. M. Long, *Powder Metall. 6*:73 (1960).
18. S. Leigh, J. E. Carless, and B. W. Burth, *J. Pharm. Sci. 56*:888 (1967).
19. E. Shotton and B. A. Obiorah, *J. Pharm. Sci. 64*:1213 (1975).
20. C. Carstensen, J. Marty, F. Puisieux, and H. Fessi, *J. Pharm. Sci. 70*:222 (1981).
21. M. Çelik, Ph.D. thesis (CNAA), Leicester Polytechnic, 1984.
22. D. N. Travers and M. P. H. Merriman, *J. Pharm. Pharmacol. 22*(suppl.):17S (1970).
23. S. Malamataris and N. Pilpel, *Powder Technol. 26*:205 (1980).
24. S. Esezobo and N. Pilpel, *J. Pharm. Pharmacol. 38*:403 (1986).
25. A. Y. K. Ho, A. Milham, L. Lockwood, and T. M. Jones, *J. Pharm. Pharmacol. 35*:114 (1983).
26. S. D. Bateman, M. H. Rubinstein, and H. S. Thacker, *Pharm. Technol. Int. 6*:30 (1990).
27. M. Çelik, *Drug Dev. Ind. Pharm. 18*:767 (1993).
28. K. Kawakita and K. H. Lüdde, *Powder Technol. 4*:61 (1970/71).

29. G. Bockstiegel, *Proceedings of the 2nd International Conference on the Compaction and Consolidation of Particulate Matter*, Brighton, East Sussex, England, 1975.
30. H. M. Macleod, in *Enlargement and Compaction of Particulate Solids* (N. G. Stanley-Wood, ed.), Butterworth, London, 1983, p. 241.
31. L. Maganti and M. Çelik, *Int. J. Pharm.*, *95*:29 (1993).
32. L. Maganti and M. Çelik, *Int. J. Pharm.*, *103*:55 (1994).
33. E. Walker, *Trans. Faraday Soc.* *19*:614 (1923).
34. M. Bal'shin, *Vestn. Mettallopromsti.* *18*(2):124 (1938).
35. K. Kawakita, *Science (Japan)* *26*(3):149 (1956).
36. P. J. James, *Powder Metall. Int.* *4*(2):82 (1972).
37. R. Ramberger and A. Burger, *Powder Technol.* *43*:1 (1985).
38. R. W. Heckel, *Trans. Met. Soc. AIME* *221*:671 (1961).
39. R. W. Heckel, *Trans. Met. Soc. AIME* *221*:100 (1961).
40. L. Athy, *Bull. Am. Assoc. Pet. Geol.* *14*:1 (1930).
41. I. Shapiro and I. Kolthoff, *J. Phys. Chem.* *51*:483 (1947).
42. K. Konopicky, *Radex-Rundschan* *3*:141 (1948).
43. J. A. Hersey and J. E. Rees, *Proceedings of the 2nd Particle Size Analysis Conference*, Society for Analytical Chemistry, Bradford, Yorkshire, England, 1970.
44. J. A. Hersey, E. T. Cole, and J. E. Rees, *Proceedings of the First International Conference on the Compaction and Consolidation of Particulate Matter*, London, 1973.
45. J. A. Hersey and J. E. Rees, *Nature Phys. Sci.* *230*:96 (1971).
46. P. York and N. Pilpel, *J. Pharm. Pharmacol.* *25*(suppl.):1P (1973).
47. J. T. Fell and J. M. Newton, *J. Pharm. Sci.* *60*:1866 (1971).
48. R. J. Roberts and R. C. Rowe, *J. Pharm. Pharmacol.* *37*:377 (1985).
49. P. J. Rue and J. E. Rees, *J. Pharm. Pharmacol.* *30*:642 (1978).
50. M. Düberg and C. Nyström, *Powder Technol.* *46*:67 (1986).
51. H. C. M. Yu, M. H. Rubinstein, I. M. Jackson, and H. M. Elsabbagh, *Drug Dev. Ind. Pharm.* *15*:801 (1989).
52. J. T. Carstensen, J. Geoffrey, and C. Dellamonica, *Powder Technol.* *62*:119 (1990).
53. A. R. Cooper and L. E. Eaton, *J. Am. Ceram. Soc.* *45*:97 (1962).
54. H. Leuenberger, *Int. J. Pharm.* *12*:41 (1982).
55. H. Leuenberger and W. Jetzer, *Powder Technol.* *37*:209 (1984).
56. H. Leuenberger and B. Rohera, *Pharm. Res.* *3*(1):12 (1986).
57. H. Leuenberger and B. Rohera, *Pharm. Res.* *3*(2):65 (1986).
58. I. Krycer, D. G. Pope, and J. A. Hersey, *Int. J. Pharm.* *12*:113 (1982).
59. J. T. Fell and J. M. Newton, *J. Pharm. Sci.* *60*:1428 (1971).
60. T. M. Jones, *Acta Pharm. Technol.* *6*:141 (1978).
61. K. Marshall, *Drug Dev. Ind. Pharm.* *15*:2153 (1989).
62. R. F. Lammens, C. J. de Blaey, and J. Polderman, *Proceedings of the 37th International Congress on Pharmaceutical Science*, F.I.P., The Hague, 1977, p. 41.
63. C. J. de Blaey and J. Polderman, *Pharm. Weekbl.* *105*:241 (1970).

64. C. J. de Blaey, M. C. B. van Oudtshoorn, and J. Polderman, *Pharm. Weekbl.* *106*:589 (1971).
65. N. A. Armstrong, N. M. A. H. Abourida, and A. M. Gough, *J. Pharm. Pharmacol. 35*:320 (1983).
66. N. A. Armstrong and L. P. Palfrey, *J. Pharm. Pharmacol. 39*:497 (1987).
67. N. A. Armstrong, *Int. J. Pharm. 49*:1 (1989).
68. K. Marshall, *Powder Technol. 16*:107 (1977).
69. J. H. Shaxby and J. C. Evans, *Trans. Faraday Soc. 19*:60 (1923).
70. H. Unckel, *Arch. Eisenhutt Wes. 18*:161 (1945).
71. H. L. Toor and S. D. Eagleton, *Ind. Eng. Chem. 48*:825 (1956).
72. E. Nelson, S. M. Naqvi, L. W. Busse, and T. Higuchi, *J. Am. Pharm. Assoc. Sci. Ed. 43*:596 (1954).
73. W. A. Strickland, T. Higuchi, and L. W. Busse, *J. Am. Pharm. Assoc. Sci. Ed. 49*:35 (1960).
74. C. J. Lewis and E. Shotton, *J. Pharm. Pharmacol. 17*(suppl.):71S (1965).
75. J. C. Guyot, A. Delacombe, C. Merle, P. Becourt, J. Ringard, and M. Traisnel, *Proceedings of the First International Conference on Pharmaceutical Technology*, Paris, 1977, Vol. 4, p. 142.
76. E. Shotton and B. A. Obiorah, *J. Pharm. Pharmacol. 25*(suppl.):37P (1973).
77. A. C. Shah and A. R. Miodozeniec, *J. Pharm. Sci. 66*:1377 (1977).
78. J. E. Rees and P. J. Rue, *J. Pharm. Pharmacol. 29*(suppl.):37P (1977).
79. P. J. Rue, P. M. R. Barkworth, P. Ridgway-Watt, P. Rough, D. C. Sharland, H. Seager, and H. Fisher, *Int. J. Pharm. Technol. Prod. Manuf. 1*(1):2 (1979).
80. I. Krycer, D. G. Pope, and J. A. Hersey, *J. Pharm. Pharmacol. 34*:802 (1982).
81. N. A. Armstrong and R. F. Haines-Nutt, *J. Pharm. Pharmacol. 22*(suppl.):8S (1970).
82. N. A. Armstrong and R. F. Haines-Nutt, *J. Pharm. Pharmacol. 24*(suppl.):135P (1972).
83. J. E. Carless and S. Leigh, *J. Pharm. Pharmacol. 26*:289 (1974).
84. P. York and E. D. Baily, *J. Pharm. Pharmacol. 29*:70 (1977).
85. M. Çelik and E. Okutgen, *Drug Dev. Ind. Pharm., 19*:2309 (1993).
86. M. P. Summers, R. P. Enever, and J. E. Carless, *J. Pharm. Pharmacol. 28*:89 (1976).
87. M. Çelik and D. N. Travers, *Drug Dev. Ind. Pharm. 11*:299 (1985).
88. O. Ejiofor, S. Esezobo, and N. Pilpel, *J. Pharm. Pharmacol. 38*:1 (1986).
89. H. C. M. Yu, M. H. Rubinstein, J. M. Jackson, and H. M. Elsabbagh, *J. Pharm. Pharmacol. 40*:669 (1988).
90. H. Leuenberger, B. D. Rohera, and Ch. Haas, *Int. J. Pharm. 38*:109 (1987).
91. L. H. Holman and H. Leuenberger, *Int. J. Pharm. 46*:35 (1988).
92. H. Leuenberger, L. H. Holman, M. Usteri, and S. Winzap, *Pharm. Acta Helv. 64*(2):34 (1989).
93. D. Stauffer, *Introduction to Percolation Theory*, Taylor & Francis, London, 1985.
94. L. H. Holman and H. Leuenberger, *Powder Technol. 60*:249 (1990).
95. D. Blattner, M. Kolb, and H. Leuenberger, *Pharm. Res. 7*(2):113 (1990).
96. D. B. Brook and K. Marshall, *J. Pharm. Sci. 57*:481 (1968).

97. J. E. Rees, J. A. Hersey, and E. T. Cole, *J. Pharm. Pharmacol.* 22(suppl.):64S (1970).
98. S. D. David and L. L. Augsburger, *J. Pharm. Sci.* 63:933 (1974).
99. B. W. Muller, K. J. Stephens, and P. H. List, *Acta Pharm. Technol.* 22:91 (1976).
100. P. L. Jarosz and E. L. Parrott, *J. Pharm. Sci.* 71:607 (1982).
101. J. T. Fell and J. M. Newton, *J. Pharm. Pharmacol.* 20:657 (1968).
102. J. T. Fell and J. M. Newton, *J. Pharm. Sci.* 59:688 (1970).
103. K. G. Pitt, J. M. Newton, and P. Stanley, *J. Pharm. Pharmacol.* 42:219 (1990).
104. K. Ridgway, M. E. Aulton, and P. H. Rosser, *J. Pharm. Pharmacol.* 22(suppl.): 70S (1970).
105. E. N. Hiestand, J. M. Bane, and E. P. Strzelinski, *J. Pharm. Sci.* 60:758 (1971).
106. M. E. Aulton, H. G. Tebby, and P. J. White, *J. Pharm. Pharmacol.* 26(suppl.): 59P (1974).
107. M. E. Aulton, *Pharm. Acta Helv.* 56:133 (1981).
108. J. E. Rees and P. J. Rue, *Drug Dev. Ind. Pharm.* 4:131 (1978).
109. G. P. Millili and J. B. Schwartz, *Drug Dev. Ind. Pharm.* 16:1411 (1990).
110. G. Alderborn and C. Nyström, *Acta Pharm. Suec.* 21:1 (1984).
111. E. N. Hiestand and D. P. Smith, *Powder Technol.* 38:145 (1984).
112. M. Juslin, L. Turakka, and P. Puumalainen, *Pharm. Ind.* 42:829 (1980).
113. A. A. Badwan, A. Abumalooh, E. Sallam, A. Abukalaf, and O. Jawan, *Drug Dev. Ind. Pharm.* 11:239 (1985).
114. W. Prapaitrakul and C. W. Whitworth, *Drug Dev. Ind. Pharm.* 15:2049 (1989).
115. R. Bodmeier and H. Chen, *J. Pharm. Sci.* 78:819 (1989).
116. S. R. Béchard and J. C. Leroux, *Drug Dev. Ind. Pharm.* 18:1927 (1992).

10

Key Factors in the Development of Modified-Release Pellets

Stuart C. Porter

Colorcon
West Point, Pennsylvania

Isaac Ghebre-Sellassie

Parke-Davis Pharmaceutical
Research Division
Warner-Lambert Company
Morris Plains, New Jersey

I. INTRODUCTION

Film coating is a process that has become widely used in the pharmaceutical industry. This process provides a great deal of flexibility to the formulator when designing high-quality pharmaceutical dosage forms. The significance of this benefit has not been lost on those designing oral, modified-release products. Indeed, Rowe [1] has considered film coating as an ideal process for the preparation of such products.

In tracing the development of modified-release dosage forms, film coating has been a key factor from the early 1950s (which saw the introduction of pellets coated with organic solvent-based solutions of wax–fat mixtures). Today, there is a growing preference to exploit the application of aqueous coating technology through the use of aqueous polymeric dispersions.

Concurrent with the emergence of newer types of coating systems has been the evolution in design of coating equipment. Previous emphasis on the use of conventional coating pans is gradually being replaced by the adoption of side-vented coating pans and fluidized-bed coating equipment as the processing methodologies of choice. Fluidized-bed technologies have gained a position of significant importance when coating multiparticulates.

Irrespective of the types of coating materials and coating process technologies used, certain key elements are critical to the success of film coating as a primary means of modifying drug release characteristics, namely:

1. Target drug release profiles must be attained in a reproducible manner.
2. Drug release characteristics should not be sensitive to minor changes in coating formulations or coating process conditions.
3. The coating formulations and coating processes should ideally be uncomplicated and facilitate scale-up from the laboratory into production.
4. The final product should be stable and, in particular, be devoid of significant time-dependent changes in drug release characteristics.
5. The coating formulations and coating process must be cost-effective.
6. The entire process (from initial development in the laboratory to ultimate product commercialization) must have a sound scientific basis so as to withstand the regulatory challenges that precede the introduction of any pharmaceutical product.

The purpose of this chapter is to review briefly the types of coating systems that are available today and to discuss in detail many of the factors that are likely to have a significant impact on the ultimate performance of film-coated pellets.

II. OVERVIEW OF VARIOUS APPROACHES TO FORMULATING MODIFIED-RELEASE COATINGS

One of the major advantages in using film-coating technology as a means of modifying the release characteristics of a drug from a dosage form is associated with the fact that the formulation flexibility afforded the product development scientist is extensive. Basic approaches to this formulation process involve the use of:

1. Film-forming materials dissolved in an appropriate solvent (usually organic in nature)
2. Aqueous polymer dispersions
3. Hot melts

Release-moderating membranes can be deposited in such a way that they are essentially continuous. Alternatively, special ingredients can be incorporated into the formulation such that pores will be created once these ingredients have been leached out. Since many of these coating systems are described in detail elsewhere in this book, a brief overview will suffice.

A. Coating Systems

Solutions of Polymers in Organic Solvents

The use of organic solvent–based solutions of polymers has provided a popular means for depositing rate-controlling membranes onto the surface of oral solid dosage forms. Many would argue that this approach provides a relatively simple but reliable means of achieving target drug release profiles. A variety of polymers has been used with such an approach. A summary (with associated references 2–14) of many of the polymers used for this purpose is given in Table 1.

The quality of coatings obtained from polymers dissolved in a solvent is defined to a large extent by:

1. The extent of the thermodynamic interaction between the solvent and the polymer
2. The relative volatility of the solvent
3. The likelihood that sufficient solvent is retained within the polymer matrix to render the coating excessively tacky during critical stages of film formation

Banker [15] has stated that optimal polymer solution will yield maximum polymer chain extension (in the dissolved state) so as to produce films with the greatest cohesive strength and thus the optimum mechanical properties.

One method for determining interactions between the polymer and solvent, and aiding in selecting the most suitable solvent, is based on the theoretical treatment of the familiar free-energy equation as proposed by Hildebrand and Scott [16] and is expressed this way:

$$\Delta H_m = V_m \left[\left(\frac{\Delta E_1}{V_1} \right)^{1/2} - \left(\frac{\Delta E_2}{V_2} \right)^{1/2} \right]^2 \phi_1 \phi_2 \tag{1}$$

where ΔH_m is the overall heat of mixing, V_m the total volume of the mixture, ΔE the energy of vaporization of either component 1 or 2, and ϕ the volume fraction of either component 1 or 2. The expression $(\Delta E)/V$ is often termed the cohesive energy density, written δ^2, where δ is the solubility parameter and is equivalent to $([\Delta E]/V)^{1/2}$ in Eq. (1). If $\delta_1 = \delta_2$, then $\Delta H_m = 0$. Thus in the free-energy equation,

$$\Delta F = \Delta H_m - T \Delta S \tag{2}$$

where ΔF is the free-energy change, T the absolute temperature, and ΔS the entropy of mixing, the free energy is now dependent on the mixing entropy. As there is a large increase in the entropy when a polymer dissolves, complete miscibility between the solvent and polymer is assured.

Table 1 Summary of Polymers Used in Preparing Modified-Release Film Coatings

Polymer	Refs.
Polymers of natural origin	
Shellac	2
Zein	3
Rosin esters	4,5
Synthetic polymers	
Ethyl cellulose	6–8
Cellulose acetate	9,10
Acrylic resins	11,12
Silicone elastomers	13
Poly(vinyl chloride)	14

Kent and Rowe [17] used the solubility parameter approach in evaluating the solubility of ethyl cellulose in various solvents for film coating. They determined the intrinsic viscosities of several grades of ethyl cellulose in a range of solvents of known solubility parameters utilizing the equations derived by Rubin and Wagner [18]. They graphically evaluated the effect of the solvent solubility parameter on intrinsic viscosity for a range of solvents classified either as (1) poorly hydrogen bonded, (2) moderately hydrogen bonded, or (3) strongly hydrogen bonded, and determined both the best class of solvent to use and the optimum solvent solubility parameter. Thus this approach can be used to determine the best solvent for the polymer from a thermodynamic standpoint. The technique can also be used for optimizing solvent blends, the individual components of which may or may not themselves be thermodynamically good solvents.

When organic solvent–based coating solutions are used, employment of solvent blends is common. Because the relative volatilities of individual solvents in the mixture often differ significantly, solvent ratios may change on evaporation. A common result is that solvent thermodynamics may change sufficiently to have a profound effect on coating mechanical strength, toughness, elasticity, and porosity, as demonstrated by Spitael & Kinget [19]. A solution to this problem involves the use of azeotropic solvent mixtures.

Use of highly volatile solvent systems, combined with employment of excessive forced drying conditions in the coating process, can cause partial spray drying of the coating. The result is a highly porous membrane with inadequate barrier properties, as described by Arwidsson [20].

Other factors may also influence the effectiveness of polymer films deposited from solution. Throughout the formation of film coatings, continual solvent loss causes contraction in volume. Ultimately, the film becomes constrained by the immobility of the polymer molecules themselves (which occurs at the so-called "solidification" point) and by adhesion of the film to its substrate. Further solvent loss creates shrinkage stresses within the film; these stresses become a contributory factor to the total internal stress that develops within the film. Internal stress has widely been described [21,22] as a major cause of membrane failure. Rowe [23] has described the importance of molecular weight (of ethyl cellulose) in helping to resist the destructive influence of internal stress, and thus enable coatings to be prepared with predictable performance characteristics. Data adapted from this work are described in Fig. 1. In addition to the influence of molecular weight, other formulation techniques (such as selection of appropriate solvents and use of effective plasticizers) may well help to minimize internal stress and thus its influence on coating structure.

Tackiness is also a factor that can create unpredictability in performance of the membrane. When depositing films from polymers dissolved in a solvent, tackiness can be the result of:

1. Excessive solvent retention
2. Overplasticization
3. Inherently low polymer glass transition temperatures

Excessive solvent retention may result from inadequate drying or as the result of a strong thermodynamic interaction (between the polymer and the

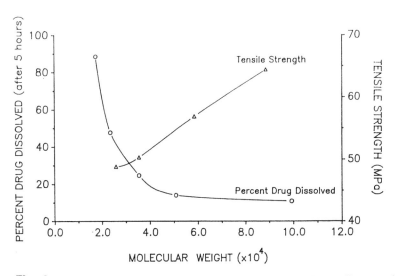

Fig. 1   Influence of molecular weight of ethyl cellulose on the tensile strength of free films and drug release from coated pellets.

solvent) which prevents effective solvent removal. Lindholm et al. [24] have described some of the factors influencing retention of organic solvents in films of ethyl cellulose.

As a result of excessive solvent retention or overplasticization, the glass transition temperature of the polymer is depressed to the point where, under normal coating process conditions, blocking of coated particles occurs. Such blocking can be predicted using the Williams–Landel–Ferry equation, as described by Wicks [25], which suggests that at product temperatures which exceed the glass transition temperature of the coating by more than 20 to 30°C, rheological conditions within the coating are sufficient to cause adhesion between adjacent coated particles should they come into contact with one another. Agglomeration of coated particles may not, per se, be catastrophic; however, agglomerated particles that become deaggregated as the result of attrition may possess coatings that are no longer continuous.

Aqueous Polymeric Dispersions

The practice of current film-coating technology shows a strong preference for the use of aqueous coating systems. When applying film coatings that are not intended to modify drug release characteristics, aqueous film coating is relatively simple, since a wide range of useful, water-soluble polymers exist.

By their very nature, polymers used in modified-release, film-coating formulations are not water soluble. Application of aqueous coating technology thus becomes more complicated. Fortunately, the use of aqueous polymeric dispersions presents the formulator with a potential solution to this challenge. The use of aqueous polymeric dispersions has been widely described by McGinity [26].

Aqueous polymeric dispersions contain polymers in the dispersed phase. Consequently, to deposit an effective coating on the surface of a pharmaceutical substrate, complete coalescence between the dispersed polymeric particles must occur. In order that such an applied coating may exhibit predictable functional characteristics, it is critical that:

1. Coalescence proceeds to an endpoint where membrane porosity is essentially eliminated.
2. Such completeness of coalescence is either achieved during the coating process or by a short curing step at the conclusion of the coating process.
3. Further gradual coalescence, over an extended period of time, is avoided.

Film formation from an aqueous polymeric dispersion is complex and, not surprisingly, many conflicting theories exist to explain the process [27]. Generally, coalescence is facilitated by the existence of capillary forces (between polymer particles) that are generated as water evaporates. Complete coalescence occurs only if there is total integration between adjacent polymer particles.

Integration occurs as the result of the interdiffusion of polymer chains at the interface between particles, a process that to some extent can be accounted for by free-volume theory. This theory presumes that sufficient intermolecular space exists in the bulk polymer to accommodate interdiffusion.

Because free volume dramatically increases at temperatures above the glass transition temperature of the polymer, the glass transition temperature of the coating system becomes a critical formulation factor when processing aqueous polymeric dispersions.

In summary, it thus becomes clear that two major factors are the keys to success when using aqueous polymeric dispersions, namely that:

1. The coating system must be appropriately formulated to ensure that its glass transition temperature is consistent with typical processing temperature that will be employed.
2. The coating process must be carefully controlled to ensure that an appropriate balance exists between the rate of removal of water (which is critical to the development of capillary forces) and processing temperatures (which are critical to ensuring coalescence is completed).

A list of aqueous coating systems that can be used in the preparation of oral, modified-release dosage forms is shown in Table 2. The existence of such an extensive list of coating systems is not intended to convey the impression that each of these systems is functionally exactly equivalent to any of the others. Significant differences exist between them:

1. The chemical nature of the polymers used
2. The molecular weight of the polymers used
3. The method by which the coating dispersion is prepared (See Table 3 for a list of common methods employed.)
4. The presence of additives that are critical to that method of preparation and, ultimately, the physical stability of the dispersion
5. The glass transition temperature of the final coating
6. The size, and size distribution, of the dispersed particles
7. The pH of the dispersion
8. The surface charge of dispersed polymer particles

These differences are likely to:

1. Influence drug release characteristics, since different chemistries are involved
2. Require the addition of critical processing additives (such as plasticizers or antitack agents)
3. Necessitate the selection of specific process conditions to ensure that appropriate coalescence occurs
4. Influence the requirement to use seal coats and overcoatings.

Table 2 Examples of Aqueous Polymeric Dispersions for Sustained-Release Film Coating

Surelease	Ethyl cellulose	Aqueous polymeric dispersion contains requisite plasticizers Addition of lake colorants should be avoided because of alkalinity of dispersion
Aquacoat	Ethyl cellulose	Pseudolatex dispersion Requires addition of plasticizers to facilitate film coalescence
Eudragit NE30D	Poly(ethyl acrylate–methyl methacrylate) 2:1	Latex dispersion No plasticizers required unless improved film flexibility is desired
Eudragit RL30D	Poly(ethyl acrylate–methyl methacrylate) triethylammonioethyl methacrylate chloride 1:2:0.2	Aqueous polymeric dispersion Requires addition of plasticizers
Eudragit RS30D	Poly(ethyl acrylate–methyl methacrylate)triethylammonioethyl methacrylate chloride 1:2:0.1	Aqueous polymeric dispersion Requires addition of plasticizer
	Silicone elastomer	Requires addition of fumed silica as a processing aid and water-soluble plasticizers [such as the poly(ethylene glycols)] to facilitate drug release
	Cellulose acetate	Requires addition of significant quantities of plasticizer; polymer stability limited unless appropriate storage conditions used

Table 3 Common Methods of Preparation of Aqueous Polymeric Dispersions

Method	Description
Emulsification polymerization	Emulsified monomer is polymerized in the presence of an initiator and surfactant to form a latex
Emulsification/solvent evaporation	A solution (of a water-insoluble polymer) in a water-immiscible organic solvent is emulsified in water; organic solvent is removed
Phase inversion technique	Water (containing a base) is added to plasticized hot melt of polymer containing a fatty acid; the initial water-in-polymer dispersion inverts to a polymer-in-water dispersion
Solvent change technique	A solution of an ionic polymer, dissolved in a water-miscible organic solvent, is mixed under agitation with water; organic solvent is removed
Dispersion of micronized polymers	A dispersion of previously micronized polymer is prepared in water at elevated temperatures

Hot-Melt Coating Systems

The process of hot-melt coating as a means of modifying drug release characteristics has received scant attention in the pharmaceutical industry, although it has been employed extensively for, inter alia, flavor encapsulation in the food industry [28]. Pharmaceutical applications for wax–fat systems include:

1. Spray congealing of molten wax–drug slurries [29,30]
2. Solvent evaporation methods [31,32]

Recently, Jozwiakowski et al. [33] have described the application of hot-melt coatings to multiparticulates using fluidized-bed processing technology. A major characteristic of hot-melt coating is that the coating system contains 100% solids. Thus deposition of the coating is rapid, a fact that compensates for the relatively low spray rates (compared to those used with other types of coating systems) that often must be employed.

Important characteristics of materials used in a hot-melt coating process relate to their:

1. Melting point
2. Melting range (a wide range complicates control of the process)
3. Melt viscosity (a relatively low viscosity is necessary if smooth, uniform coatings are desired)

Because of the practical temperature limitations of common pharmaceutical coating processes, coating materials used in the hot-melt process are limited to those having a melting point in the range 40 to 80°C. Consequently, hydrogenated cottonseed oils (having melting ranges of 64 to 67°C) have been found to be suitable for pharmaceutical hot-melt coating processes [33].

Unlike common pharmaceutical film-coating processes (which rely on the formation of a coating as volatile materials are removed), solid coatings derived from hot melts are formed as the result of congealing of molten materials. With this process, control of processing parameters is critical. Especially, processing air temperatures must be controlled to allow molten material to spread over the surface of the material being coated, yet congeal fast enough to ensure that agglomeration does not occur. Coating quality is usually best when the processing temperature is close to the congealing temperature of the coating material.

If the hot melt is congealed too rapidly, a porous coating will result; conversely, if the coating is congealed too slowly, the product bed becomes viscous, sticky, and ultimately agglomeration will occur.

B. Additional Critical Ingredients

While modified-release coating systems may be relatively simple and consist essentially of a primary polymer contained in an appropriate vehicle, rarely are such coating systems used without the inclusion of other additives.

Such additives include materials designed to:

1. Modify film structure in a predictable manner.
2. Overcome some of the deficiencies of the primary polymer.
3. Facilitate coalescence of aqueous polymeric dispersions.
4. Minimize tackiness of the coating system during application.
5. Impart color.

An extensive (but not necessarily all-inclusive) list of specific additives might include:

1. Secondary polymers
2. Pore-forming agents
3. Plasticizers
4. Antitack agents
5. Pigments

Because modified-release film coatings are, by their very nature, highly functional coatings, the potential influence of additives (deliberate or otherwise) on that functionality must be recognized and well understood. In particular, careful consideration must be given to the:

1. Compatibility between the additive and the main polymer (or polymer dispersion for latex systems)

2. Influence of additives on the film-forming characteristics of aqueous polymeric dispersions
3. Influence of additives on the mechanical strength of the coating
4. Uniformity of distribution (within the film) of additives that are insoluble in the coating solvent/vehicle

Influence of Secondary Polymers

It is not unusual to discover, when formulating modified-release products, that no one polymer possesses the ideal characteristics to meet the objectives set for the final product. Using mixtures of polymers often:

1. Allows coatings with optimal barrier properties to be created
2. Facilitates the release of large drug molecules through membranes that would otherwise be too effective a barrier

The use of polymer mixtures has thus been a common practice in the preparation of film coatings designed to modify drug release characteristics from a pharmaceutical oral solid dosage form. By way of example, Vasilevska et al. [34] have described the use of mixtures of acrylic copolymers on the release characteristics of diltiazem from coated pellets. McAinsh and Rowe [35] have used mixtures of ethyl cellulose and hydroxypropyl methylcellulose to modify drug release from propranolol-laden pellets. In both of these examples, the polymers were dissolved in organic solvents. When using organic solvent–based solutions of polymers Rowe [36] and Sakellariou et al. [37] have expressed concern over the compatibility of polymers used in such mixtures. It has been established [36], for example, that mixtures of ethyl cellulose with either hydroxypropyl methylcellulose or hydroxypropyl cellulose are only partially compatible, and phase separation is likely to occur. This phenomenon has been confirmed using thermomechanical analysis, by Sakellariou [38]. While the use of secondary polymers as release-modifying agents has been quite common with organic solvent-based polymer solutions, similar approaches have begun to be employed with aqueous polymeric dispersions. Using chlorpheniramine maleate and acetaminophen as model drugs, Zhang et al. [39] have described the influence on drug release of adding hydroxypropyl methylcellulose to an aqueous dispersion of ethyl cellulose. Of particular interest in this case was the demonstration that the addition of a water-soluble polymer need not always increase the rate at which a drug is released through the membrane. The addition of hydroxypropyl methylcellulose increased the release rate of chlorpheniramine, but decreased that of acetaminophen, through the membrane. The differences observed were related to the fact that the addition of hydroxypropyl methylcellulose:

1. Increased both the drug solubility within the film and the drug diffusion coefficient through the film in the case of chlorpheniramine

2.   Increased the drug diffusion coefficient through the film, but decreased drug solubility within the film, in the case of acetaminophen

These findings were confirmed by conducting permeability studies using an appropriate free-film, diffusion-cell technique, as well as conducting drug solubility studies in isolated film segments.

When considering the addition of high-molecular-weight additives (such as secondary polymers) to aqueous polymeric dispersions, one always has to consider the influence of the additive on the film-forming characteristics of the dispersions. Such additives will typically reside outside the latex particles, and if used in sufficiently high concentrations, may inhibit to some degree the coalescence of the latex.

Influence of Plasticizers

Plasticizers are typically added to coating systems to:

1.   Improve the mechanical properties of the coating by reducing brittleness, increasing flexibility and reducing the elastic modulus of the polymer.
2.   Reduce the glass transition temperature, and thus the minimum film-forming temperature, of aqueous polymeric dispersions.

When considering the use of plasticizers several factors are of prime importance:

1.   Plasticizer compatibility with the polymer
2.   Plasticizer efficiency
3.   Plasticizer permanence

Plasticizer compatibility is often determined by analysis of thermodynamic interactions, as predicted by assessment of solubility parameters [40]. As an extension of this type of evaluation, several authors [41,42] have assessed the compatibility of a series of phthalate esters with various grades of ethyl cellulose using intrinsic viscosity measurements.

Plasticizer efficiency is equally important because:

1.   It has been shown that plasticizers do influence the rate at which a drug is released through the membrane [43].
2.   Overzealous use of plasticizers can cause excessive tackiness.

Plasticizer efficiency is often considered a measure of the magnitude of the decrease in the glass transition temperature of the polymer caused by the addition of the plasticizer. Using this as a basis, Rowe et al. [44] determined that for a series of phthalate esters, diethyl phthalate not only showed the greatest compatibility with ethyl cellulose but was also the most efficient plasticizer of the series. The importance of appropriate plasticizer selection and the ability of this type of additive to improve the performance characteristics of polymers that would be deemed otherwise inadequate have been ably demonstrated by Rowe [23].

The use of mixtures of polymers may require that more than one type of plasticizer may have to be used, especially if total lack of polymer compatibility causes phase separation [37]. When dealing with aqueous polymeric dispersions, a critical need for plasticizers occurs when the glass transition temperature of the polymer is substantially above the normal coating processing temperatures used. As shown previously (see Table 2), many aqueous polymeric dispersions do require the presence of a plasticizer to facilitate film formation. While some polymeric dispersions are plasticized during the manufacture of the dispersion, others require that an appropriate quantity of plasticizer be added by the pharmaceutical formulator. In such cases, the performance of the final coating is very much affected by the nature and concentration of the plasticizers used. In this regard, Arwidsson et al. [45] and Goodhart et al. [46] have described the importance of plasticizers when using aqueous dispersions of ethyl cellulose. Additionally, Iyer et al. [47] have described the importance of mixing time when incorporating plasticizers into an aqueous dispersion of ethyl cellulose. When using dibutyl sebacate as the plasticizer, they found that a minimum of 30 min mixing time was required. Even so, the efficiency of plasticization did not appear to match that obtained when the plasticizer is incorporated during manufacture of such an aqueous dispersion. In contrast to these findings, Lippold et al. [48] discovered that when adding dibutyl sebacate or diethyl phthalate to an aqueous dispersion of ethyl cellulose, approximately 5 to 10 h of standing time was required after addition of the plasticizer and prior to commencement of coating to ensure effective interaction (as measured by influence on the minimum film-forming temperature of the coating system) between the polymer and plasticizer. Under these circumstances they found dibutyl sebacate to be a more effective plasticizer than diethyl phthalate, which has been established as an effective plasticizer for ethyl cellulose when using organic solvent–based solutions of this polymer [44].

While the importance of plasticizers to the performance characteristics of aqueous dispersions of ethyl cellulose has been a major discussion point in the pharmaceutical literature, similar attention has recently been focused on aqueous dispersions of cellulose acetate. Since this polymer is commonly used with osmotic delivery systems, it is critical that strong films are obtained. Bindshaedler et al. [49] have evaluated the effectiveness of various plasticizers for this polymer system. Their findings suggest that the levels of plasticizer required are substantially higher than those required for aqueous ethyl cellulose dispersions.

Influence of Pore-Forming Agents

Although it can be argued that when water-soluble polymers (such as hydroxypropyl methylcellulose) are incorporated into coatings formed from water-insoluble polymers they may function as pore-forming agents, these additives are not usually regarded as true pore formers. The distinction may be based on the contention that water-soluble polymers, being essentially high-molecular-weight materials, may be precluded from being totally leached out from the coating to form a well-defined pore structure.

Thus many of the pore-forming agents that are commonly used are typically low-molecular-weight materials. Often they are dispersed as discrete particles in a coating system in which they remain insoluble. Classic use of this approach has been described by Källstrand and Ekman [14] who have used solutions of poly(vinyl chloride) in acetone into which has been dispersed micronized sucrose. With such a system, the particles of sucrose are dissolved out on exposure to an aqueous fluid, leaving a membrane with a well-defined pore structure. Similar approaches have been used by Lindholm et al. [50] and Hennig and Kala [51]. An interesting concept was proposed by van Bommel et al. [52], who used drug (dispersed in the coating) as the pore-forming agent. As expected, these additives (in the range 0 to 50%) substantially reduced the tensile strength of the ethyl cellulose films and, at the same time, significantly increased both the permeability coefficients and diffusion coefficients for the drug through the membrane.

When pore-forming agents are used with aqueous coating systems, although similar trends to those obtained with coatings deposited from organic solvent–based solutions occur, there are some different issues to be considered since the additives reside outside the latex polymer particles:

1. The distribution of the additive is likely to be more heterogeneous.
2. Film formation may be affected when high concentrations of additive are used.
3. The mechanical strength of the coatings can potentially be affected to a greater degree than when using films prepared from similar polymers applied from organic solvent–based solutions.

Nonetheless, the use of pore-forming additives with aqueous latex coating systems has proven effective. Li and Peck [53] have described the important of poly(ethylene glycol) when dispersed in water-based silicone elastomer coatings. These authors found that at high loadings of the pore-forming additive, transpore diffusion became a dominant factor in determining drug release from products coated with the silicone elastomer.

When pore-forming agents are employed with aqueous coating systems, and the final coating is applied to a core material capable (on interaction with a dissolution fluid) of generating a significant osmotic pressure, the mechanical properties of that coating become a critical factor if membrane rupture is to be avoided. Appel and Zentner [54] have shown that when evaluating ethyl cellulose coatings modified with urea, a significant quantity of coating may have to be applied to create a functional membrane of acceptable mechanical strength.

Since many pore-forming additives are, out of necessity, water soluble, they are likely to be dispersed on a molecular scale in the coating (since they will dissolve in the vehicle of the coating system). Consequently, the pores that are ultimately formed will be extremely small in size. This is in sharp contrast to the effects described earlier [14] when pore-forming agents were dispersed in solutions of polymer dissolved in organic solvents. Larger pores can be created with aqueous coating systems when using pore-forming agents having selective water solubility. Such an effect was achieved by Bodmeier and Paeratakul [55] when using calcium phosphate dispersed in latex coating systems. In this case, the pore-forming agents were leached out when the coated products were exposed to gastric juice.

Influence of Insoluble Additives
Insoluble additives that might be included in modified-release, film-coating formulations are those that:

1. Impart color (e.g., pigments)
2. Act as antitack/antiadhesive agents (e.g., talc)

The presence of such additives will almost certainly influence the barrier properties of the applied coating. Parker et al. [56] determined that many factors can influence the permeability characteristics of coatings containing dispersed, insoluble additives, including:

1. Concentration of the additive (where concentration effects are usually linked to the binding capacity of the film former used)
2. Packing and distribution of the additive
3. The relative hydrophilicity/hydrophobicity of the additive
4. Influence of the additive on the microstructure of the coating

Their findings appeared to fit the Chatfield theory [57], which predicts that the presence of dispersed particles decreases permeability of a coating (by effectively increasing the length of the diffusion pathway through increased tortuosity) until the critical pigment volume concentration (CPVC) is exceeded. This effect is represented schematically in Fig. 2.

The Chatfield theory is more predictive of membrane behavior when the insoluble additives are dispersed in organic solvent–based solutions of polymers. Aqueous polymeric systems can behave somewhat differently owing to the fact that the additives tend not to be dispersed uniformly throughout the coating, but rather become concentrated in the interstices between latex particles. Goodhart et al. [46] have described the influence of additives such as talc or magnesium stearate when dispersed in ethyl cellulose pseudolattices. In this case, the additives were intended to function as antiadhesive agents.

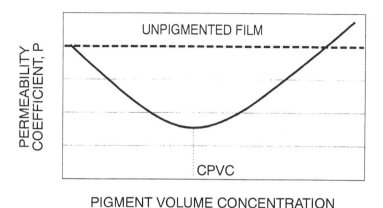

PIGMENT VOLUME CONCENTRATION

Fig. 2   Influence of pigments on the moisture permeability of polymer films.

III.   KEY FACTORS INFLUENCING DRUG RELEASE FROM PELLETS COATED WITH MODIFIED-RELEASE COATINGS

Although utilization of pellets in the design and development of modified-release dosage forms has been practiced since the mid-1950s [58], it is only in the last 10 to 15 years that the technology has become the focus of intensive research both in terms of processing equipment and dosage-form development. Introduction of new and specialized equipment, which allows optimum control of process parameters, and the availability of literature replete with information about the physical and mechanical forces that govern pellet formation and growth, have transformed pellet manufacture from an art practiced by a handful of experienced operators to a discipline characterized by scientific rationale and reasoning.

These developments, coupled with advances made in film-coating technology and the documented biopharmaceutical and therapeutic advantages of pellets over single-unit dosage forms such as tablets [59], permit the full potential of pelletized dosage forms to be realized. Not only do pellets offer flexibility during formulation development, but they also possess characteristics that maximize the ultimate objective of the dosage form (i.e., optimum delivery of the drug to its site of absorption). Because pellets are distributed freely throughout the gastro-intestinal tract, they tend to reduce variations in gastric emptying and intestinal transit times as well as minimize inter- and intrasubject variability. As a result, pelletized multiparticulate modified-release dosage forms generally maximize drug absorption, minimize potential side effects, and provide relatively stable plasma profiles characterized by fewer peaks and valleys. Since the dose is divided into several units, failure of the coating material on a few pellets does not significantly impair the overall performance of the dosage form as would be the case with single-unit products. This allows pelletized dosage forms to have a wide, built-in margin of safety against dose dumping and thus increases their reliability as a dosage form.

Development of modified-release pellets involves a series of phases that need to be characterized and evaluated separately prior to the formulation of the final dosage forms. Detailed analysis and understanding of these phases up-front not only save a tremendous amount of time and resources during the course of development, but also ensure batch-to-batch reproducibility during scale-up and production.

The objective of this discussion is to highlight those key factors that must be considered during the development of modified-release pellets. In this regard consideration will be given to those factors that relate to:

1. The pellets (as a substrate suitable for coating)
2. The coating formulation
3. The coating process
4. Aging or time-dependent changes that may occur once the coating has been applied

The influence of many of these factors on drug release from film-coated, modified-release dosage forms has been discussed by Mehta [60].

A. Physicochemical Properties of Core Pellets

The term *pellets* generally conjures an image of spheroidal particles that contain drug either layered on the surface or distributed throughout the mass of that pellet. The term can be expanded, however, to include all forms of multiparticulates, including:

1. Drug-containing granules
2. Drug/ion-exchange resin complexes
3. Drug crystals
4. Minitablets

Pellets (i.e., spheroidal particles) can be prepared using a number of techniques including:

1. The application of a solution, suspension, or powder (containing the drug and often excipients) onto the surface of nonpareils
2. The preparation of spheroids, consisting of drug mixed with appropriate excipients, using a combination of blending, agglomerating, and spheronizing techniques

Spheronization is often a multistep process that is employed primarily in the preparation of pellets where high drug loadings are required. Layering techniques in which solutions or suspensions are applied to the surfaces of nonpareils are typically employed only when low-to-medium drug loadings are required; for high drug loadings, processing times tend to be long. Such a layering process is often relatively simple. In contrast, layering of powders can be more complex, since there is a need to meter powders accurately if acceptable uniformity is to be achieved. The layering process may utilize a wide range of coating equipment, although fluidized-bed processors, with their greatly enhanced drying capabilities, are often preferred.

Depending on the type of process used in their preparation and the physicochemical properties of the drug and excipients, pellets differ widely in chemical, physical, and mechanical properties. Thus it is critical that prior to the application of the modified-release coating, pellets are systematically evaluated and characterized. Failure to do so will increase the likelihood that consistent results will be difficult to achieve. Some of the key pellet variables that need to be considered are summarized in Table 4.

Table 4 Summary of Substrate Variables and Their Potential Effects

Variable	Has influence on:
Substrate size	Batch surface area
	Quantity of coating to be applied
Size distribution	Batch surface area
	Batchwise variability in surface area
Surface roughness	Batch surface area
Density	Uniformity of distribution of coating
Friability	Drug erosion during application of coating
Chemical properties of drug	Interaction with coating (and thus diffusivity and partitioning effects)
	Osmotic effects
Solubility of drug	Drug release rate
	Potential for drug to be leached from core during application of coating
Molecular weight of drug	Ability of drug to penetrate through membrane
Nature of excipients	Ability of core to absorb liquid during coating process
	Ability of core to absorb fluids in dissolution test or after ingestion
	Osmotic effects

Physical Characteristics of Substrate

Surface Area. Major characteristics of pellets (and many other forms of multiparticulates) include the fact that they:

1. Exhibit relatively small particle sizes
2. Feature a wide range of particle sizes within a batch
3. Often possess ill-defined surface topographical characteristics

The result is that the surface area of a batch is potentially large and variable. Successive batches of product may thus demonstrate fluctuating surface areas that require the application of differing levels of coating if consistent drug release characteristics are to be obtained. The influence of particle size (of pellets) on batch surface area and requirements for the amount of coating that needs to be applied to maintain a constant coating thickness has been discussed by Lehmann and Dreher [11]. From their studies it becomes apparent that the influence of particle size becomes dramatic for pellets below 0.5 to 1.0 mm in size.

The data highlighted in Fig. 3 describe how dramatically the particle size of the pellets being coated can influence drug release characteristics. Variations of this magnitude, however, should rarely be experienced when the process for preparing pellets has been well defined and controlled. It is more likely that the particle size variation will be manifested as variability in particle size distribution from lot to lot, even though the average particle size may be reasonably consistent from lot to lot. Li et al. [61] have described how such variability can affect drug release characteristics. In attempting to control the effect of variation in surface area, these authors proposed that the surface area of each lot of material should be determined, and coating level be adjusted to maintain a constant quantity of coating per unit surface area of material.

The implication of variation in pellet particle size and particle size distribution described so far is that such variation affects the total surface area of the batch to be coated, and hence influences how much coating must be applied for consistent results.

Particle size distribution is also critical when pellets are coated in fluidized-bed equipment. During the coating process, the pellets are fluidized randomly within the processing chamber. The distance that the pellets travel during fluidization is partly related to the size and density of the particles. As a result, the residence times of the pellets in the coating zone may vary and lead to pellets that are coated to different levels as shown by Wesdyk et al. [62]. This potential problem is routinely overcome by charging into the coating machine core pellets that have a narrow mesh fraction, and hence a narrow particle size distribution.

Pellet topography is also a factor that can influence batch surface area. Depending on the physicochemical properties of the formulation components and

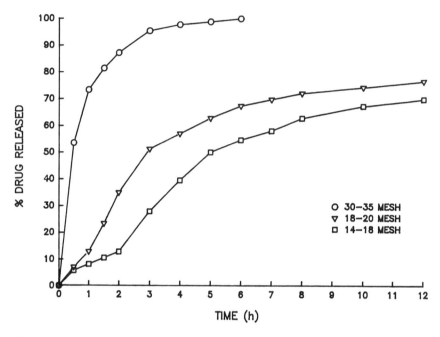

Fig. 3 Release of chlorpheniramine maleate from pellets coated with Surelease (10% theoretical weight gain) in the Wurster process.

the method of pellet preparation and conditions, pellets that have different surface topographies are obtained. Some tend to have smooth surfaces while others may feature irregular and rough surfaces that are aesthetically undesirable. Surface topography is also critical in that it literally dictates the amount of coating material that needs to be applied to achieve a target drug release rate. At a given coating level, smoother pellets generally provide lower release rates than rough pellets, as shown in Fig. 4. The observed differences in release rates are attributed to the higher surface area of the rougher pellets and the relatively shorter diffusional path length at the ridges where the coating can be thinner.

Multiparticulate Morphology and Porosity. The method used to prepare pellets may influence other factors that can have an impact on drug release characteristics. The data depicted in Fig. 5 compare the results obtained when coating pellets (containing propranolol hydrochloride) prepared by different techniques. In one case, the pellets were prepared by layering a drug suspension onto the surface of nonpareils; in the other, the pellets were prepared by extrusion spheronization. In each case, the pellets produced were of equivalent particle sizes and were coated with the same amount of coating material. Differ-

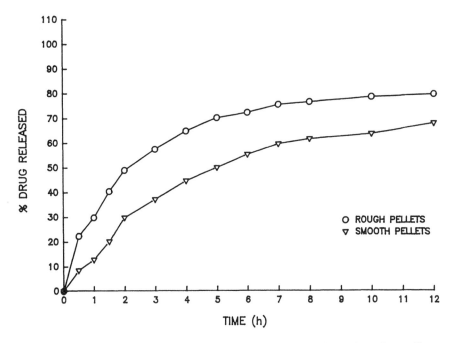

Fig. 4 Effect of pellet morphology on release of chlorpheniramine maleate from pellets coated with Surelease (10% theoretical weight gain) using the Wurster process.

ences in drug release characteristics could be attributed to differences in smoothness and porosity of the pellets used.

These results are similar to those obtained by Zhang et al. [63], who evaluated pellets produced by two different spheronization techniques. These authors found that the method of pellet preparation influenced the critical coating level (which is defined as that level of applied coating where drug release changes from pore moderated to membrane moderated).

An additional factor that must be considered is the influence of the method of pellet preparation on pellet porosity. Porosity has been shown to influence drug release characteristics by interfering with the mechanism of film formation of aqueous polymeric dispersions [64]. During the coating process, water evaporates; water is also removed from the coating formulation, at least initially, by penetration into the core. If the rate of water absorption or penetration into the core during coating is slow, it may assist the drying process and allow the development of capillary forces that are sufficient for polymer packing, subsequent deformation, and fusion, as indicated earlier. This leads to the formation of films that have optimum mechanical properties. If, however, the rate of water

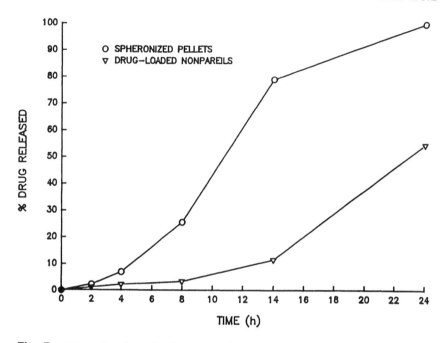

Fig. 5 Effect of method of pellet preparation on release of propranolol hydrochloride from pellets coated with Surelease (15% theoretical weight gain) using the Wurster process.

penetration is too fast, it produces, at the interface, films that have poorly co-alesced polymeric spheres. The rate and extent of water penetration are directly proportional to the pore diameter in the pellets and increase as the size of the pores increases. It is important, therefore, that the diameter of the pores, if present in the pellets, be very small and less than the diameter of the colloidal particles of the dispersion.

These effects have been confirmed by Iley [65], who found that application of a latex coating dispersion to a highly porous substrate produced, in turn, a coating that was itself highly porous. This result contrasted dramatically with that obtained when the same coating material was applied to a nonporous core material.

Friability. During tablet manufacture, friability is considered so critical to the development of the dosage form that it is employed routinely as an in-process control. Tablets that are out of specification with respect of friability are not processed further. Friability also plays a crucial role in the development of modified-release pellets. During the coating process, pellets are subjected to appreciable particle-to-particle and particle-to-wall frictional forces. Friable pel-

lets will generate a significant amount of fines that become suspended temporarily in the expansion chamber. Some of these fines return to the product
chamber (due to gravitational forces), where they run the risk of becoming
embedded in the film being deposited on the pellet. Others are trapped in the
filter bags and get dislodged under their own weight or during the intermittent
shaking of the filters. Once dislodged, drug particles can also become embedded
in the film as the coating process progresses. As a result, during dissolution
testing, the embedded particles can be leached from the coating and create pores.
The presence of such pores will not only lead to faster release rates than
expected, but also, due to the randomness of the distribution of the pores, make
the release rates variable. In some cases, extremely friable pellets may fragment
during fluidization, increase the particle population in the bed and hence the
surface area of the core particles, and once again, provide faster release rates
than desired.

Summarizing, for a given coating level, friable pellets produce drug
release rates that are faster than those of nonfriable pellets (see Fig. 6). It is,
therefore, imperative to produce core pellets that withstand the vigorous agitation
that occurs in the coating chamber.

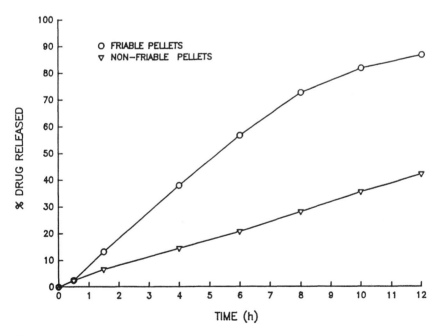

Fig. 6 Effect of pellet friability on drug release from pellets coated with Aquacoat.

Chemical Characteristics of the Substrate

During any film-coating process, opportunity exists for strong interaction between the coating materials and ingredients contained in the core. This interaction can be increased when ingredients in the core are soluble in the solvent/vehicle of the coating system. Chemistry plays an important part in defining drug release characteristics, with some of the important issues being:

1. The influence of drug/excipient solubility on drug release rate
2. The likelihood that the drug will partition into the coating
3. The influence of drug molecular weight on drug release rate
4. The influence of core ingredients on generation of osmotic pressure within the reservoir
5. The influence of core surface counterions on the stability of aqueous polymeric dispersions and hence ultimate membrane structure

Influence of Solubility.   The solubility characteristics of core ingredients affect the development of modified-release pellets in several ways. Core pellets containing highly water-soluble drugs usually require, all other factors being equal, higher coating levels to achieve a predetermined release rate compared to those containing poorly water-soluble drugs [66,67], as shown by the data highlighted in Fig. 7. When the release of the drug results essentially from transport of dissolved drug through water-filled pores or channels within the coating, the solubility of the drug in water will be a significant factor in determining drug release rate. The influence of drug solubility in water, however, will be severely diminished when dealing with large drug molecules, when pore-wall effects become a factor in determining drug release rate. In this case, unless pores in the coating are relatively large and numerous, drug release will be slower than would otherwise be predicted by drug solubility in water.

Conversely, drugs of low water solubility may be released more rapidly than expected. Such a result has been described by Wald et al. [68], who discovered that the unexpected behavior could be attributed to a high solubility of the drug in the polymer membrane (hence true diffusion becomes the major mechanism controlling drug release).

The aqueous solubility of core ingredients can be a major contributory factor influencing the generation of osmotic pressure within the coated pellet once the pellet interacts with the dissolution fluid. Osmotic effects are a predominant factor in mediating drug release from reservoir systems having a delivery orifice. Ghebre-Sellassie et al. [69] have shown that osmotic pressure may well be an important factor contributing to drug release from coated pellets without such an orifice.

Solubility also plays a major role if pellets are coated under inadequate drying conditions. During the application of the coating material, surface dissolution and subsequent migration of drug molecules into the film coat may occur.

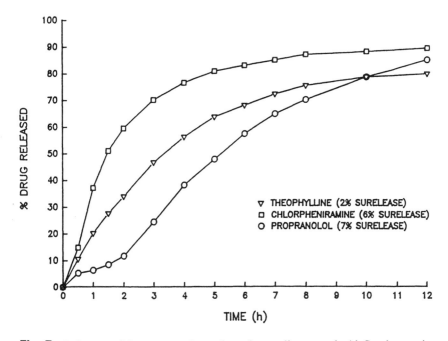

Fig. 7   Influence of drug type on drug release from pellets coated with Surelease using the Wurster process.

The degree of migration depends on the intrinsic dissolution rate and solubility of the active ingredient, which, in turn, dictates the rate and extent of drug release by altering the porosity and hence permeability of the film. Migration could be prevented by spraying the coating formulation initially at low application rates, by applying a seal coat prior to application of the modified-release coating or by optimizing the drying conditions of the coating process. A schematic representation of drug migration during the coating process is given in Fig. 8. The data shown in Fig. 9 indicate the usefulness of employing seal coats (which are applied to the pellets before application of the modified-release coating).

While much has been written about the influence of drug solubility on ultimate drug release rate, there has been a paucity of information regarding the influence of excipients (with the exception of that relating to the influence of excipient solubility on the generation of osmotic pressure within the reservoir). Recently, however, Eerikäinen et al. [70] have discussed the influence of the solubility of excipients on drug release rate. Using indomethacin-sodium as a model water-soluble drug, they found that using excipients of low water solubility (such as calcium hydrogen phosphate or cornstarch) had a retardant effect on drug release characteristics. These results contrasted with those obtained

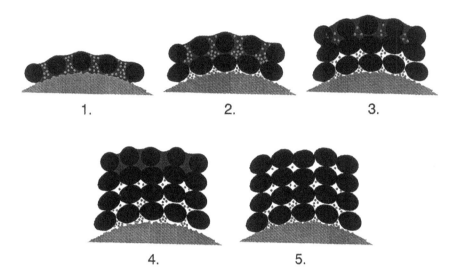

Fig. 8   Schematic illustration of drug leaching (from core pellets) during the application of an aqueous polymeric dispersion.

when using a water-insoluble, but hydrophilic, excipient such as microcrystalline cellulose. In this case, the highly water-absorbant properties of microcrystalline cellulose tend to increase the rate at which water is absorbed by the coated pellets, thus enhancing drug dissolution rate within the pellets.

Influence of the Chemical Composition of the Core.   It was described earlier how core characteristics, particularly porosity, can affect the quality of a modified-release coating, especially when such a coating is applied as an aqueous polymeric dispersion. The structure of these types of coatings can also be influenced by interaction with core ingredients that are ionic in nature. It is widely known that aqueous polymeric dispersions tend to coagulate irreversibly when exposed to ionic solutions. The extent depends on the type and concentration of the ions. Therefore, when dispersion-based coating formulations are applied to pellet substrates composed of inorganic salts of active ingredients, they may partially dissolve the substrate and allow the ions to come in contact with the uncoalesced colloidal particles. This process may then lead to spot coagulation and impede, at least initially, the formation of a continuous film. As a result, the first layers of coating become discontinuous and porous and require the application of more layers of coating to achieve the desired release profiles than otherwise would be the case. The problem, which appears to be commonly observed with sodium salts of active ingredients, can be overcome by the

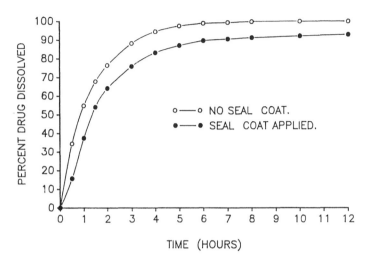

Fig. 9 Influence of a seal coating on release of chlorpheniramine maleate from pellets (coated with Surelease, 10% theoretical weight gain) using the Wurster process.

application of a water-soluble seal coat between the substrate and the controlled-release coating (see Fig. 10). The seal coat is used solely to prevent the ions from interacting with the dispersion coating formulation. It is interesting to note that the effect of chloride ions on film structure is negligible, as indicated.

Stability Issues. The chemical characteristics of coating systems, particularly those based on aqueous polymeric dispersions, may vary substantially. In particular, the pH may differ significantly for different coating systems. It is, therefore, essential that active ingredients that are sensitive to pH are coated only with those dispersions where the pH ensures maximum stability. Thus the development scientist has to have a good understanding of the chemistry and stability of the active ingredient if proper selection of the coating dispersion is to be made. Chemical instability may result from the in situ formation of salts and complexes, ester hydrolysis, or ring opening and cyclization. When the coating dispersion has an inappropriate pH or chemical composition, chemical interaction between the substrate and the coating formulation can potentially be prevented by the application of a seal-coat formulation where the pH does not lead to chemical instability of the active ingredient. Again, the seal coat should be composed of water-soluble materials that do not impede drug release or have any influence on the rate-controlling properties of the modified-release coating.

Many drug substances exist in more than one polymorphic form and, under the right conditions, including exposure to solubilizing liquids, could be converted into forms having physical properties that are different from those of the

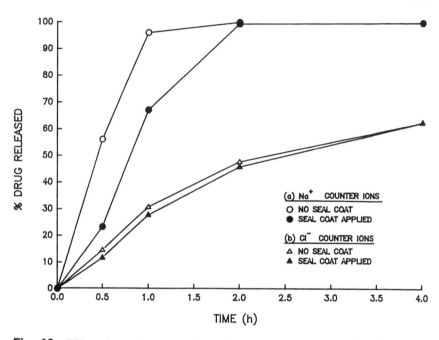

Fig. 10   Effect of counterions on drug release from pellets coated with Aquacoat.

original form. It is conceivable, therefore, that during the coating process, drug substances on the surface of the pellets may dissolve and recrystallize into new forms that may have different intrinsic dissolution rates. When and if this occurs, the release kinetics of the coated pellets may vary with the processing conditions and consequently become difficult to control. If the product bed temperature is controlled to maximize drying and prevent high-moisture buildup, the effect may become negligible. If the potential for polymorphic conversion exists, however, it is advisable to consider the incorporation of a seal-coating step in the process. The solvent that is used to prepare the seal-coat formulation should not solubilize the active ingredient or, if it does, should not promote recrystallization of the drug into a different polymorphic form during the coating process.

B.   Composition of the Coating Formulations

Earlier in this chapter, a comprehensive review was given of the various types of film-coating ingredients that can be used in modified-release film coating and how these ingredients might influence the performance of the coating. When contemplating the development of a coating suitable for producing modified-release pellets, the basic choices involve using either organic solvent–based polymeric solutions, aqueous polymer dispersions, or hot melts. Organic solvent–based solutions of polymers still retain a place of significance for these types of functional coatings. Key ingredients (such as polymers and plasticizers) are "dispersed" on the molecular scale and hence interaction between them is maximized. This type of system facilitates the inclusion of water-soluble, pore-forming agents as relatively large particles. However, appropriate solvent selection is paramount to achieving coatings that have the requisite structural and functional characteristics. Additionally, use of organic solvents in film-coating operations continues to attract regulatory scrutiny (particularly from the environmental standpoint).

Aqueous polymeric dispersions provide a means of offsetting the problems associated with the use of organic solvents. These coating systems do, however, have unique film-forming characteristics that require optimized coating procedures to be employed. This requirement ensures that acceptable coating performance is achieved and time-dependent changes in drug release characteristics are avoided. Finally, use of hot melts totally eliminates the need to consider drying during the coating process. Nonetheless, coating conditions must be carefully controlled if optimal product performance is to occur.

Regardless of the type of coating system selected, the presence of key ingredients in the coating formulation is likely to have a major influence on the results obtained. The information in Table 5 summarizes the effects of some of the key formulation variables. During the formulation development process, key questions that must be answered include:

1. What impact does polymer chemistry have on the results to be obtained?
2. In a similar vein, what is the impact of selecting an aqueous polymeric dispersion instead of an organic solution of the same polymer?
3. What effect will additives such as plasticizers, pigments, antitack agents, surfactants, and other secondary polymers have?
4. What influence does the method of incorporation of all the various ingredients have?

Although aqueous polymeric dispersions have yet to gain prominence as coating systems for preparing modified-release pellets, their use is attracting substantial interest in the pharmaceutical industry. Some major issues that must be confronted when selecting a coating system of this type include a desire to ensure that:

1. The coating formulation enables appropriate drug release characteristics to be achieved in a consistent manner.
2. The membrane obtained is structurally sound.
3. The coating systems are readily adaptable to existing coating process technology.

Table 5 Summary of Coating-Formulation Variables and Their Potential Effects

Variable	Has influence on:
Polymer type	Permeability characteristics of membrane
	Film formation
	T_g of membrane
Molecular weight of polymer	Film mechanical strength and elasticity
Plasticizers (type and concentration)	Film flexibility
	T_g of membrane
	Coalescence of latex coatings
Insoluble additives	Permeability of membrane
	Film mechanical strength and elasticity
	Tackiness of coating
Soluble additives	Permeability of membrane
	Changes in membrane characteristics with time and pH
Concentration of coating ingredients in coating liquid	Uniformity of distribution of coating
	Drying process
	Processing time
Physicochemical properties of solvent/vehicle	Drying rate
	Leaching of drug from substrate
	Porosity of membrane
	Physicomechanical properties of membrane
Method of preparation	Interaction of coating ingredients
	Variability in membrane performance

4. The coating formulations and processes are sufficiently optimized to prevent the occurrence of time-dependent changes in drug release characteristics.

The specialized nature of aqueous polymeric dispersions also dictates that additional care be exercised when incorporating additives. In particular, it is necessary to ensure that:

1. The method of incorporation is reproducible so that variable behavior (batch to batch) is avoided.
2. The method of incorporation or the presence of the additive does not compromise the stability of the dispersion.
3. The presence of the additive does not compromise the film-forming characteristics of the dispersion.

In addition to the influence of the coating formulation (including additives) on drug release characteristics, a major variable to be encountered (and controlled) is the quantity of coating applied. Typical results are shown in Fig. 11.

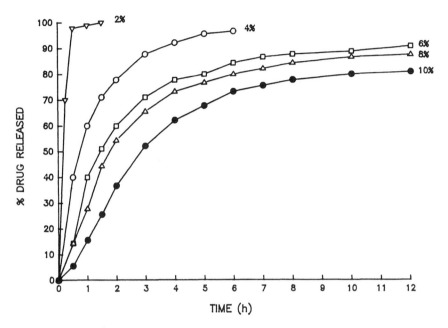

Fig. 11 Influence of quantity of Surelease coating applied on release of chlor-pheniramine maleate from coated pellets.

Similar results when using organic solvent–based coating systems have been described by Kawashima et al. [71]. Not unexpectedly, as the coating level is increased, coating thickness and length of the diffusion pathway also increase, with the result that drug release rate is reduced. This explanation is somewhat simplistic since recent data suggest that as coating progresses, a fundamental change in membrane structure occurs, as shown schematically in Fig. 12. In describing this phenomenon, Zhang et al. [72] have suggested that:

1. At low levels of applied coating, flaws (or pores) exist which, being contiguous between the pellet–coating interface and the surface of the coating, permit drug to be released readily through a water-filled porous network. In this case, cumulative drug release is linear with respect to the square root of time.
2. At higher levels of applied coating, the pores become sufficiently sealed so as to ensure that drug release occurs through an intact membrane and follows zero-order kinetics.
3. The transition point, where drug release begins to be defined by zero-order kinetics, is called the critical coating level.

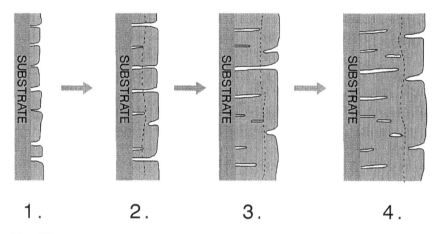

Fig. 12 Modified-release film coatings: model defining membrane structure as coating progresses.

These authors have developed a model that characterizes drug release under these conditions. This model predicts that the quantity of drug released (per unit surface area), Q_t, in time t is given by

$$Q_t = \left[\left(2D_a \; \frac{\epsilon}{\tau} \; C_s A\right)^{1/2} - kL\right]t^{1/2} + \frac{D_m K_m \omega c_s}{L} \; t \tag{3}$$

where L is the thickness of the coating, D_a and D_m are drug diffusion coefficients in water and the membrane, respectively; C_s the drug solubility in water; K_m the drug partition coefficient between water and the polymer phase; A the drug concentration in the core material, which has a porosity ϵ and tortuosity τ; and k and ω are constants.

Using this model, it is apparent that the mechanism for drug release and the critical coating level are dependent on:

1. The core material (ingredients, formulation, and method of manufacture)
2. The composition and structure of the membrane

Influence of Coating Formulation on Drug Release Characteristics

Comparison of Coating Systems.   Much has been said about the relative merits of organic solvent–based coating systems (organosols) compared to those based on aqueous polymeric dispersions [73,74]. Generally, recent evidence suggests [47] that aqueous systems can equal the performance of their organic solvent–based counterparts as long as appropriate coating process conditions are employed. Results to some extent are, however, dependent on the type of aqueous coating system used. From the product development standpoint, it would be prudent to consider that no two coating formulations will perform identically, as shown by the results outlined in Fig. 13.

Influence of Release-Modifying Additives.   Release-modifying additives are incorporated in coating formulations for a variety of reasons. Spray application of film coatings to pellets, irrespective of whether pan-coating or fluidized-bed processing techniques are used, is a process that can have serious limitations. Inability to achieve absolute uniformity of distribution of the coating and batch-to-batch fluctuations in coating process efficiency (i.e., a measure of how much coating is actually deposited in comparison to theoretical quantities based on mass-balance determinations) can be problematic if drug release characteristics are very sensitive to even minor variances in coating level. The results depicted in Fig. 14 are a good case in point.

Ideally, a robust process will seek to minimize these variances. Variability in the process can, however, often be compensated for by decreasing the sensitivity of the coating formulation to such variation in coating distribution. For example, increasing the permeability of the coating to the drug, and at the same time increasing the quantity of coating applied, can produce acceptable results, as shown in Fig. 15.

The general effects of adding water-soluble polymers to ethyl cellulose films derived from aqueous polymeric dispersions are shown in Fig. 16. The effects of such additives are not always predictable, however, as discussed earlier in this chapter and described by Zhang et al. [39].

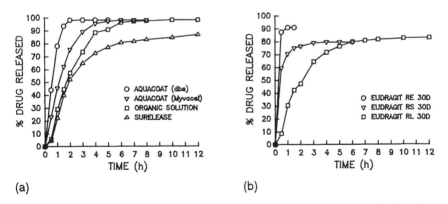

Fig. 13 Influence of type of coating system used on release of chlorpheniramine maleate from coated pellets (10% theoretical weight gain): (a) ethyl cellulose coatings; (b) acrylic coatings.

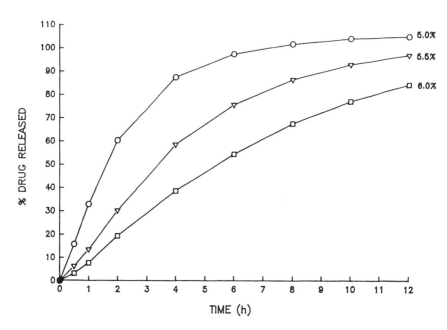

Fig. 14 Effect of coating level on drug release from pellets coated with Aquacoat.

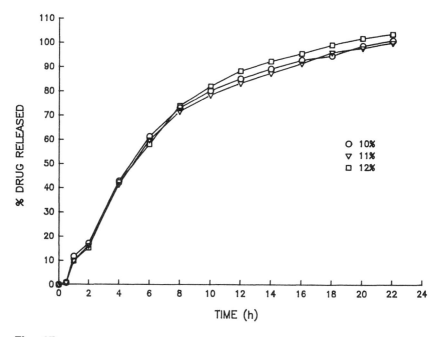

Fig. 15 Effect of coating level on drug release from pellets coated with Aquacoat modified by the inclusion of a water-soluble additive.

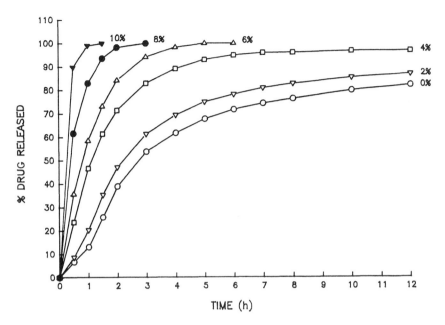

Fig. 16 Effect of adding methylcellulose to Surelease on the release of chlorpheniramine maleate from coated pellets (10% theoretical weight gain).

Since nonuniformity in distribution of the coating is likely to be at its greatest in the early stages of the coating process, the use of release-modifying additives is of paramount importance when target drug release profiles are normally achieved with unmodified coatings at very low levels of applied coating.

Finally, the benefits of including release-modifying additives become well recognized when upgrading coating technologies. Many early modified-release pellet formulations were coated with shellac-based coating systems, the behavior of which is normally difficult to match with the newer aqueous polymeric dispersions. Shellac coating systems especially exhibit release characteristics that are highly pH dependent (a fact that results from the increasing aqueous solubility of shellac as the pH of a dissolution fluid is increased).

Although some aqueous coating systems have been described to have pH-dependent properties [75], most are little influenced by the pH of the dissolution fluid [47,69]. Consequently, the properties of aqueous polymeric dispersions must be modified if they are to mimic those obtained with shellac-based coating systems. Such pH-dependent behavior can be induced by incorporating enteric polymers into the aqueous coating system. Typical results are shown in Fig. 17.

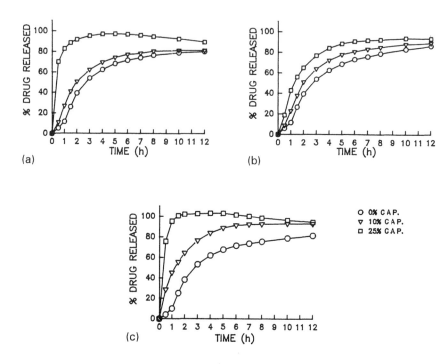

Fig. 17 Effect of addition of cellulose acetate phthalate to Surelease on release of chlorpheniramine maleate from coated pellets (10% theoretical weight gain applied): (a) distilled water; (b) 0.1N HCl solution; (c) buffer solution, pH 6.8.

Case Study.   This case study describes the preparation of modified-release pellets containing propranolol hydrochloride and illustrates the benefits of using a release-modifying ingredient [in this case, hydroxypropyl methylcellulose (HPMC)] to achieve the target release profile. The basic processing conditions are detailed in Table 6.

Initial results for an unmodified coating system are shown in Fig. 18. While these results bracket the target release profile, they demonstrate that drug release rate is highly sensitive to small fluctuations in coating level. This condition could pose serious problems on scale-up into manufacturing. In this study, the target release profile was attained and sensitivity to coating level minimized by incorporating a small quantity of water-soluble polymer (HPMC) into the coating formulation (see Fig. 19).

Table 6 Coating Process Conditions Used in the Preparation of Modified-Release Pellets Containing Propranolol Hydrochloride

Processing equipment:	Glatt GPCG-3 unit fitted with a Wurster insert and a Schlick Model 970 air spray gun with a 1.0-mm liquid orifice
Coating formulation:	Surelease E-7-7050 diluted to 15% solids
Pellets:	2.3 kg of 18 to 20 mesh nonpareils, layered with propranolol hydrochloride and sealed with a water-soluble polymer coating
Coating process conditions	
Partition height	25 mm
Fluidizing air volume	60–90 m³/h
Inlet air temperature	59–61°C
Product temperature	38–43°C
Exhaust air temperature	37–41°C
Atomizing air pressure	3 bar
Spray rate	24–28 g/min

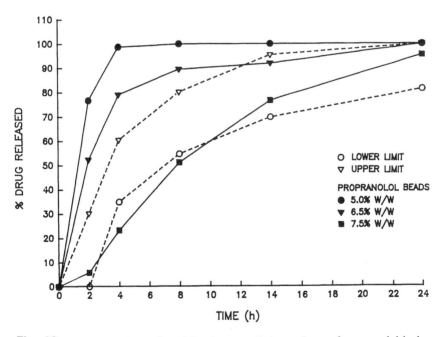

Fig. 18 Influence of quantity of Surelease applied on release of propranolol hydrochloride from coated pellets (produced in the Wurster process).

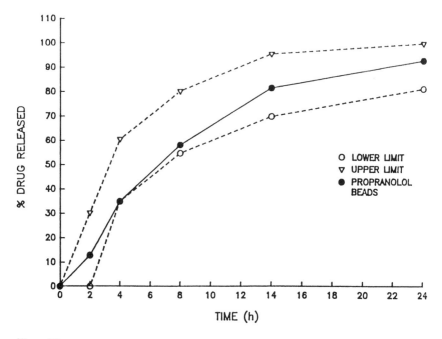

Fig. 19 Influence of addition of Opadry to Surelease on release of propranolol hydrochloride from coated pellets (8% theoretical weight gain, produced in Wurster process).

C. Influence of Coating Process Conditions

Although those factors relating to both the formulation of the core and the coating can, as suggested, have a significant impact on the behavior of modified-release products, many would consider that the influence of the coating process should not be underemphasized. It is possible to argue, in fact, that all our good intentions when formulating the core and the coating are for naught if a major emphasis is not placed on designing an effective, robust process. A modified-release coating, when applied to a pellet, is the major factor that influences drug release characteristics. Thus when attempting to develop an appropriate coating process, factors that are of paramount importance are those that:

1. Affect the quality of the applied coating
2. Influence the uniformity of distribution of that coating across the surface of all the pellets in the batch

Quality issues when dealing with the influence on drug release characteristics are essentially structural in nature and include those relating to:

1. Porosity of the coating that may be caused by excessive "spray drying" of coatings derived from organosols or incomplete coalescence of coatings deposited from aqueous polymeric dispersions.
2. Presence of stress cracks that might be induced by use of overvigorous drying conditions in the coating process or excessive agitation of the pellets during the coating process.
3. "Picking" of the coating as a result of tackiness caused by excessive spray application rates with organosols or excessively high temperatures when using aqueous polymeric dispersions.

The pharmaceutical literature is replete with examples that clearly indicate how the identification and control of key processing parameters is a major prerequisite to achieving appropriate product quality (particularly in the functional sense) when using aqueous polymeric dispersions as coating systems. Specific examples are discussed later in this section.

Uniformity of distribution of the coating can be affected by many processing parameters and can be confounded when coating pellets which can vary (on both an intra- and interbatch basis) with respect to particle size and density, and batchwise with respect to surface area.

Obtaining a product that meets certain functional guidelines, especially when dealing with modified-release products, is of paramount importance. To do so, certain operational conditions must be met. At the same time, capital and operating costs must also be factored in. Thus, from an operational standpoint, it is important when designing a process to:

1. Keep the process simple.
2. Ensure that the process is reproducible from one coating run to another.
3. Make the process robust so that any variation within the control limits of that process does not impact final product performance.
4. Be cognizant of a common desire (on the part of manufacturing departments) to utilize processing equipment that already exists in the manufacturing plant, unless introduction of new processing technology is a key factor in product performance.

In summary, the coating process can be as critical (if not more so) as both pellet and coating formulation considerations in defining the performance of the final coated product. An overview of important processing factors (many of which are discussed in more detail later) and their potential effects is provided in Table 7.

Selection of Processing Equipment
A wide variety of coating equipment exists today and any one type of equipment may be eminently suitable for the production of coated, modified-release pellets.

Table 7 Summary of Processing Variables and Their Potential Effects

Variable	Has influence on:
Coating equipment	Quality and functionality of coating
	Economics
Coating dispersion solids	Membrane structure
content	Uniformity of distribution of coating
	Processing costs
Spray rate	Membrane structure
	Uniformity of distribution of coating
	Processing costs
Atomizing air pressure/volume	Membrane structure
	Uniformity of distribution of coating
Quantity of coating applied	Drug release rate
	Uniformity of distribution of coating
	Processing costs
Drying conditions (air volume, temperature, and humidity)	
With organic solvent– based solutions	Membrane structure
	Drug release rate
With aqueous polymeric dispersions	Coalescence of film
	Drug leaching
	Tackiness of the coating
	Drug release rate

Conventional coating pans have long been used to prepare coated pellets, especially when using coating systems based on shellac that are applied by manual ladling techniques. Conventional coating pans do suffer some key disadvantages, irrespective of whether the coating material is applied by ladling or spray techniques. Many of these disadvantages relate to the fact that:

1. Mixing is extremely poor even when aided by baffles, such that achieving a uniform distribution of the coating may be extremely difficult.
2. Drying is also poor, such that the chances for excessive tackiness (with its associated risk of picking) and leaching of drug (if soluble in the solvent used in the coating system) from the core are increased.
3. The hazards of dealing with flammable or toxic organic solvents are more difficult to contain.

Despite these difficulties, use of these types of coating pans persists as described, for example, by Bianchini and Vecchio [76]. Use of modified coating pans, especially the side-vented (or perforated) variety, has become commonplace in the pharmaceutical industry for coating conventional solid dosage forms, especially when employing aqueous coating technology. Coating pans of this type may well be eminently suited to the production of coated pellets; modifications to pan design, however, may be necessary to ensure that pellets are contained inside the coating chamber.

Of particular interest when coating pellets is the availability of various types of fluidized-bed coating equipment. Fluidized-bed processing has evolved substantially in the pharmaceutical industry. As a result, the multiprocessor concept (where a basic processing unit can be modified by insertion of one of several types of processing chambers) has become a standard. Equipment of this type has proven to be particularly suitable for coating pellets, where any one of three basic processing approaches may be used:

1. Top spray
2. Bottom spray
3. Tangential spray

The basic features of these processes and their relative advantages and disadvantages are described in Table 8. In rationalizing the suitability of any one type of coating process compared to another, many factors may well have to be considered during the selection process:

1. How representative is the processing equipment of the various processes already in place at the many facilities worldwide that may also be called upon to manufacture the product in question?
2. What is the nature of the product being coated and the type of coating system that will be used?
3. What are the functional characteristics of the coating being applied?
4. How critical to the ultimate performance of the final product is the type of processing equipment that will be employed?
5. Can the equipment be operated in such a way as to guarantee reproducibility in performance of the final product from batch to batch?
6. What is the probable preferred batch size and ultimate product volume to be coated?
7. How well does a particular type of processing equipment facilitate scale-up from the product development laboratory through the pilot plant and, ultimately, into full-scale manufacturing?
8. What are the relative costs of installing and operating the processing equipment that is being considered?

When considering the performance of the final coated product, the quality of the coating that is to be deposited will be a major key to success. Mehta and Jones [77] have provided a general discussion of the influence of various types of coating process on the quality of applied modified-release film coatings. Their

Table 8 Features of Three Basic Approaches to Fluidized-Bed Coating of Modified-Release Pellets

Method	Description	Advantages	Disadvantages
Top spray	Adaptation of the typical fluidized-bed granulation process, with the spray nozzle typically positioned lower in the product container	Easy access to spray nozzle(s) Simple to set up Facilitates processing of large batches Good mixing characteristics Less sensitive to impact of interparticle size/density differences on coating uniformity	Impractical for high weight gains Tendency to produce highly porous coatings, particularly with organosols More likely to be associated with lower coating efficiencies
Bottom spray	Wurster process	Facilitates uniform distribution of coating Adaptable to a wide range of coating applications Minimizies spray-drying potential	More sensitive to impact of interparticle size/density differences on coating uniformity Nozzles not readily accessible, causing interruption of process when nozzle blockages occur Requires tallest expansion chamber
Tangential spray	Adaptation of the rotor-granulation, spheronization process	Relatively simple to set up Easy access to spray nozzle(s) Shortest machine High application rates possible Greater flexibility with respect to batch size and batch-size changes during processing Less sensitive to impact of interparticle size/density differences on coating uniformity	Subjects product to high mechanical stress Generally accommodates smaller batch sizes Relatively expensive capital item

findings generally conclude with the observation that the quality of the coatings follows this order (of decreasing quality) with respect to type of process chosen:

$$\text{Wurster} = \frac{\text{tangential}}{\text{spray}} > \frac{\text{side-vented}}{\text{pan}} >> \frac{\text{conventional}}{\text{pan}}$$

There is no doubt that the fluidized-bed processes possess the best drying characteristics and that the proximity of the spray nozzle to the product being coated when using the Wurster or tangential-spray process helps to control deposition of the coating fluid and evaporation of the solvent/vehicle in that fluid and thus maximize quality of the final coating.

When comparing two fluidized-bed processes used in the application of organic solvent–based polymer solutions, Li et al. [78] have shown how the Wurster process was more efficient (in terms of actual amount of coating deposited) than the top-spray process. When attempting to analyze these observed differences, consideration must be given to the relative volatilities of the solvent/vehicles used in the coating formulations and how well that volatility is suited to the particular process/process conditions employed. For comparison, the latent heats of vaporization for solvents/vehicles commonly employed in film coating are summarized in Table 9. These values suggest, for example, that:

1. Use of the more volatile solvents may be problematic in a process where the spray nozzle is remote from the product being coated (i.e., in a coating pan) or in a process that uses countercurrent spraying (i.e., in a top-spray fluidized-bed process) (see Fig. 20).
2. Use of slow-drying liquids (such as water) will be problematic in processes where drying is poor (i.e., conventional coating pans).

This is not to say, however, that the difficulties suggested are insurmountable. It is interesting to note that while the results obtained may, as shown in Fig. 20, be quite sensitive to the type of process used when employing organic

Table 9 Latent Heat of Vaporization of Solvents Used in Film Coating

Solvent	Latent heat (kJ/kg)
Ethyl acetate	386.9
Acetone	524.0
Methylene chloride	556.7
Isopropyl alcohol	604.3
Ethyl alcohol	774.6
Methyl alcohol	1078.1
Water	2260.4

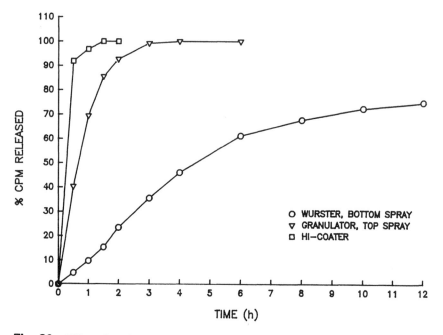

Fig. 20   Effect of coating method on release of chlorpheniramine maleate from pellets coated with an organic solvent–based solution of ethyl cellulose (10% theoretical weight gain).

solvents, that sensitivity may be reduced when aqueous polymeric dispersions are used as the coating system. For example, the data shown in Fig. 21 represent a comparison of results obtained when using three different fluidized-bed processes. The data in Fig. 22 compare results obtained when using either the Wurster process or side-vented coating pan.

Notwithstanding the importance of the particular coating process used, the way in which a particular coating process is set up may well have a significant impact on the results obtained. For example, when setting up the Wurster process, the positioning of the inner partition can be extremely important. The height of the inner partition controls to a large degree the rate at which pellets enter the coating zone. The data in Fig. 23 indicate how the positioning of this inner partition can influence the results obtained. In this case it is likely that at the lower partition height, the rate at which pellets entered the coating zone was insufficient to prevent coating material being deposited on the sides of the inner partition or from being prematurely dried and thus trapped in the filters. The result in either case is a lower coating efficiency with an attendant lower quantity of coating deposited on the pellets.

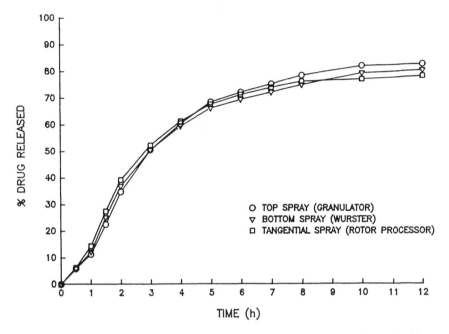

Fig. 21   Effect of type of fluidized-bed coating process used on release of chlor-pheniramine maleate from pellets coated with Surelease (10% theoretical weight gain).

Finally, the influence of the nozzle height setting in a top-spray fluidized-bed coating unit on ultimate drug release characteristics is shown in Fig. 24. In this case it is expected that coating efficiency and coating quality are decreased when the nozzle is positioned further away from the product bed.

Influence of Specific Processing Conditions

As discussed in the preceding section, selection of appropriate processing equipment for coating pellets is extremely important. This selection process, however, is only the beginning of the story when considering the impact of processing as a whole. The effort involved in preparing robust formulations and selecting appropriate processing equipment can be wasted if the same emphasis is not placed on identifying critical processing parameters and establishing processing conditions based on sound specific evaluation. This requirement is a fundamental issue for the success of any film-coating operation; it is critical to the success of any coating operation where a modified-release film coating is to be applied to a substrate as potentially complex as drug-loaded pellets.

Some key processing parameters that may have to be considered when film-coating pellets are:

1. Product batch size
2. Pellet size (and size distribution)
3. Pellet density (and variability within the batch)
4. Coating liquid: organosol or aqueous polymer dispersion?
5. Rheological characteristics of coating liquid
6. Solids content of coating liquids
7. Quantity of coating liquid that is to be applied
8. Spray application rate
9. Drying air temperature, volume, and humidity
10. Product temperature (or exhaust temperature if measurement of product temperature proves difficult)
11. Atomizing air pressure/volume
12. Partition height (Wurster process)
13. Pan speed (coating pan process)
14. Baffle design (coating pan process)
15. Spray nozzle design (fluid orifice size, fluid orifice design, air cap design)
16. Disk speed (tangential spray process)
17. Slit opening (tangential spray process)

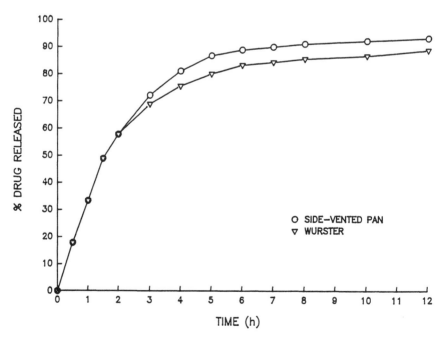

Fig. 22   Effect of coating method on release of chlorpheniramine maleate from pellets coated with Surelease (10% theoretical weight gain).

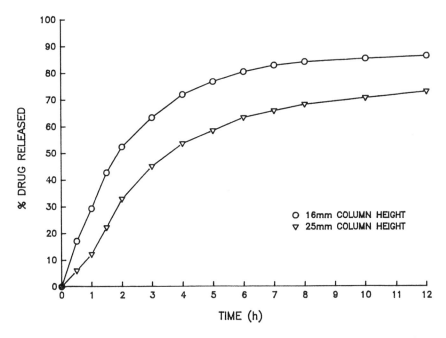

Fig. 23 Influence of partition height on release of chlorpheniramine maleate from pellets coated with Surelease (10% theoretical weight gain) in a Glatt GPCG-3 Wurster process.

Although this list is extensive (but not necessarily all inclusive), in general terms it is important to consider how these processing parameters will influence product behavior, both during and after the coating process. An overview of some important considerations is given in Table 10. The special nature of aqueous polymeric dispersions and the relative complexity of film formation with these systems necessitate that the influence of processing conditions on coating performance be well characterized.

The drying conditions employed when coating pellets potentially have a major influence on the performance characteristics of any modified-release coating. In any coating process, drying conditions collectively involve consideration of these independent variables:

1. Drying air volume
2. Drying air temperature
3. Drying air humidity

Although a dependent variable, product temperature (or exhaust temperature if measurement of product temperature is impractical) is usually a key

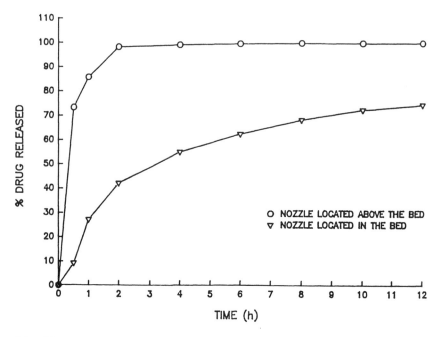

Fig. 24 Influence of nozzle height setting (top spray process) on drug release characteristics from coated pellets.

parameter that has a major influence on the quality of the coating, especially when using aqueous polymeric dispersions.

Drying conditions alone have less meaning, however, unless considered in the context of spray application rate and solids content of the coating liquid. When considering the impact of drying conditions on the behavior of applied coatings, the emphasis may well be different when dealing with organic solvent–based solutions of polymers (organosols) than when using aqueous polymeric dispersions.

With organosols, the impact of drying conditions needs to focus on:

1. The potential for creation of unacceptably porous coatings that result from use of overzealous drying conditions (spray drying)
2. The opportunities for increased tackiness, a phenomenon that occurs as a result of inadequate drying or exposure of coated products to high temperatures
3. The risk of "picking" as a result of inadequate drying

Thus, when applying organosols, drying conditions must be carefully regulated to avoid both inadequate drying and overdrying. In discussing the

Table 10 Important Product Characteristics That May Be Influenced by Processing Conditions

Issue to be considered	Processing conditions that may have impact
Film formation (and structural quality of the coating)	Drying conditions (processing our temperature, volume, and humidity)
	Spray application rate
	Atomizing air pressure/volume
Tackiness of coating	Drying conditions
	Spray application rate
	Solids content of coating liquid
Uniformity of distribution of coating	Mixing of pellets in process
	Number of spray nozzles used
	Design of spray nozzles
	Atomizing air pressure/volume
	Spray application rate
	Solids content of coating liquid
	Quantity of coating applied
Leaching of drug from pellet	Nature of coating liquid (organosol or aqueous polymeric dispersion)
	Drying conditions
	Spray application rate
	Atomizing air pressure/volume
	Solids content of coating liquid

influence of processing factors on the properties of ethyl cellulose coatings deposited from organic solvent–based solutions, Arwidsson [20] has described how drying air temperature has a significant influence on drug release rate (with increasing temperature causing an increase in release rate).

In contrast, when contemplating the application of aqueous polymeric dispersions, emphasis is placed on the influence of drying conditions on:

1. Coalescence of the coating
2. Tackiness due to the exposure of the coating to temperatures significantly in excess (>20 to 30°C) of its glass transition temperature
3. Leaching of the drug from the pellet (a potential problem when the drug is water soluble and drying conditions are inadequate)

Using aqueous polymeric dispersions of ethyl cellulose, Yang and Ghebre-Sellassie [79] have shown how product temperature, although a dependent variable, is the critical factor (related to drying conditions) which influences drug release characteristics. This finding is supported by the data highlighted in Fig. 25.

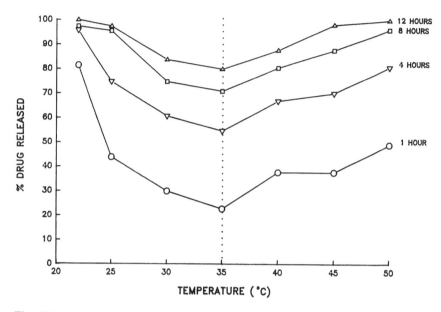

Fig. 25 Influence of product temperature on release of diphenhydramine hydrochloride from pellets coated with an aqueous ethyl cellulose dispersion (15% theoretical weight gain).

Pellets coated at a product bed temperature of 22°C released more than 90% of the drug after only 2 h, implying that the coating was highly porous and permeable. The porous nature of the coating is related to the high-water solubility of the drug (diphenydramine hydrochloride in this case) in the core pellets as well as the low product bed temperature used.

During the early stages of the coating process, drug molecules can dissolve in the coating liquid and migrate from the pellet into the film. If this occurs, the presence of drug in the film can adversely affect the mechanism of film formation of the coating dispersion. First, a significant amount of drug may dissolve in the coating formulation and thus reduce the surface tension of the liquid, thereby lowering the capillary forces needed for the deformation of the polymeric spheres. Second, during the coating process, drug molecules may become dispersed within the polymer coating. Drug may then diffuse out during the dissolution phase to generate a porous and hence more permeable coating. Finally, the product bed temperatures may be lower than the minimum film-formation temperature of the plasticized polymer and hence may not be conducive to optimum film formation. In this case, the polymeric spheres would remain hard and resist deformation. This means that the capillary forces generated during water evaporation may not be strong enough to cause significant polymer deformation and coalescence. As a result, a discontinuous and porous film that provides relatively fast release rates is obtained.

Coating at too high product temperatures may also produce less-than-optimum films, again causing fast release rates [as demonstrated by the release profile of pellets coated at 50°C (Fig. 25)]. The evaporation rate of water at such elevated temperatures may be so high that while the migration of drug from the pellet into the coating is minimized, coating quality is again compromised. At high temperatures, the rate of water evaporation overwhelms the normal diffusion process (of water through the capillary network formed between adjacent polymer particles as the coating dries) that is critical to effective consolidation and subsequent coalescence of the polymer particles. Consequently, water evaporates long before sufficient capillary pressure is exerted to deform the polymer particles. Again, the result is a porous film containing poorly coalesced polymer particles (even though processing temperatures are substantially in excess of the minimum film-forming temperature of the coating system). The results shown in Fig. 25 indicate that processing at product temperatures in excess of 35°C does not guarantee, per se, that a suitably coalesced film having optimal properties will be obtained.

Once the optimal drying conditions are deduced, it should be possible to produce reproducible and stable films. In addition, it may be possible to keep to a minimum the amount of applied coating that is needed to achieve a target release profile. This has implications with respect to both processing times and coating raw material costs. From the standpoint of process control, it may well be possible that once a optimum product temperature has been established, the process may be robust enough to allow spray rates to be varied over quite a wide range without compromising the drug release rate. This conclusion is supported to some extent by the results shown in Fig. 26, where essentially constant release rates are obtained over spray application rates in the range 12 to 24 g/min. Such a result not only provides manufacturing flexibility but also facilitates better control over operating costs.

Leaching of drug from the pellet, particularly during application of aqueous coating systems, is another matter that often has to be resolved. This problem needs to be contained by effectively controlling the evaporation of water without compromising the quality of the coating, a result that once again requires a thorough understanding of the interaction of key operating parameters.

Processing deficiencies that contribute to leaching include:

1. Poor drying conditions (poor airflow, low operating temperatures, excessive humidities)
2. Low solids content in coating fluid
3. Employment of excessively high spray rates
4. Equipment failures

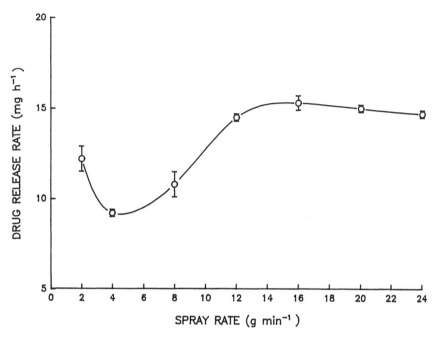

Fig. 26   Effect of spray application rate on release of chlorpheniramine maleate from pellets coated with Surelease (10% theoretical weight gain) using the Wurster process.

An example of what can happen when any of those conditions apply is shown in Fig. 27. The faster release rates seen with the 40-mg dose of propranolol hydrochloride occurred because drug was leached from the pellet during coating, a result that was confirmed by examination of samples by means of scanning electron microscopy. Although inadequate drying conditions contributed to the problem in this case, what would otherwise be considered to be acceptable drying conditions may not guarantee success if the drying environment is overwhelmed by the quantity of solvent/vehicle that must be removed. The issue in this case is the solids content of the coating liquid.

When using an aqueous coating system, the solids content of the coating dispersion can have a significant impact on product performance, as indicated in Fig. 28. These results suggest that when coating at low dispersion solids, excess water can penetrate the pellet and then be removed (leaching drug with it) when the pellets are dried afterward at elevated temperatures (60°C). Thus for pellet samples coated with low solids dispersions, the quantity of drug release after a given time is greater after extended drying of the sample than before such drying.

The purpose of employing dilute coating dispersions is to improve uniformity of distribution. The data in Fig. 28 suggest that under these circumstances,

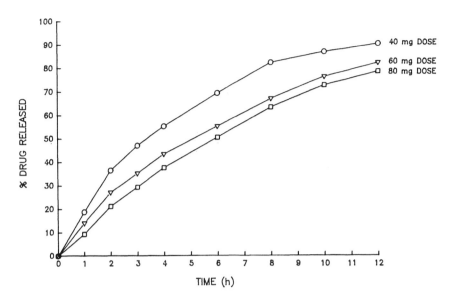

Fig. 27 Release of propranolol hydrochloride from pellets coated with Surelease (8% theoretical coating level).

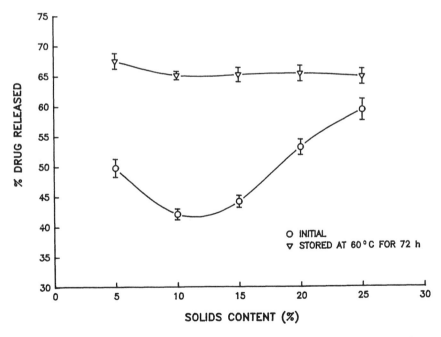

Fig. 28 Effect of dispersion solids content on release of chlorpheniramine maleate from pellets coated with Surelease (10% theoretical coating level).

269

improved drying conditions must be employed if drug leaching is not to become a problem. For example, without changing processing temperatures, the drying conditions can be improved by increasing fluidizing air velocity (within certain limits to keep attrition at a minimum), as suggested by the results highlighted in Fig. 29.

It may be apparent from the discussion so far that film coating with aqueous polymeric dispersions requires considerable attention to employment of appropriate processing conditions if target objectives are to be met. A key factor is that product temperature must exceed the minimum film-forming temperature (MFFT) of the coating system to ensure complete coalescence of that coating. However, if the product temperature exceeds the glass transition temperature (usually closely related, numerically, to MFFT) by more than 20 to 30°C, the polymer film will become soft enough that tackiness becomes a major problem. Of particular concern is the approach where, at the conclusion of the coating process, coated product may be dried (cured) at elevated temperatures to facilitate complete removal of moisture and completion of the coalescence process. Once this additional drying step has been completed, if the product so treated is not allowed to cool before discharge into a bulk container, tackiness of the coating will simply bond pellets together during bulk storage. This problem, which occurs when hot bulk product remains stationary for some time, is called blocking. A simple solution to the blocking problem involves the application of

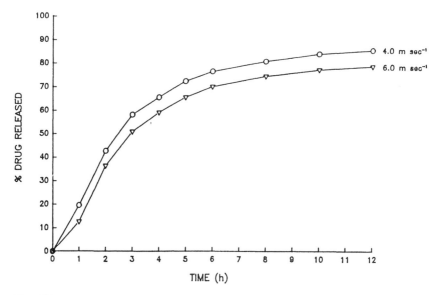

Fig. 29   Effect of fluidizing air velocity on release of chlorpheniramine maleate from pellets coated with Surelease (10% theoretical coating level) using the Wurster process.

an overcoat that is applied as an aqueous solution of a polymer such as hydrox-ypropyl methylcellulose.

Importance of Process Optimization

In the preceding discussion, one overriding issue that should be readily apparent is that to produce a quality modified-release product in the form of film-coated pellets, the number of variables that needs to be evaluated is quite extensive. The common practice, often called a process of successive approximation (or the trial-and-error approach), whereby potentially key variables are identified on the basis of past experience and then investigated one at a time, is a time-consuming process that rarely produces all the information needed to develop an optimal product.

For this reason, employment of one of the many statistical approaches to product and process optimization is becoming commonplace. These approaches allow several variables to be changed during each coating run and the responses to be examined by regression analysis and response surface analysis. A significant advantage to such an approach is that interactions between key variables can often be determined. For example, the influence on drug release rate of changing spray rate, changing solids content of coating liquid, and changing drying air temperature can be evaluated simultaneously and the combined impact determined.

The really important benefits of utilizing such approaches for optimization include the fact that information generated in this way can be used to:

1. Support the validation process
2. Provide appropriate documentation for review in the NDA preapproval inspection process

Using a suitable experimental design procedure, Dietrich and Brausse [80] have described an approach for validating the pellet coating process to manufacture a sustained-release theophylline product. Their procedure employed a combination of factorial and central composite designs, with results being analyzed statistically using a method involving both qualitative evaluation by analysis of variance (ANOVA) and quantitative evaluation of multilinear and multisquare regression. These techniques enable empiricism to be kept out of the validation process.

Arwidsson [20] and Arwidsson et al. [81] have employed statistical experimental design techniques to study the influence of process factors on the properties of ethyl cellulose films deposited from organic solvent–based solutions and aqueous polymeric dispersions, respectively. In each case, the spraying techniques used enabled samples of free film to be obtained that could be subjected to tensile stress analysis; limitations of the procedure resulted from the fact that samples used in drug release studies lacked the characteristics of material produced under the more dynamic conditions encountered in a typical coating process.

Johansson et al. [82] studied the influence of spray rate, fluidizing air velocity, inlet air temperature, and distance of the spray nozzle from the pellet bed on coating efficiency and absence of agglomeration of pellets coated with an organic solvent–based solution of ethyl cellulose in a top-spray fluidized-bed coating process (Glatt WSG-5). Their approach to evaluation of the effect of these parameters was based on use of a reduced factional design. Their results suggest that with this type of process and coating system, coating efficiency cannot be increased beyond 75% without risking some agglomeration of pellets.

Finally, Jozwiakowski et al. [33] have characterized a fluidized-bed process (top spray, Glatt GPCG-5) for the application of hot melts to a multiparticulate product. Using the XSTAT software package [83] and employing the simplex technique, these researchers found that to achieve a target drug release of approximately 60% in 60 min, maximize reproducibility in the process operating parameters, and achieve a cost-effective process, the optimal operating parameters for the hot-melt coating employed were:

Temperature of molten wax: 120°C
Atomizing air temperature: 120°C
Atomizing air pressure: 5 bar
Spray rate: 30 g/min

D. Scaling Up the Pellet-Coating Process

Without trying to oversimplify what can often be a complex process, the major object in scaling up a product is to achieve product performance characteristics that duplicate (within certain tolerances) those obtained during the product development process. Duplication must also be achievable on a consistent basis. Certainly, the special performance requirements of modified-release pellets add a level of complexity to scale-up issues for this type of product. When scaling up the pellet-coating process, there are some factors that are often considered predictable:

1. Droplet size of the coating fluid (impacted by coating liquid rheology, spray rates, nozzle sizes, and atomizing air pressure/volume)
2. Evaporation rate (impacted by spray rate, fineness of atomization of coating liquid, and drying-air volume/temperature/humidity)
3. Spray rate (impacted by nozzle dynamics, coating liquid rheology, pump settings)

Other factors are less predictable and include:

1. Influence of increasing batch sizes on mass effects such as mixing and attrition

2. Coating efficiency (i.e., the actual quantity of coating deposited compared to what is delivered in theory)
3. Behavior of both coating liquid and pellets in the coating zone as a result of changes in equipment configuration (nozzles, Wurster partition height settings) on scale-up

It is always important to be aware of the fact that when a product is introduced into manufacturing, other potential variables may become more prominent. In particular, there may be several manufacturing sites to consider where key environmental factors such as humidity are variable.

To improve the chances of success on scale-up, these issues should be attended to:

1. Both the pellet and coating formulations should be robust.
2. The development-scale process should be well characterized.
3. Detailed pilot-scale studies should have been completed.

The implications of scale-up to the pellet-coating process has received only scant attention in the pharmaceutical literature. However, both Mehta [84] and Jones [85] have provided useful overviews in this area with respect to the fluidized-bed process. Using the approach recommended by Mehta (where prediction of spray rate on scale-up can be determined from the respective fluidizing air volumes used in the developmental-scale process and the pilot/manufacturing process), scale-up procedures have been developed for the application of an aqueous ethyl cellulose dispersion to pellets using the Wurster process. The fixed and specific process operating parameters used in this study are shown in Table 11.

Although no attempt was made in this study to compensate for changes in fineness of atomization of coating liquid as the spray rates increased on scale-up (an issue that should really be addressed in a process dealing with small-particle coating), the results obtained for drug release characteristics are remarkably consistent (as shown in Fig. 30). When applying functional film coatings to pellets, it is not unusual to find that the quantity of coating needed to achieve a target drug release profile may have to be reduced when going to the larger manufacturing scales. Process efficiencies change (and may be improved) on scale-up, as does the influence of increased mass of pellets on coating morphology, two factors that may well contribute to the need for reduced levels of coating. The higher processing temperatures required and the relative softness of coating under such temperature conditions may be contributory factors when applying aqueous polymeric dispersions as modified-release coatings. The need for reduced levels of coating is clearly demonstrated in Fig. 31, where results are shown for drug release characteristics from pellets coated in a tangential-spray process. These results are supported by similar data obtained in a Wurster process (Fig. 32).

Table 11 Operating Parameters Used in Scale-up of Wurster Process for Applying Surelease to Pellets

	Fixed operating parameters
Substrate	18/20 mesh nonpareils containing chlorpheniramine maleate (37.5 mg/g)
Inlet air temperature	60°C
Outlet temperature	30°C
Atomizing air pressure	2.5 bar
Coating formulation	Surelease (diluted to 15% w/w solids)
Theoretical quantity of coating applied	10% w/w (dry solids basis)

			Spray rate (g/min)	
Process	Batch size (kg)	Air volume (m³/h)	Predicted	Actual
GPCG-1 (6-in. Wurster)	1.0	60	—	20
GPCG-3 (7-in. Wurster)	3.0	70	23	25
GPCG-5/9 (7-in. Wurster)	5.0	100	33	36
GPCG-60 (18-in. Wurster)	40.0	550	183	197

When examining the pellet-coating process, a major consideration is to ensure that the morphological characteristics of the coating remain essentially unchanged on scale-up. The drying conditions experienced in the process will have a major influence in this regard. Of the three major variables that contribute to drying, processing temperature (particularly as it relates to product temperature) and humidity should ideally remain consistent with what was achieved on the development scale. Drying-air volume, a parameter that will certainly be changed on scale-up, is not only a drying factor but one that influences pellet movement, especially in a fluidized-bed process. In the fluidized-bed process, air velocity (through the distributor plate) should remain essentially constant on scale-up. Total air volume will obviously increase because the area of the air distributor plate will increase. Despite the need to increase total air volume, it is quite common to find that the quantity of air required to fluidize 1 kg of pellets decreases as the size of the machine increases.

Scaling-up spray rate requirements is often overcomplicated. Essentially, the change in spray rate should be in direct relationship to the drying (fluidizing)

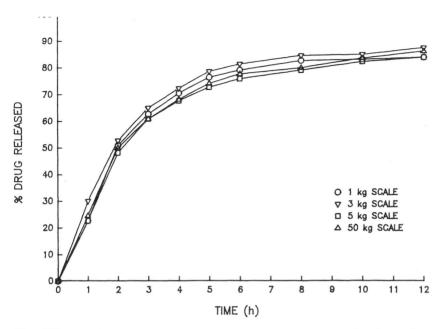

Fig. 30 Comparison of drug release characteristics for chlorpheniramine maleate coated with Surelease (theoretical 10% weight gain) on various processing scales.

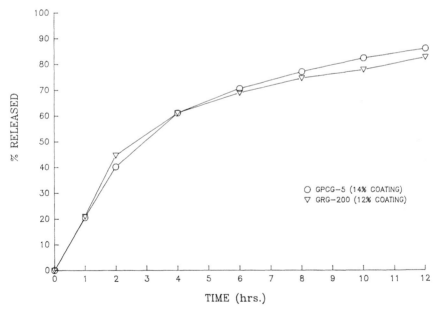

Fig. 31 Comparative dissolution profiles for pellets coated with an aqueous ethyl cellulose dispersion using the rotor process in either a GPCG-5 or GRG-200.

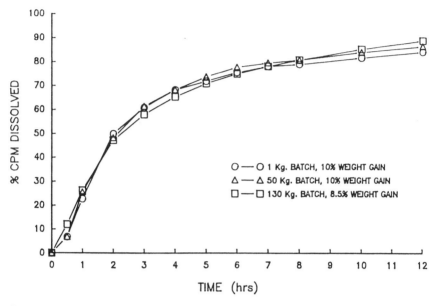

Fig. 32   Influence of processing scale (Wurster process) on release of chlorpheniramine maleate from pellets coated with Surelease.

air volume, not the change in batch size. There are many ways to increase liquid delivery rates (in addition, of course, to just increasing pumping speed):

1. Increasing nozzle orifice size
2. Switching to multiple-head nozzles
3. Increasing the number of nozzles

Each approach has its merits, and choice may well be dictated by the type of processing machine used. However, as spray rates are increased, it is extremely important to keep fineness of atomization (i.e., droplet size) consistent with what was achieved on the development scale. The usual approach to maintaining the same degree of atomization as spray rates are increased is to increase atomizing air pressure/volume. There are limitations to this approach, especially when increasing spray rate from 20 g/min (lab scale) to 2000 g/min (production scale). Generally, the slower a liquid passes through a nozzle and intermixes with atomizing air, the finer the atomization that is achieved. Thus the objective on scale-up is, of course, to maintain fineness of atomization by increasing atomizing air pressure; additionally, flow rate through a nozzle can be held in check by increasing the number of nozzles used, a result that is achieved by either using multinozzle spray gun or by increasing the number of spray guns used.

Finally, as scale-up progresses, it is often necessary to deal with a change in equipment configuration. Such changes may involve:

1. Changing from a single-headed nozzle to a multiple-headed nozzle (typically in the top-spray fluidized-bed process)
2. Increasing the number of spray guns used (usually occurs in a coating pan or the Wurster and tangential-spray fluidized-bed processes)
3. Changing to multiple inserts in the Wurster process
4. Changing the inner-partition height in the Wurster
5. Changing pan speeds (coating pan process) or rotor speeds (tangential-spray process)
6. Changing orifice plate configurations (particularly in the Wurster process where it is necessary to maintain equivalent airflow through each coating zone of a multiple insert unit without dramatically increasing pellet attrition)

E. Controlling Time-Dependent Changes in Drug Release Characteristics

A primary concern during the development of modified-release pellets is to meet a particular objective in a reproducible manner with respect to drug release characteristics. Such consistency, however, must be extended to the behavior of the product subsequent to the time of manufacture and until that product is consumed by the patient.

It should thus be apparent that changes in drug release characteristics that occur with time are going to be of great concern. Ideally, no such changes should occur. Realistically, we might expect that some change in behavior might be observed but that the magnitude of the change will not be so substantial that performance of the product is compromised.

When dealing with possible time-dependent changes, three distinct possibilities exist:

1. No significant changes occur (either the product is stable or opposing mechanisms for change cancel each other out).
2. Drug release rate increases with time.
3. Drug release rate decreases with time.

Factors Causing an Increase in Drug Release Rate
Time-dependent increases in drug release rate are likely to be associated with:

1. Membrane failure
2. Partitioning of drug into the membrane with associated gradual diffusion of drug through that membrane
3. Gradual release of a water-soluble drug as the result of egress of entrapped moisture

Membrane failure is likely to result from inadequate coating strength, which can often be corrected by utilizing a higher-molecular-weight grade of polymer or by effective plasticization as described by Rowe [23].

Wald et al. [68] have described how an increase in drug release rate can occur unexpectedly when the drug shows a significant affinity for the coating, resulting in substantial partitioning of that drug in the membrane. Under these circumstances, the problem can be minimized by applying a seal coating over the pellet (prior to application of the release-modifying coating). In this case, the seal coating should be one for which the drug has low affinity.

An example of a water-soluble drug being released gradually with time as the result of egress of entrapped moisture is provided by the data highlighted in Fig. 33. This problem is likely to be more prevalent when coating with aqueous polymeric dispersions, since moisture can readily penetrate the pellet during coating if drying conditions are inadequate for the process.

Whether drug transfers into the coating via partitioning or as the result of moisture transfer, storage of the product at accelerated temperatures is likely to cause an increase in drug release rate with time. This situation is caused by the fact that at temperatures in excess of the glass transition temperature (T_g) of the

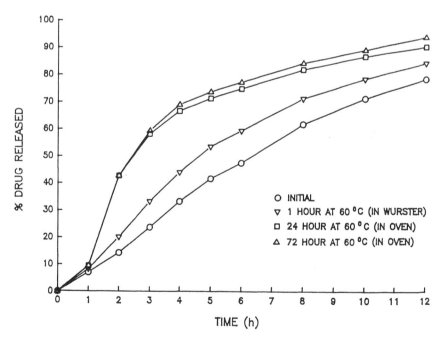

Fig. 33   Effect of drying time (at 60°C) on release of chlorpheniramine maleate from pellets coated with Aquacoat (10% theoretical weight gain).

coating, free volume increases with increasing temperature. Thus the membrane becomes more "open" to the passage of the drug. It has also been suggested by Lippold et al. [86] that the activation energy for diffusion is inversely proportional to the fractional free volume of the polymer above its T_g. That is, an increase in free volume facilitates increased diffusion through the film as a direct result of the fact that the activation energy for diffusion is decreased.

Factors Causing a Decrease in Drug Release Rate
There are two factors that might cause a decrease in drug release rate with time:

1. Gradual evolution of entrapped organic solvent
2. Further gradual coalescence (curing) of a coating applied as an aqueous polymeric dispersion

   In the first case, continued loss of organic solvent will cause a reduction in free volume within the coating and a concurrent increase in the glass transition temperature of that coating. Thus the activation energy for diffusion through the coating will increase, and the barrier properties of the membrane will improve (unless coating structure is compromised by the cracking induced by increasing internal stress as solvent is evolved).

   Much has been talked about "curing" effects that can occur when product coated with an aqueous polymeric dispersion is stored at elevated temperatures. Such behavior has been described by Goodhart et al. [46], Ghebre-Sellassie et al. [66], and Li et al. [87]. In contrast, Porter [6] has shown that such curing effects do not always occur. This disparity in results suggests that a postcoating curing step may or may not be necessary, depending on the type of pseudolatex employed.

   Utilization of appropriate processing conditions may be an important approach when attempting to eliminate potential curing effects with aqueous polymeric dispersions. Success in this regard will also be contingent on using a suitably formulated coating system. For example, when using an aqueous ethyl cellulose dispersion, the drug release characteristics shown in Fig. 34 were obtained. These results clearly indicate that curing effects occurred in one case but not the other. The coating process conditions used in each case are summarized in Table 12. It is apparent that the only significant difference between the two procedures relates to the solids content of the coating dispersion used. Under other circumstances, we would expect that applying a dilute coating dispersion might increase drug-leaching problems and cause time-dependent increases in drug release rate. In this case, however, no drug leaching or curing effects are evident. Worthy of note, however, is the duration of the coating run in each case. Stable results were obtained with the longer process, while curing effects were observed with the shorter process. This finding suggests that to minimize curing effects:

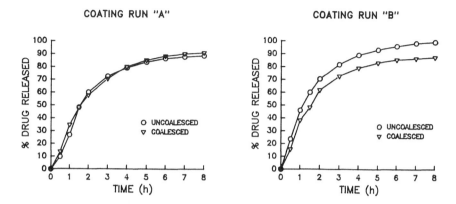

Fig. 34   Effect of process variation on release of chlorpheniramine maleate from pellets coated with Surelease.

1.   Optimal processing conditions are required to ensure complete coalescence of the aqueous polymeric dispersion occurs during the coating process.
2.   The pellets being coated must be exposed to those optimal conditions for a minimum amount of time to ensure that complete coalescence will occur.

   To explore the time issue a little further, similar coating trials have been completed where processing conditions have been kept constant, and the only variable was the amount of coating applied (under this premise, increasing the amount of coating applied automatically lengthens the process). Of course, as the quantity of coating is increased, drug release characteristics change; the impor-

Table 12   Coating Process Conditions That Have Been Shown
to Influence Time-Dependent Decreases in Drug Release Rate
When Applying Surelease in a Uni-Glatt Wurster

	Specific process conditions	
Process parameters	Run A	Run B
Inlet air temperature (°C)	60	60
Exhaust air temperature (°C)	28	30
Flap setting (deg)	30	30
Atomizing air pressure (bar)	2.5	2.5
Coating dispersion solids (% w/w)	5.0	25.0
Spray rate (g/min)	15	15
Theoretical weight gain (%)	6.0	6.0
Coating time (min)	80	16

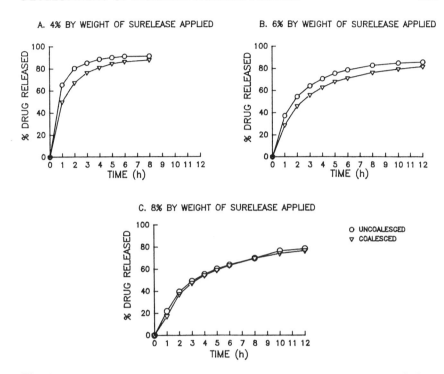

Fig. 35 Potential curing effects: influence of quantity of Surelease coating applied.

tant question, however, is whether the need for curing also changes. The results shown in Fig. 35 suggest that this is the case.

Although these findings may not be entirely conclusive, they do suggest that the need to conduct subsequent curing trials can often be avoided if optimized process conditions are employed (with the utilization of an appropriate time element to be considered as part of the optimization process).

Summarizing, when dealing with time-dependent changes that can occur in the drug release characteristics of modified-release pellets, some critical issues that must be considered involve:

1. Recognizing how key formulation and process variables influence the morphology of the coating
2. Optimizing the coating process so that curing of coatings derived from aqueous polymeric dispersions is completed, as much as possible, during the time frame of the coating process
3. Always conducting curing studies (especially when the aqueous polymeric dispersion formulations are modified by inclusion of other additives)

4. Ensuring that coating process conditions are adequate to eliminate the risk of drug leaching from pellets during application of the coating
5. Being cognizant of potential changes in coating behavior as the process is scaled up

REFERENCES

1. R. C. Rowe, *Pharm. Int.*, January, pp. 14–17 (1985).
2. R.-K. Chang, G. Iturrioz, and C.-W. Luo, *Int. J. Pharm. 60*(2):171 (1990).
3. J. A. Butler and J. J. Vance, U.S. Patent 3,365,365 (1968).
4. Y. V. Pathak, R. L. Nikore, and A. K. Dorle, *Int. J. Pharm. 24*:351 (1985).
5. Y. V. Pathak and A. K. Dorle, *J. Controlled Release 5*:63 (1987).
6. S. C. Porter, *Drug Dev. Ind. Pharm. 15*(10):1495 (1989).
7. L. P. Amsel, O. N. Hinsvark, K. Rotenberg, and J. L. Sheumaker, *Pharm. Technol. 8*(4):28 (1984).
8. H. Arwidsson, *Acta Pharm. Nord. 3*(1):25 (1991).
9. F. Theeuwes and T. Higuchi, U.S. Patent 3,845,770 (1974).
10. F. Theeuwes, D. Swanson, P. Wong, P. Bonsen, V. Place, K. Heimlich, and K. C. Kwan, *J. Pharm. Sci. 72*(3):253 (1983).
11. K. Lehman and D. Dreher, *Pharm. Technol. 3*(3):52 (1979).
12. S. P. Li, K. M. Feld, and C. R. Kowarski, *Drug Dev. Ind. Pharm. 17*(12): 1655 (1991).
13. R. J. Kostelnik, *Polymeric Delivery Systems*, Gordon & Breach, New York, 1978.
14. G. Källstrand and B. Ekman, *J. Pharm. Sci. 72*(7):772 (1983).
15. G. S. Banker, *J. Pharm. Sci. 58*:81 (1966).
16. J. Hildebrand and R. Scott, *The Solubility of Non-electrolytes*, 3rd ed., Reinhold, New York, 1949.
17. D. J. Kent and R. C. Rowe, *J. Pharm. Pharmacol. 31*(5):269 (1979).
18. A. Rubin and R. A. Wagner, *J. Appl. Polym. Sci. 19*:3361 (1975).
19. J. Spitael and R. Kinget, *Pharm. Acta Helv. 55*(6) (1980).
20. H. Arwidsson, *Acta Pharm. Nord. 3*(1):25 (1991).
21. R. C. Rowe, *J. Pharm. Pharmacol. 33*:423 (1981).
22. R. C. Rowe, *J. Pharm. Pharmacol. 35*:112 (1982).
23. R. C. Rowe, *Int. J. Pharm. 29*:37 (1986).
24. T. Lindholm, A. Huhtikangas, and P. Saarikivi, *Int. J. Pharm. 21*: 119 (1984).
25. Z. W. Wicks, Jr., *J. Coatings Technol. 58*(743):23 (1986).
26. J. W. McGinity, *Aqueous Polymeric Coatings for Pharmaceutical Dosage Forms*, Marcel Dekker, New York, 1988.
27. C. Bindschaedler, R. Gurny, and E. Doelker, *Labo-Pharma-Probl. Tech. 31*(331):3869 (1983).
28. W. Wizerkaniuk, U.S. Patent 4,129,666 (1978).
29. I. S. Hamid and C. H. Becker, *J. Pharm. Sci. 59*:511 (1970).
30. S. Motycka and J. G. Nairn, *J. Pharm. Sci. 67*:500 (1978).
31. S. P. Patel and C. I. Jarowski, *J. Pharm. Sci. 64*:869 (1975).
32. C. Igwilo and N. Pilpel, *J. Pharm. Pharmacol. 39*:301 (1987).

33. M. J. Jozwiakowski, D. M. Jones, and R. M. Franz, *Pharm. Res.* 7(11):1119 (1990).
34. K. Vasilevska, Z. Djuric, M. Jovanovic, M. Strupar, and A. Simov, *Pharmazie* 46:54 (1991).
35. J. McAinsh and R. C. Rowe, U.S. Patent 4,138,475 (1979).
36. R. C. Rowe, *J. Pharm. Pharmacol.* 38:214 (1986).
37. P. Sakellariou, R. C. Rowe, and E. F. T. White, *J. Appl. Polym. Sci.* 34:2507 (1987).
38. P. Sakellariou, Ph.D. thesis, University of Manchester Institute of Science and Technology (1984).
39. G. H. Zhang, J. B. Schwartz, R. L. Schnaare, R. J. Wigent, and E. J. Sugita, *Proceedings of the International Symposium on Controlled Release of Bioactive Materials,* 17:194 (1990).
40. H. Burrell, in *Polymer Handbook,* 2nd ed. (J. Brandrup and E. H. Immergut, eds.), Wiley-Interscience, New York, 1975, pp. 337–360.
41. D. J. Kent and R. C. Rowe, *J. Pharm. Pharmacol.* 30:808 (1978).
42. C. A. Entwistle and R. C. Rowe, *J. Pharm. Pharmacol.* 31:269 (1979).
43. R. C. Rowe, *Int. J. Pharm.* 29:37 (1986).
44. R. C. Rowe, A. D. Koteras, and E. F. T. White, *Int. J. Pharm.* 22:57 (1984).
45. H. Arwidsson, O. Hjelstuen, D. Ingason, and C. Graffner, *Acta Pharm. Nord.* 3(2):65 (1991).
46. F. W. Goodhart, M. R. Harvis, K. S. Murthy, and R. U. Nesbitt, *Pharm. Technol.* 8(4):64 (1984).
47. V. Iyer, W.-H. Hong, N. Das, and J. Ghebre-Sellassie, *Pharm. Technol.* 14(9):68 (1990).
48. B. C. Lippold, B. H. Lippold, B. Sutter, and W. Gunder, *Drug Dev. Ind. Pharm.* 16(11):1725 (1990).
49. C. Bindschaedler, R. Gurny, and E. Doelker, *J. Pharm. Pharmacol.* 39:335 (1987).
50. T. Lindholm, M. Juslin, B. A. Lindholm, M. Poikala, S. Tiilikainen, and H. Varis, *Pharm. Ind.* 49(7):740–746 (1987).
51. D. Hennig and H. Kala, *Pharmazie* 42:26 (1987).
52. E. M. G. van Bommel, J. G. Fokkens, and D. J. A. Crommalin, *Acta Pharm. Technol.* 35(4):232 (1989).
53. L. C. Li and G. E. Peck, *Drug Dev. Ind. Pharm.* 15(12):1943 (1989).
54. L. E. Appel and G. M. Zentner, *Pharm. Res.* 8(5):600 (1991).
55. R. Bodmeier and O. Paeratakul, *Pharm. Res.* 8(3):355 (1991).
56. J. W. Parker, G. E. Peck, and G. S. Banker, *J. Pharm. Sci.* 63(1):119 (1974).
57. H. W. Chatfield, *Science of Surface Coatings,* Van Nostrand, New York, 1962, p. 453.
58. I. Ghebre-Sellassie, Pellets: a general overview, in *Pharmaceutical Pelletization Technology* (I. Ghebre-Sellassie, ed.), Marcel Dekker, New York, 1989.
59. H. Bechgaard and G. H. Nielson, *Drug Dev. Ind. Pharm.* 4:53 (1978).
60. A. M. Mehta, *Pharm. Manuf.* 3(1) (1986).
61. S. P. Li, G. N. Mehta, J. D. Buehler, W. M. Grim, and R. J. Harwood, *Drug Dev. Ind. Pharm.* 14(4):573 (1988).

62. R. Wesdyk, Y. M. Joshi, N. B. Jain, K. Morvis, and A. Newman, *Int. J. Pharm.* *65*:69 (1990).
63. G. Zhang, J. B. Schwartz, R. L. Schnaare, R. J. Wigent, and E. T. Sugita, *Drug Dev. Ind. Pharm. 17*(6):817 (1991).
64. Emulsion properties 3: film formation, in *Surface Coatings*, Vol. 1 (OCCAA, Eds.), Chapman & Hall, New York, 1983, pp. 175–183.
65. W. J. Iley, *Powder Technol. 65*:441 (1991).
66. I. Ghebre-Sellassie, V. Iyer, D. Kubert, and M. B. Fawzi, *Pharm. Technol. 12*(9):96 (1988).
67. S. C. Porter, *Drug Dev. Ind. Pharm. 15*:1495 (1989).
68. R. J. Wald, S. L. Saddler, and G. F. Amidon, *Pharm. Res. (Suppl.)*, *5*(10):S-115 (1988).
69. J. Ghebre-Sellassie, R. H. Gordon, R. U. Nesbitt, and M. G. Fawzi, *Int. J. Pharm. 37*:211 (1987).
70. S. Eerikäinen, J. Yliruusi, and R. Laakso, *Int. J. Pharm. 71*:201 (1991).
71. Y. Kawashima, T. Handa, A. Kasai, H. Takenaka, and S. Y. Lin, *Chem. Pharm Bull. 33*(6):2469 (1985).
72. G. Zhang, J. B. Schwartz, and R. L. Schnaare, *Pharm. Res. 8*(3): 331 (1991).
73. E. Horvath and Z. Ormos, *Acta Pharm. Technol. 35*(2):90 (1989).
74. D. Hennig and H. Kala, *Pharmazie 40*:554 (1985).
75. R. K. Chang, C. H. Hsiao, and J. R. Robinson, *Pharm. Technol. 8*(3):56 (1987).
76. R. Bianchini and C. Vecchio, *Boll. Chim. Farm. 126*(11):441 (1987).
77. A. M. Mehta and D. M. Jones, *Pharm. Technol. 9*(6):52 (1985).
78. S. P. Li, K. M. Feld, and C. R. Kowarski, *Drug Dev. Ind. Pharm. 15*(8):1137 (1989).
79. S. T. Yang and J. Ghebre-Sellassie, *Int. J. Pharm. 60*:109 (1990).
80. R. Dietrich and R. Brausse, *Arzneim. Forsch./Drug Res. 38*(11):1210 (1988).
81. H. Arwidsson, O. Hjelstuen, D. Ingason, and C. Graffner, *Acta Pharm. Nord. 3*(4):223 (1991).
82. M. E. Johansson, A. Ringberg, and M. Nicklasson, *J. Microencapsulation 4*(3):217 (1987).
83. J. S. Murray, *X-Stat: Statistical Experimental Design/Data Analysis/Nonlinear Optimization*, Wiley, New York, 1984.
84. A. M. Mehta, *Pharm. Technol.* (2):46 (1988).
85. D. Jones, *Controlled-Release Society Workshop on Scale-up Processes for the Manufacture of Controlled-Release Systems*, Reno, Nev., July 26–27, 1990.
86. B. H. Lippold, B. K. Sutler, and B. C. Lippold, *Int. J. Pharm. 54*:15 (1989).
87. S. P. Li, R. Jhawar, G. M. Mehta, R. J. Harwood, and W. M. Grim, *Drug Dev. Ind. Pharm. 15*(8):1231 (1989).

11

Mechanisms of Release from Coated Pellets

Jennifer B. Dressman* and Bernhard Ø. Palsson

The University of Michigan
Ann Arbor, Michigan

Asuman Ozturk

Chiron Corporation
Emeryville, California

Sadettin Ozturk

Miles, Inc.
Berkeley, California

I. INTRODUCTION

It would simplify the development process considerably if one were able to predict the in vivo performance of a pelleted dosage form a priori from its formulation. To do this, one must bear in mind that there are several processes after ingestion of the dosage form which may affect the temporal profile of the drug's blood level. These include the rate at which the pellets move through the gastrointestinal (GI) tract, the rate at which they release drug in each region of the GI tract, how efficiently the drug is absorbed once it is released, and whether it is subject to metabolism in the gut wall or during its first pass through the liver.

A primary advantage of pelleted dosage forms is that they tend to have more reproducible upper GI transit patterns than monolithic dosage forms, especially if one compares dosing in the fed and fasted states. The transit behavior of dosage forms has been the subject of several conferences and many publications. In addition to Chapters 12 and 13 herein, the reader is referred to Hardy et al. [1] for a thorough discussion of transit behavior. The rate at which

---

Current affiliation: Institute for Pharmaceutical Technology, J.W. Goethe University, Frankfurt, Germany.

the drug is released at any given location within the GI tract may depend on the local environment as well as the mechanism by which the drug is released from the pellets. Following release, the drug is available for absorption. The rate of uptake may be linear with concentration, or if absorbed by an active mechanism, it may be saturable. The kinetics of uptake are compound specific and need to be tested for each drug, especially if one intends to modify the release rate and hence the drug's concentration at the absorptive sites. The permeability of the gut mucosa to the drug may also vary among regions of the GI tract. The stomach appears to be relatively impermeable to most drugs. The small intestine, with its large surface area and specialized transport mechanisms is the main site of absorption for the majority of compounds, when they are formulated in conventional dosage forms. Colonic uptake rate may be comparable to that in the small intestine and should be determined if released from the dosage form is intended to continue for more than 6 to 8 h. During and after uptake there may be metabolic limitations to bioavailability: namely, gut wall and hepatic metabolism. If these processes are saturable, the fraction of drug reaching the general circulation may be altered when it is formulated in a slow-release dosage form. Again, this behavior is case specific. Wagner [2] has developed models for determining whether formulation into a slow-release dosage form will affect the overall bioavailability of a drug that undergoes first-pass metabolism.

This chapter is devoted to the mechanisms by which drug is released, under GI conditions, from pelleted dosage forms. Together with the physiological considerations outlined above, a knowledge of the release mechanism will help the pharmaceutical scientist to predict the in vivo behavior of the dosage form. Additionally, if the current dosage form does not provide an optimal delivery pattern, a knowledge of the release mechanism can serve as a guide as to which aspects of the formulation should be modified to improve the release profile. Finally, the release mechanism may influence the degree to which variations in physiological conditions will translate into variability in dosage-form performance.

The major mechanism by which the drug is released from a pelleted dosage form will naturally depend on the type of coating and the method by which it is applied. An important determinant of the kinetics of release is the solubility behavior of the coating material under GI conditions. Behavior may be classified according to three general types: (1) the coating is insoluble under all physiologically relevant conditions, (2) the solubility changes dramatically at some point the GI tract, and (3) the coating is slowly erodible under GI conditions. Less obvious but also important to the kinetics of release are the influences of the core formulation, in terms of both the physical properties and amounts of the drug and excipient materials present, and the physiological environment into which the drug is released.

II. PELLETS COATED WITH A POLYMER NOT SOLUBLE UNDER GI CONDITIONS

The aim of coating pellets with an insoluble polymer is to retard the rate of drug release, so that blood levels are sustained over a prolonged period. The most commonly used materials are the insoluble ethers of cellulose, such as ethyl cellulose. These may be applied from organic solution, or preferably, from an aqueous dispersion. The method of application and processing conditions may influence the porosity of the coating and consequently the release mechanism. Cellulose esters such as acetates and butyrates can also be applied to produce insoluble films. Some methylacrylate polyesters, such as Eudragit S, which are insoluble over the pH range 1 to 7.5, may also be considered as coatings that are essentially insoluble under GI conditions. Although many other insoluble materials have developed for sustaining release (silastics, polylactic acids, etc.), the release rates are usually much too slow for application to GI delivery, where complete depletion of the drug reservoir should occur within the transit time from the mouth to the midcolon, a time that averages about 12 to 16 h [1]. A further restriction as to the choice of coating agent is the ease with which it can be applied to pellets as a coating, vis-à-vis making a matrix-type system. Finally, coating materials are required to be nontoxic, a property that has not yet been verified for some of the more recently developed materials.

There are several possible mechanisms by which release from pellet dosage forms coated with GI-insoluble polymers may occur: solution/diffusion through the continuous plasticized polymer phase, solution/diffusion through plasticized channels, diffusion through aqueous pores, and osmotically driven release [3].

A. Solution/Diffusion Through the Continuous Plasticized Polymer Phase

This mechanism assumes that the polymer forms a continuous phase in which the plasticizer and other additives are homogeneously dispersed. The diffusion of a solute molecule within an amorphous polymer phase is an activated process involving the cooperative movements of the penetrant (drug) and the polymer chain segments around it [4]. In effect, thermal fluctuations of chain segments allow sufficient local separation of adjacent chains to permit the passage of a penetrant. It is by this stepwise process that hindered molecular diffusion occurs.

Another, less likely mechanism of release is the movement of the drug on the polymer chains, known as configurational diffusion. Release by diffusion/ solution through the plasticized polymer phase is depicted in Fig. 1. The release rate when the solution/diffusion mechanism is operating can be described by

$$J = \frac{P_m}{\delta} (C_s - C_b) \tag{1}$$

where J is the flux (release rate per unit surface area of coating), C_s and C_b are the concentration of drug at the drug–coating interface and the bulk, respectively, and δ is the coating thickness. The permeability coefficient, P_m, of the coating polymer can be written as

$$P_m = \frac{D\epsilon}{\tau\beta} K = D'K \tag{2}$$

where D is the molecular diffusivity of the drug, K the distribution coefficient of the drug between the polymer membrane and fluid in the core (imbibed water), ϵ the volume fraction of the chain openings, β a chain immobilization factor, and τ the tortuosity factor [5].

The apparent diffusivities (D') of compounds of varying molecular weights in a permeable polymer above its glass transition temperature (natural rubber), and in a stiff polymer below its glass transition temperature (polystyrene), are compared in Fig. 2 with their diffusivities in water. The frequency with which a diffusion step occurs depends on (1) the size and shape of the drug, (2) how tightly packed and how much force of attraction there is between adjacent polymer chains, and (3) the stiffness of the polymer chains. In general, the farther below its glass transition temperature (T_g), the less permeable will be the polymer. Plasticizers lower the T_g, increase free volume, and promote polymer–segment mobility, thus increasing the diffusivity. Coatings likely to be applied to sustain release from pelleted dosage forms will usually exhibit behavior intermediate between the two extreme cases shown in Fig. 2.

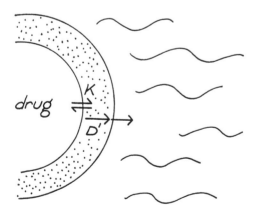

Fig. 1 Drug release from coated pellets via solution/diffusion through the polymer film. (From Ref. 3.)

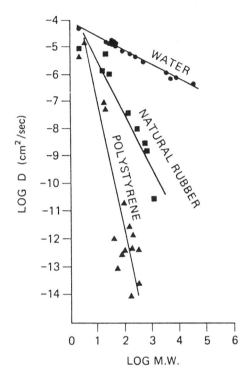

Fig. 2 Variation of diffusion coefficient with solute molecular weight in water, natural rubber, and polystyrene. (From Ref. 4.)

From Eq. (2) it can be seen that the overall permeability of the polymer to the drug will depend on the ability of the drug to partition into the polymer as well as its ability to diffuse through the polymer. In general, this can be estimated from the solubility parameters of the drug and polymer [6].

Taking a hypothetical case where one desires a release rate of 10 mg/h of a drug with a solubility of 10 mg/mL from a dose of pellets having 50 cm^2 surface area, coated by a polymer 20 μm thick, the permeability must be

$$P = \text{desired release rate} \times \frac{1}{\text{SA}} \times \frac{\text{coating thickness (cm)}}{\text{solubility (mg/cm}^3)}$$
$$\text{(mg/s)}$$

$$\approx 10^{-8} \text{ cm}^2/\text{s}$$

For P_m to be the dominant (fastest) mechanism of release, the partition coefficient needs to be about 1:1, in the case of a rubbery polymer where $D \approx 10^{-8}$ cm^2/s, while if the polymer is stiff, with $D \approx 10^{-12}$ cm^2/s, K will need to be greater than 10^4.

The solution/diffusion mechanism has been demonstrated for many polymer films prepared from organic solvents (e.g., Refs. 7–11), which tend to form complete films. In general, it will be dominant only in those cases where the film is continuous (lacks pores) and flexible, and where the drug has a high affinity for the polymer relative to water.

B. Solution/Diffusion Through Plasticizer Channels

When the plasticizer is not uniformly distributed in the coating polymer, and if its content is high, the plasticizer could conceivably take the form of a continuous phase in the form of patched channels [3]. This mechanism is shown in Fig. 3. The release rate for this model can be described by Eq. (1), with P_{pl}, the permeability of the plasticizer, replacing P_m, the permeability of the coating polymer [3]:

$$P_{pl} = \frac{D_{pl}\epsilon_{pl}}{\tau_{pl}} K_{pl} \tag{3}$$

In this case, K_{pl} is the distribution coefficient of the drug between plasticizer and the core fluid (imbibed water), τ_{pl} the tortuosity of the plasticizer channels, and ϵ_{pl} the volume fraction of plasticized channels. For this mechanism to be dominant,

$$P \approx 10^{-8} \text{ cm}^2/\text{s} = \frac{D_{pl}\epsilon_{pl}}{\tau_{pl}} K_{pl}$$

Diffusivity in the plasticizer will generally be lower than in water since plasticizers tend to be relatively viscous. Assuming that $D_{pl} \cong 10^{-6}$ cm^2/s, a

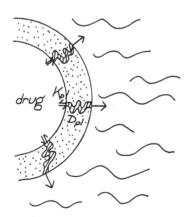

Fig. 3 Drug release from coated pellets via solution/diffusion through plasticizer channels. (From Ref. 3.)

plasticizer load of 40% with half forming channels ($\epsilon = 0.2$) and a low tortuosity of $\tau = 2$, the ability of the drug partition into the plasticizer should be at least

$$K_{pl} = \frac{10^{-8} \times 2}{10^{-6} \times 0.2} = 0.1$$

Using the solubility ratio as a rough guide to the distribution coefficient for phenylpropanolamine HCl between water and four plasticizers, Ozturk et al. [3] estimated K values ranging from 1.694/40 for triacetin to 3.954/40 for Myvacet, suggesting that this mechanism would be a little too slow to explain the release rates observed.

In fact, there has not been any study in the literature surveyed which observed this mechanism for coatings applied either from organic solvents or aqueous dispersions, most likely because a localization of the plasticizer phase to form continuous channels represents an extreme condition [3]. Under most circumstances one anticipates that the plasticizer will be distributed more or less uniformly through the polymer.

C. Diffusion Through Aqueous Pores

In this model, the coating is not homogeneous and continuous but is punctuated with pores. These pores fill with solution when the pellets come in contact with an aqueous medium, thus facilitating diffusion of the drug, as illustrated in Fig. 4. This mechanism is more likely to be operative for coatings formed from aqueous dispersions of pseudolatexes than when the coating is applied from an organic solvent. During the coating and curing processes, the pseudolatex particles often do not fuse completely, thereby creating pores in the coating. These pores may be on the order of 1 μm in diameter, as shown in Fig. 5.

Fig. 4 Drug release from coated pellets via diffusion through aqueous channels. (From Ref. 3.)

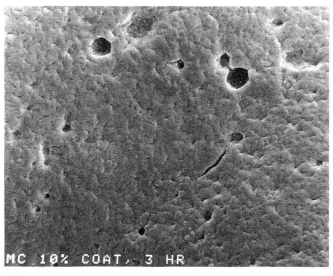

(a)

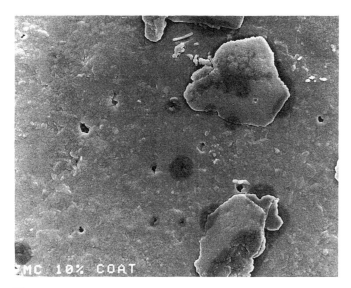

(b)

Fig. 5 Scanning electron micrographs of the porous coating formed when ethyl cellulose is applied to pellets containing phenylpropanolamine HCl from an aqueous dispersion. (a) Coating after 3 h dissolution testing. (b) Coating before dissolution testing. (From Ref. 3.)

For diffusion through aqueous pores, the permeability coefficient, P_p, is given by

$$P_p = \frac{D_p \epsilon_p}{\tau_p} \tag{4}$$

where D_p is the aqueous diffusivity of the drug, ϵ_p the volume fraction of the aqueous channels, and τ_p the tortuosity of the aqueous channels. The partition coefficient, K, will be unity in this case.

For diffusion through aqueous pores to be the mechanism driving the release rate, consider the example above, where P should be on the order of 10^{-8} cm^2/s. At $\tau = 10$,

$$\epsilon = \frac{10^{-8} \times 10}{5 \times 10^{-6}} = 0.02$$

indicating that $>2\%$ of the surface area should consist of pores, while at $\tau = 2.5$, $\epsilon = 0.005$, indicating that $>0.5\%$ of the surface area should consist of pores. From these calculations one may conclude that if SEM consistently indicates the presence of pores in the coating, it is likely that diffusion through the pores will contribute significantly to the overall release rate.

D. Osmotically Driven Release

When the coating is porous, there is also the possibility of release being driven by an osmotic pressure difference between the core materials and the release environment. Sources of osmotic pressure in the core formulation include low-molecular-weight excipients (e.g., the sugar constituting Nu-Pareil seeds) and the drug. At saturation, the concentration of sucrose is $5.85\ M$, corresponding to an osmotic pressure of 150 bar. For the drug to contribute significantly to the osmotic pressure, it should be highly water soluble, be of low molecular weight, and be present in a substantial dose (i.e., capable of achieving saturation concentration in the core). For example, phenylpropanolamine HCl, with a solubility of $2.12\ M$, can generate an osmotic pressure of 54.4 bar at saturation. By comparison, the proximal small intestine maintains essentially isoosmotic conditions, that is, about 8 bar.

When the pellets come into contact with an aqueous environment, water is imbibed through the coating, creating a solution in the core. The excipients and/or drug dissolve in the imbibed water, generating the interior osmotic pressure. The osmotic pressure difference between the core and the external medium then provides the driving force for efflux through pores in the coating (see Fig. 6). The release rate for this process can be described by

$$J = K\sigma\ \Delta\pi\ (C_i - C_m) \tag{5}$$

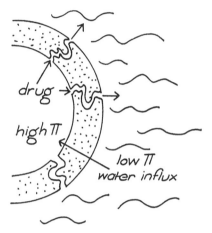

Fig. 6 Drug release from coated pellets, driven by an osmotic pressure difference across the coating. (From Ref. 3.)

where K is the filtration coefficient, σ the reflection coefficient of the coating, $\Delta\pi$ the osmotic pressure difference across the coating, and C_i and C_m the interior and media drug concentrations, respectively [12].

As mentioned above, the osmotic pressure in the small intestine is low and practically invariant. In the stomach, by contrast, the osmotic pressure will vary with the type of fluid ingested, ranging from water (0 bar) to soft drinks [typically, 10 to 12% sucrose (i.e., about 30 bar)]. Ingestion of these types of dosage forms under hypertonic conditions (e.g., with sugary drinks or dense meals) may therefore result in initially slow release rates or perhaps even a delay in release.

Assuming that drug is released under sink conditions (i.e., $C_m \approx 0$) in the intestine, zero-order release is achieved while (1) the materials responsible for maintaining the osmotic pressure difference are present in the core above their solubilities (hence constant $\Delta\pi$ is maintained), and (2) the drug is present at a level greater than its solubility (constant C_i). Achieving osmotically driven release requires a semipermeable coating, the existence of aqueous pores within the coating, and the presence of core materials that are capable of generating sufficient osmotic pressure. Bindschaedler et al. [13] demonstrated that even if the coating is applied from an aqueous dispersion, its semipermeable properties can be retained. Aqueous pores can be deliberately created by adding water-soluble excipients such as poly(ethylene glycol) (PEG), poly(vinyl acetate) (PVA), or sorbitol to the coating formulation [12,14,15]. The choice of core material will influence the degree of osmotic pressure generated. Because of their high sugar content, use of Nu-Pareil seeds is more likely to result in osmotically driven release than are granules in which the drug is spheronized with high-molecular-weight materials such as Avicel.

The usual method to check for osmotically driven release is to add various amounts of urea to the dissolution media and observe whether the release rate is inhibited [12,14,15]. Sodium chloride is less preferable, as in this case both osmotic pressure and ionic strength effects can contribute to changes in the release profile. One should also check that the drug solubility (and hence the driving force for release by the diffusion through an aqueous pore mechanism) is not affected by the presence of large concentrations of the osmotic agent used.

III. COATINGS THAT UNDERGO A DRAMATIC CHANGE IN SOLUBILITY WITH LOCATION IN THE GI TRACT

Traditionally, the coatings that have been employed because of their large increase in solubility at some point in the GI tract have been those that are pH sensitive. This sensitivity has been utilized to prevent release in the stomach while affording complete release in the intestine. The pH differential between stomach and small intestine is greatest in the fasted state, about 5 pH units [16]. During and soon after meal ingestion this differential is reduced because the gastric contents are buffered by ingested food (Fig. 7). Gastric pH may temporarily reach pH values of 4 to 5, thus considerably narrowing the range over which the coating solubility must change. Table 1 lists some commonly used enteric coating materials and the pH values at which they become rapidly soluble.

Achieving selective release to the colon via a pH-sensitive change in the coating solubility is more problematic than for the small intestine, as the pH differential between the ileum and proximal colon is much less distinct than that between the stomach and small intestine. Work by Evans and co-workers [17] showed that the mean pH in the proximal colon is 6.4 ± 0.6, whereas the pH in the intestine tends to range from pH 5 (upper intestine, fed state) to pH 6.5 (lower small intestine, fasted state) [18]. Ingestion of carbohydrates not digested by pancreatic enzymes may result in more acidic pH values in the colon, due to fermentation by the colonic bacteria to small volatile fatty acids such as acetic, propionic, and butyric acids. Thus, for reproducible and selective delivery to the colon, one must look to other region-specific mechanisms of release, such as azoreductase activity [19] and other reactions catalyzed by enzymes derived from the colonic bacteria.

Another application of pH-sensitive coating materials in oral dosage forms is in the design of sustained-release formulations of weakly basic drugs that are not very soluble in their free-base forms. As the dosage form moves into the intestine and the pH value nears the pK_a value of the drug, its solubility dimin-

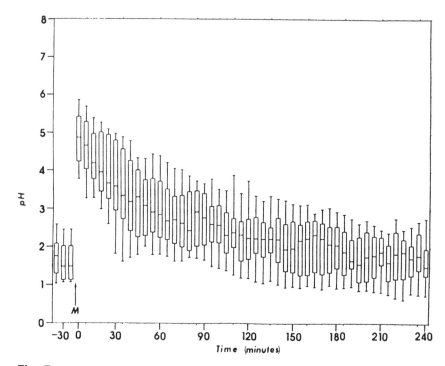

Fig. 7 Pre- and postprandial gastric pH in 24 healthy young men and women. Bars represent 25th and 75th percentiles of pooled data. (From Ref. 16.)

ishes, thus reducing the driving force for release. If a pH-sensitive polymer is incorporated in the barrier coating, the surface area available for release (pore surface area) will increase when the polymer dissolves, offsetting the reduction in concentration gradient across the coating and maintaining the release rate of drug from the dosage form.

As a pellet coated with a pH-sensitive polymer moves through the GI tract, there are three stages in the drug release process, each with a distinct mechanism: the stage at which the coating is insoluble, the stage at which the coating is dissolving, and the stage at which the coating has dissolved.

A. Mechanism of Release while the Coating is Insoluble

At this stage, release will occur according to one or more of the mechanisms described in Section II. Usually, one wants to limit the rate of release as much as possible in this region (e.g., enteric-coated pellet formulated to prevent release

Table 1 pH-Sensitive Polymers Used as Enteric Coating Materials

Polymer	pH at which the polymer starts to dissolve rapidly
Poly(vinyl acetate/phthalate) (PVAP)	4.7
Hydroxypropyl methylcellulose phthalate	
HP50[a]	5.0
HP55[a]	5.5
Cellulose acetate phthalate (CAP)	6.0
Methacrylic acid/methacrylic acid methyl ester	
Eudragit L100[b]	5.5
Eudragit L[b]	6.0

[a]Shinetsu.
[b]Rohm Pharma.

of an acid-labile drug in the stomach). In this case one would aim for a complete film and a solution/diffusion mechanism to minimize release.

B. Mechanism of Release while the Coating is Dissolving

As the pellet reaches an environment with a higher pH, a pH gradient, and hence a solubility gradient for the polymer, develop across the film. The coating begins to dissolve at the bulk edge but remains insoluble near the core.

Essentially, there is an insoluble coating layer, the thickness of which is diminishing with time. As a result, the rate of drug release will increase with time. This process continues until the coating becomes so thin that it ruptures mechanically, after which release of drug will occur via the usual mechanisms associated with immediate-release dosage forms. Two parameters are of interest: (1) how fast the drug is being released while the coating is dissolving, and (2) how long it will take for the coating to rupture. Comparison of the coating rupture time to rates at which the pellet moves through the intestine will provide an estimate of the region in the small intestine in which most of the drug will be released. A schematic for the processes described above is shown in Fig. 8.

In addition to the polymer pK_a and intestinal pH values, other parameters important to the rate of dissolution of the polymer include polymer solubility, the buffer capacity of the intestinal contents, and the concentration and pK_a of any acids or bases (including the drug) present in the pellet core. All these factors combine to determine the pH profile within the coating layer. The time at which the coating will rupture can be estimated by assuming that a 90 to 95% reduction

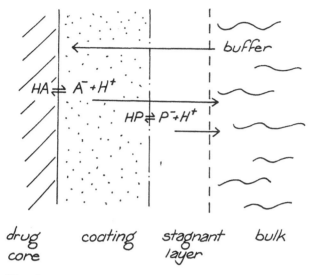

Fig. 8 Schematic representation of polymer (HP,P⁻) dissolution and drug release (HA,A⁻) from an enteric-coated product. H^+ represents hydrogen ion concentration. (From Ref. 20.)

in the coating thickness must occur before the coating ruptures. The rate at which the polymer dissolves is given in general terms by

$$J_{HP} = \rho_m \frac{dR}{dt} \tag{6}$$

where J_{HP} is the polymer flux, ρ_m the molal density, and dR/dt the rate at which the coating thickness decreases. For the coating, one may also write

$$J_{HP} = \frac{D_{HP}}{\delta} \left([HP]_{T,p} - [HP]_{T,b} \right) \left(1 + \frac{\delta}{R} \right) \tag{7}$$

where D_{HP} is the polymer diffusivity, δ the boundary layer thickness, $[HP]_{T,p}$ and $[HP]_{T,b}$ the concentrations of polymer of the coating surface and in the bulk, respectively, and R the radius of the coated pellet. Assuming that $\delta \ll R$ and sink conditions,

$$\rho_m \frac{dR}{dt} = - \frac{D_{HP}}{\delta} [HP]_{T,p} \tag{8}$$

integrating between $t = 0$ $(R = R_2)$ and $t = t$ $(R = R)$ yields

$$t = \frac{\rho_m \delta}{D_{HP}} \int_R^{R_2} \frac{dR}{[HP]_{T,p}} \tag{9}$$

Setting $R = R_1 + 0.05(R_2 - R_1)$ and integrating numerically will yield the time at which the coating has undergone a 95% reduction in thickness. Note that if the pellets are not perfectly spherical, rupture of the coating may occur earlier due to mechanical failure of a part of the film that is initially considerably thinner than the average thickness.

While the polymer is dissolving, the rate of drug release can be described by

$$J_{HA} = \frac{D_{HA}}{\delta} \left\{ [HA]_{T,p} - [HA]_{T,b} \left(1 + \frac{\delta}{R_1} \right) \right\} \tag{10}$$

for a spherical pellet. Symbols have the same meanings as above, with HA representing an acidic drug rather than the polymer (HP). As before, under sink conditions and for the usual case where $\delta \ll R_1$,

$$J_{HA} = \frac{D_{HA}}{\delta} [HA]_{T,p} \tag{11}$$

To calculate $[HA]_{T,p}$, it is necessary to know the pH at the dissolving edge of the polymer. A method for calculating this H^+ concentration, based on the pK_a and solubility of the drug and polymer and the pK_a and concentration of buffers in the release medium is given by Ozturk et al. [20].

The effect of including acid and basic materials in the pellet formulation is shown in Fig. 9. As the pH of the core is increased, the ability of the polymer to buffer the pH in the coating layer to an acidic pH is reduced and this facilitates coating dissolution. This buffering ability naturally also depends on the pK_a and intrinsic solubility of the polymer: the lower the pK_a and the higher the intrinsic solubility, the more powerful the ability of the polymer to keep the pH of the coating layer low. Model predictions for the effect of polymer solubility and pK_a on coating dissolution time are shown in Fig. 10. The third important determinant of the pH gradient in the coating layer is the buffer capacity of the release medium, illustrated in Fig. 11. In Table 2 data for pH and buffer capacity in the canine GI tract are compared with solutions commonly used for in vitro dissolution testing.

In summary, the model for enteric-coated dosage-form performance presented above suggests three possible contributing factors toward achieving release at the desired location within the GI tract: (1) choice of coating material with respect to pK_a and solubility, knowing the pH and buffer capacity profile of the small intestine; (2) choice of coating thickness, since rate of dissolution is inversely proportional to this factor; and (3) manipulation of the formulation of the pellet core by appropriate addition of acidic or basic excipients. Finally, it is worth reiterating that in order for in vitro tests to be predictive of in vivo performance, one should adjust both the pH and buffer capacity of the medium to values that are as close to physiologic as possible.

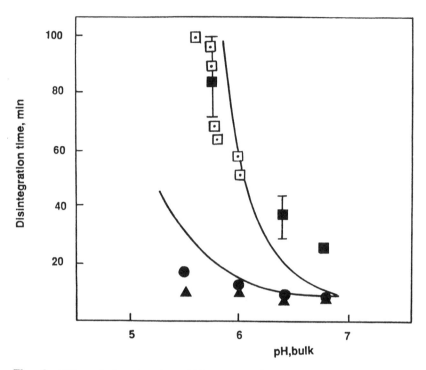

Fig. 9 Effect of changing the acid-base properties of the core formulation on the disintegration time of enteric-coated tablets. Simulated (solid lines) and experimental data for time of onset of disintegration of PVAP-coated aspirin (■), placebo (●) and citrate (▲) tablets, as a function of bulk pH. (From Ref. 20.)

IV. BIOERODIBLE COATINGS

Several features of GI physiology may be capitalized upon to achieve slow and/ or site-specific release. These include the pH profile, discussed above, the pancreatic enzymes and bile output in the duodenum, and the presence of exo-enzymes of the colonic bacteria. To achieve prolonged release while utilizing the pH profile to prevent gastric release, one can employ a heterogeneous coating such as shellac. This resin yields a complex mixture of aliphatic hydroxy acids (the major component of which is aleuritic acid) and alicyclic hydroxy acids (the major component of which is shellolic acid) and their polyesters [21]. The acidic nature of the hydrolysate explains the enteric properties of shellac coatings, while the heterogeneity of the composition explains why release occurs over a prolonged period. Triglycerides and other wax/fat substances have also been utilized as either coatings or in matrix formulations to provide prolonged drug

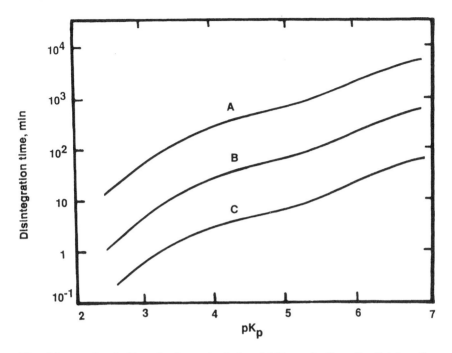

Fig. 10 Predicted effect of polymer intrinsic solubility and pK_a on the disintegration time of an enteric coating. Curve A corresponds to an intrinsic solubility of $10^{-7} M$, curve B to $10^{-6} M$, and curve C to $10^{-5} M$. (From Ref. 20.)

release. Their digestion and dissolution is facilitated by lipases released in the pancreatic juice and by bile salts. Thus one can expect drug release from pellets incorporating or coated with waxy and fatty substances to occur primarily in the small intestine. To predict the rate of in vivo release from in vitro experiments, one should incorporate physiological amounts of pancreatin and bile salts (or some other surfactant with equivalent wetting and solubilization properties) in the release media.

As already mentioned, the high degree of colonization of the colon by bacteria compared with other regions of the GI tract [22] affords an opportunity for designing site-specific release to this region. This would be particularly useful for treating local problems such as inflammatory bowel disease, since one could achieve high concentrations in the local tissue without overburdening systemic levels. To date, the azoreductases have been utilized to achieve colon-specific delivery [19] and the use of glycosidic linkages is also under study [23]. Table 3 lists the relative concentrations of bacteria at various locations in the GI tract. The wide range of bacteria present in large quantity in the colon suggests

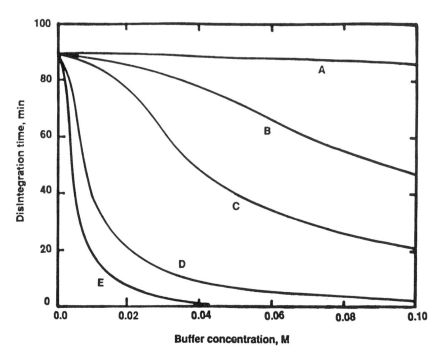

Fig. 11   Predicted effect of the pH of the dissolution medium on the disintegration time of enteric-coated tablets. In this simulation the buffer pK_a was 7.1, and pH values simulated were pH 5 (curve A), pH 5.5 (curve B), pH 6 (curve C), pH 6.5 (curve D), and pH 7 (curve E). The polymer pK_a was set at $pK_a = 4.5$. (From Ref. 20.)

Table 2   Buffer Capacity of Chyme Recovered from the Midgut in Two Fistulated Dogs, and of Simulated Intestinal Fluid USP (Sans Pancreatin)

	Recovery of fluid	Initial pH	Buffer capacity (mEq/liter/pH unit)
Dog 1	Early chyme	6.75	28
	Mid	6.82	32
	Late	6.88	48
Dog 2	Early	6.50	24
	Mid	6.59	35
	Late	6.79	44
Simulated intestinal fluid, USP		6.8	26

Source: Ref. 20.

Table 3   Bacterial Types and Counts in Various Regions of the GI Tract

Bacteria type	Number/mL			
	Stomach	Jejunum	Ileum	Feces
Aerobic/facultative				
Enterobacteria	$0-10^2$	$0-10^5$	10^3-10^7	10^7-10^{12}
Streptococci	$0-10^3$	$0-10^4$	10^2-10^6	10^5-10^{10}
Staphylococci	$0-10^2$	$0-10^3$	10^2-10^5	10^4-10^7
Lactobacilli	$0-10^3$	$0-10^4$	10^2-10^5	10^6-10^{10}
Anaerobic				
Bactrooides	Rare	$0-10^2$	10^3-10^7	$10^{10}-10^{12}$
Bifidobacteria	Rare	$0-10^3$	10^3-10^5	10^8-10^{12}
Streptococci	Rare	$0-10^3$	10^2-10^4	10^8-10^{11}
Clostridia	Rare	Rare	10^2-10^4	10^6-10^{11}
Eurobacteria	Rare	Rare	Rare	10^9-10^{12}

Source: Ref. 22.

that numerous opportunities exist for the development of polymers that would be selectively degraded in the colon. More sophisticated yet would be a prolonged release dosage form for which the rate of release was dictated by some ingredient within the formulation itself, rather than relying on the external environment. Such is the case for bioerodible polymers, in which catalysts for the erosion process are included in the formulation. The topic of bioerodible drug delivery systems has been reviewed thoroughly by Chasin and Langer [24]. Most of the bioerodible polymers that have been developed for implants erode too slowly to be useful for GI delivery; they degrade on a time sale of weeks to months as opposed to hours. However, for some polymers, such as the polyorthoesters, one can incorporate a catalyst to promote the rate of erosion (Fig. 12). By adding phthalic anhydride in the polyorthoester matrix, for example, a surface erosion mechanism (as opposed to bulk erosion) is achieved and zero-order release results, because the rate of water penetration into the polymer becomes the rate-limiting step to its erosion. For a cyclobenzaprine HCl formulation, this approach has been shown to produce release over a period of 10 to 15 h. The design considerations for such short-term release formulations have been discussed by Heller et al. [25]. The advantages of this type of design are threefold: (1) release should be relatively independent of GI conditions, (2) one can manipulate the release rate by modifying the level of catalyst used, and (3) complete release should be possible provided that the polymer degrades completely with the GI transit time.

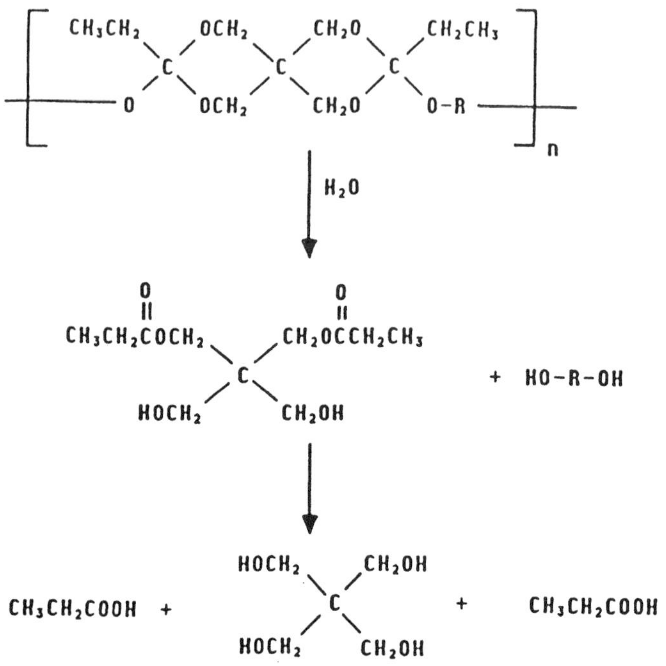

Fig. 12 Schematic for the decomposition of polyorthoesters. (From Ref. 25.)

V. SUMMARY

There are a wide variety of coating materials available today which can be used to modify the rate at which drug is released from pelletized dosage forms in the gastrointestinal tract. As would be expected from their diverse chemical and physical properties, the release mechanisms of these coatings are numerous. A knowledge of their release mechanisms facilitates selection of the best polymer for the job, depending on what release profile is desired. It also helps to design appropriate in vitro test conditions for the dosage form.

The future developments in controlled-release dosage forms will probably lie in systems that are less sensitive to the vagaries of the GI environment: that is, dosage forms from which the release rate is determined principally by the formulation itself. The use of erodible polymers, where the rate of degradation is controlled by an inclusive catalyst, and osmotically driven release devices represent two examples. On the other hand, dosage forms intended to provide site-specific release will need to rely on changes in GI parameters that are very clear cut. The pH changes along the GI tract, apart from gastric to duodenal in the fasted state, are not distinct enough for accurate regional delivery. Parameters that exhibit more promising differentials are the levels of pancreatic enzymes in the small intestine and the levels of bacteria in the colon. These two areas merit further investigation.

REFERENCES

1. J. G. Hardy, S. S. Davis, and C. G. Wilson, eds., *Drug Delivery to the Gastrointestinal Tract*, Ellis Horwood, Chichester, West Sussex, England, 1989, Chapters 3–6.
2. J. G. Wagner, *Clin. Pharmacol. Ther. 37*:481 (1985).
3. A. G. Ozturk, S. S. Ozturk, B. O. Palsson, T. A. Wheatley, and J. B. Dressman, *J. Controlled Release, 14*:203 (1990).
4. J. O'Neill, in *Controlled Release Technologies* (A. F. Kydonieus, ed.), CRC Press, Boca Raton, Fla., 1980, pp. 130–139.
5. T. K. Sherwood, R. L. Pigford, and C. R. Wilke, *Mass Transfer*, McGraw-Hill, New York, 1975, pp. 43–148.
6. A. S. Michaels, P. S. L. Wong, R. Prather, and R. M. Gale, *AIChEJ. 21*:1073 (1975).
7. V. Vidmar, I. Jalsenjak, and T. Kondo, *J. Pharm. Pharmacol. 34*:411 (1982).
8. S. Benita and M. Donbrow, *Int. J. Pharm. 12*:251 (1982).
9. K. Uno, M. Arakawa, T. Kondo, and M. Donbrow, *J. Microencapsulation 1*:335 (1984).
10. F. M. Sakr, E.-D. Zim, E. Esmat, and F. M. Hasheim, *Acta Pharm. Technol. 33*:31 (1987).
11. C. D. Melia, I. R. Wilding, and K. A. Khan, *Pharm. Technol. Int. 3*:24 (1991).
12. G. M. Zentner, G. S. Rork, and K. J. Himmelstein, *J. Controlled Release 1*:269 (1985).
13. C. Bindschaedler, R. Gurny, and E. Doeller, *J. Controlled Release 4*:203 (1986).
14. G. S. Rekhi, S. C. Porter, and S. S. Jambhekar, Mechanism and some factors affecting the release of propanolol hydrochloride from beads coated with aqueous polymeric dispersions, *Proceedings of the International Symposium on Controlled Release of Bioactive Materials*, Basel, 1988, pp. 372–373.
15. R. U. Nesbitt, M. Mahjour, N. L. Mills, and M. B. Fawzi, Membrane controlled release dosage forms, *Arden House Conference Proceedings*, Harriman, New York, 1986 (handout).
16. J. B. Dressman, R. R. Berardi, T. L. Russell, L. C. Dermentzoglou, K. M. Jarvenpaa, S. Schmaltz, and J. L. Barnett, *Pharm. Res. 7*: 756 (1990).
17. D. F. Evans, G. Pye, R. Bramley, A. G. Clark, T. J. Dyson, and J. D. Hardcastle, *Gut 29*:1035 (1988).
18. C. A. Youngberg, R. R. Berardi, W. F. Howatt, M. L. Hyneck, G. L. Amidon, J. H. Meyer, and J. B. Dressman, *Dig. Dis. Sci. 32*:472 (1986).
19. M. Saffran, G. S. Kumar, C. Savariar, J. C. Burnham, F. Williams, and D. C. Neckers, *Science 233*:1081 (1986).
20. S. S. Ozturk, B. O. Palsson, B. Donohoe, and J. B. Dressman, *Pharm. Res. 5*:550 (1988).

21. S. Budavari, ed., *Merck Index*, 11th ed., Merck & Co., Rahway, N.J., 1989, p. 1343.

22. G. L. Simon and S. L. Gorbach, Intestinal flora and gastrointestinal function, in *Physiology of the Gastrointestinal Tract* (L. R. Johnson, ed.), Raven Press, New York, 1987, pp. 1729–1747.

23. D. Friend, Delivery of glycoside prodrugs of steroids to the colon, *Proceedings of the 2nd Jerusalem Conference on Pharmaceutical Sciences and Clinical Pharmacology*, Jerusalem, 1992.

24. M. Chasin and R. L. Langer, eds., *Biodegradable Polymers as Drug Delivery Systems*, Marcel Dekker, New York, 1990.

25. J. Heller, R. V. Sparer, and G. M. Zentner, in *Biodegradable Polymers as Drug Delivery Systems* (M. Chasin and R. L. Langer, eds.), Marcel Dekker, New York, 1990, pp. 121–161.

12

Biopharmaceutical Aspects of Multiparticulates

Johannes Krämer and Henning Blume

Deutsches Arzneiprüfungsinstitut
Eschborn, Germany

I. MODIFIED-RELEASE ORAL DOSAGE FORMS

Multiparticulates in pharmaceutics are of special interest in the case of controlled- and delayed-release oral pharmaceuticals summarized under the term *modified-release dosage forms*. Medicinal products with controlled release of the active substance are developed for several purposes, which specifically include the following:

1. The maintenance of constant therapeutic drug concentrations in plasma, which are mostly equivalent to an extension of the drug's pharmacodynamic effects
2. The avoidance of excessively high plasma concentration peaks, which leads to a reduction of undesired side effects
3. The extension of dosage intervals, whereby patient compliance may be improved

In Fig. 1, the plasma concentration versus time profiles are depicted, one of an oral solution of the drug and the other of a controlled-release preparation. With application of the controlled-release preparation not only was a reduction of peak concentration achieved, but a considerable extension of the desired duration

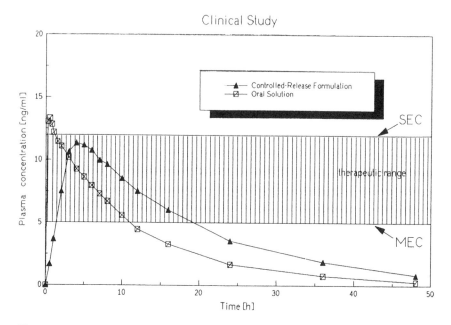

Fig. 1 Plasma concentration–time plots: controlled-release formulation in comparison with oral solution.

of the drug effects. This can be seen in the figure by comparing the time spans of both curves above the minimum effective concentration (MEC). The shaded area marks the therapeutic range to be above MEC and below the minimal side-effect concentration (SEC), which is narrow for some drugs (e.g., theophylline). A narrow therapeutic range requires a very careful adjustment of the dosage regimen above all for the use of immediate-release oral dosage forms. The application of a dosage form with prolonged release of the drug consequently allows us to extend the dosage interval. This may be of great importance, especially in long-term treatment, where poor fluctuation from peak to trough in the concentration versus time profiles is desired during the entire dosage interval. In addition to this general advantage of all controlled-release dosage forms there is a special therapeutic necessity for controlled-release formulations of, for instance, theophylline. Due to circadian rhythms, asthmatic attacks occur predominantly in the early morning hours between 4 and 5 o'clock. Therefore, the evening dose must be high enough to provide therapeutically effective concentrations during the entire night. Treatment with prolonged-release dosage forms may be unavoidable.

A. Need for Controlled-Release Formulations

The percentage of oral controlled-release pharmaceuticals toward all solid oral dosage forms in the German market increased during the last decade. Market analyses of the Central Laboratory of German Pharmacists, ZL, prove that there is an increasing need for controlled-release oral products. In 1980 there were 203 controlled-release dosage forms [2]. In 1990 their total number grew to 345 [3]. Multiparticulate dosage forms' portion of all controlled-release dosage forms increased from 34% to 53% (Fig. 2). This shows that multiparticulates are gaining greater importance within this galenical group. Although these data characterize only a small segment of the international market, they should be representative of development in the field of all solid oral dosage forms. One reason for the permanently increasing number of multiparticulates may be the progress in manufacturing, and a second may be their biopharmaceutical advantages.

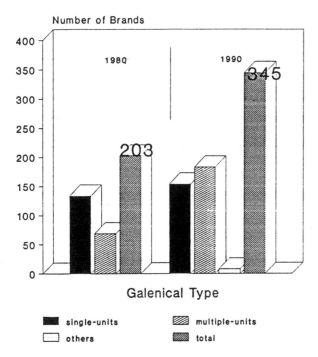

Fig. 2 Development of controlled-release medicinal products for the German market: comparison of galenical types; single unit versus multiparticulates. (From Ref. 1.)

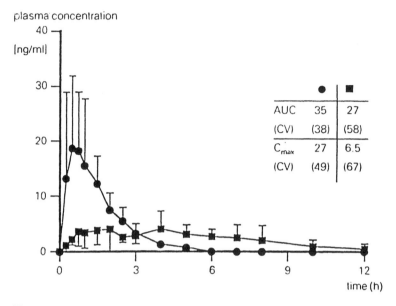

Fig. 3 Effect of the pharmaceutical form on the bioavailability of ISDN. ●, Immediate-release tablet; ■, controlled-release tablet.

In the development of controlled-release formulations, a certain loss of bioavailability must be accepted. This is illustrated in Fig. 3 with reference to the example of ISDN preparations. Using the same quantity of processed drug substance (20 mg in each case), the extent of bioavailability is approximately 30% higher after application of the immediate-release preparation than after administration of the controlled-release product. At the same time, controlled-release preparations often produce markedly higher individual variabilities of plasma profiles. This is also documented in Fig. 3 with reference to the values of relative standard deviation listed in the table as coefficients of variation (CV). For the controlled-release tablets for which the mean profiles are displayed, about a 50% higher variability was determined. But the application of controlled-release formulations does not necessarily cause a higher variability. A comparison of the results of isosorbide-5-mononitrate immediate-release tablets with a controlled-release formulation of the same drug substance (Fig. 4) shows an even lower variability of about 25%. The bioavailability is again lower in the case of the controlled-release formulation.

In addition to the pharmacokinetic appearance, it is necessary to consider more fundamental questions, such as whether the general goal of getting smoothly oscillating plasma levels, often described as "the flatter the better," over long periods generally provides pharmacodynamic advantages. ISDN is a particularly

plasma concentration

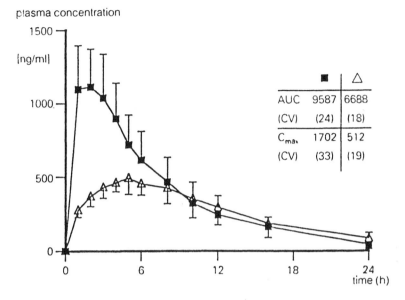

Fig. 4 Effect of the pharmaceutical form on the bioavailability of 5-ISMN. ■, Immediate-release tablet; △, controlled-release tablet.

relevant example in this respect, since tolerance development must be taken into consideration. Therapeutic tolerance toward controlled-release ISDN can be developed if the concentration of ISDN or its 5-mononitrate metabolite are constantly maintained at therapeutic levels. According to the present stage of knowledge, a higher fluctuation in plasma levels is somewhat desirable. For this purpose the dosage scheme has to assure that for a certain period of time, plasma levels are constantly below the threshold of tolerance.

Fluctuations in the analytically determined plasma concentration versus time plots of each subject participating in a clinical trial depend not only on the type of galenical formulation under investigation but also on other factors. Among these, the drug-dependent individually changing variability of the analytical method, particularly, has to be taken into consideration. In addition, a very important aspect is the homogeneity of the drug product lot as well as the lot-to-lot homogeneity. Distinct problems can arise in relation to the required content uniformity of multiparticulates in hard gelatine capsules, especially when various types of pellets or granules containing different active substances are combined. In such cases either stepwise filling of the various pellets into capsules (sequential process) or, alternatively, synchronic filling of a prefabricated mixture can be conducted. In the latter case, importance is ascribed not only to the homogeneity of the primary mixture but particularly, to possible segregation of various components. Examples of controlled-release nifedipine preparations of the German market illustrate that homogeneity is not guaranteed in all cases. In Fig. 5 almost superimposible mean plasma concentration profiles of three batches of one product prove constant lot-to-lot quality; the rate and extent of bioavailability do not differ significantly. In another case, considerable differences in results were found for three batches of another nifedipine controlled-release pellet formulation; the rate and extent of bioavailability differ significantly (Fig. 6). To achieve constant pharmacodynamic efficacy in therapy, uniform, regular pharmaceutical quality within lots and from lot to lot are of prime importance.

B. Gastric Transit Time: An Essential Parameter Relevant to the Variability of Plasma Concentration Versus Time Curves

Physiology of the GI Tract

A decisive influence on possible intrasubject and intersubject fluctuations of plasma concentration versus time profiles is that of alterations in the individual gastrointestinal (GI) transit times. The transit of particles measured, for instance, as a single stool transit or by gamma-camera imaging depends additionally on the transit rate through various segments of the large intestine [4]. This is commonly

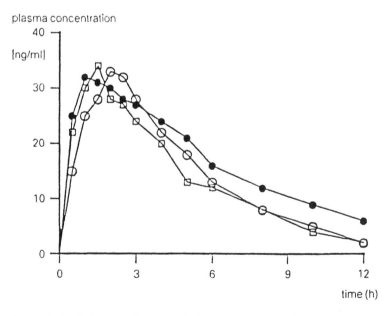

Fig. 5 Sufficient conformity of different batches of one particular nifedipine controlled-release product.

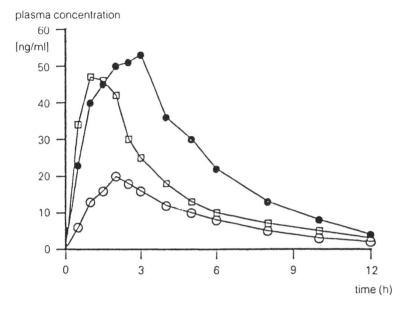

Fig. 6 Nonconformity of different batches of one particular nifedipine controlled release product.

of minor interest because due to its small active surface, absorption in distal parts of the digestive tract may be reduced compared to the proximal part of the small intestine. Therefore, the arrival time and stay at the site of absorption are the most relevant factors. Transit through the stomach and intestine, measured as the cecum arrival time, is modified mainly by the gastric emptying time [5].

The effect of the rate of gastric transit on the onset and slope of plasma concentration versus time curves after administration of an immediate-release glibenclamide tablet is shown in Fig. 7. In a bioequivalence study a commercial glibenclamide product of the German market was given in a repetitive trial. Normally, glibenclamide is absorbed rapidly, leading to a steep slope in the plasma curve soon after ingestion of the tablets. Peak concentrations are reached within 1 to 2 h. One of the participating volunteers showed remarkable differences in the plasma profiles in the two blocks of the trial. After one application a considerable delay in absorption was detected. As usual, the rise of concentration was steep, but only after a remarkable lag time of 3 h. This delay could be explained by an unusual long stay of the dosage form in the stomach, where only poor dissolution, and hence poor absorption, occurred before the dosage form progressed to the duodenum, where the active compound was then rapidly dissolved and absorbed from the dosage form. The pronounced differences in the

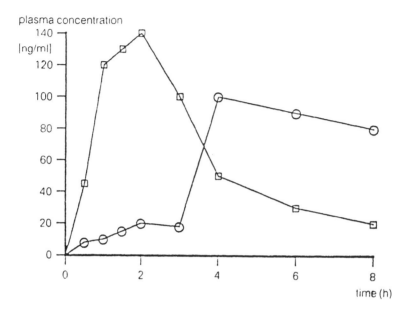

Fig. 7 Effect of emptying of the stomach on the progress of plasma concentration–time plots (e.g., intake of one tablet of the same glibenclamide product on two different days of the study).

results cannot be explained by different pharmaceutical qualities of the drug product since the identical product was administered in both phases of the study.

The bioavailability, above all, of modified-release oral dosage forms depends on their size and their ability to move together with solid or liquid compounds of the food through the narrow valves that regulate the propulsion of any intraluminal content. It must be mentioned that the movement of intraluminal content is poorly influenced by gravity. In the case of multiparticulates, the gastric transit of pellets of different densities was shown to be unaffected by gravity [6].

The GI tract is divided in three major parts. From proximal to distal any food first contacts the stomach, a storage and grinding organ, with its pyloric region as the junction to the following small intestine (duodenum, jejunum, ileum). The small intestine is connected to the colon, where remarkable absorption can occur [7] via the ileocecal valve. The transit rate of solid particles such as food or oral dosage forms throughout the entire GI tract as a tube 5 to 7 m long with a variable diameter is commonly measured by gamma-camera imaging with radiopaque markers. Transit through the GI tract is normally influenced by active propulsionary processes and varies in healthy subjects. The factors of influence can be attributed to both physiological variables caused by body posture, emotional status, circadian rhythms, and the very important nutritional status; and to the impacts of drugs (e.g., metoclopramide, loperamide) and the geometries of dosage forms.

Transit to absorption sites is controlled primarily via the pyloric region of the distal stomach, which works as a sieving gate, classifying particles of different sizes. The distal stomach is concerned primarily with retention and trituration of food and the pretention of duodenal-gastric reflux. The characteristic contract of the distal stomach is the peristaltic wave. This wave causes changes in intraluminal pressure and usually lasts for a period of 1 to 4 s. The pattern and force of the peristaltic movements vary depending on the nutritional status.

Importance of Nutritional Status

In the fasted state the interdigestive pattern of the myoelectric cycles has four phases [8]:

1. Phase I, the basal phase, which lasts 45 to 60 min, shows few or no potentials in the gastric wall.
2. Phase II, the preburst phase, has intermittent action potentials and lasts for about 30 to 45 min.
3. In phase III, called the burst, powerful distal gastric bursts occur for 5 to 15 min. One possible function may be to sweep large indigestible particles out of the stomach; therefore, it is called "housekeeper wave."
4. Phase IV, the terminal phase of the myoelectric cycle, leads back to phase I.

Within a migration time of 2 h, each phase of the cycle migrates distally, via the small intestine, to the colon. In the fed state there is only one phase with regular, frequent contractions (four or five per minute) at an amplitude lower than that of phase III. It lasts as long as food is present in the stomach.

C. Progression of Drug Particles with Intraluminal Content

Gastric emptying in the fed state is regulated by feedback receptors located in the duodenum [9]. Its rate depends on the caloric content of the chymus. Carbohydrates empty faster than proteins, which in turn empty faster than fats. Since isocaloric amounts of carbohydrates, proteins, and fats empty at the same rate [10], the caloric density of food determines the rate of emptying such that the energy that arrives at the duodenum is nearly constant. Moreover, food with an osmolality higher than the physiological value of about 280 mOsm/kg stays longer in the stomach than does isoosmotic food. Boluses of ingested food are stored in the proximal stomach. Solids form a mass in the stomach, whereas swallowed liquids as well as gastric juice flow readily outside the mass through the pyloric junction to the duodenum. Driven by the difference in intestinal and gastric pressure, liquids easily pass through the pylorus, whereas solids are retained.

The gastric emptying of solid particles is influenced largely by the resistance of the small diameter of the stomach at the gastroduodenal junction. As solids are swept into the distal stomach by the peristaltic wave the pylorus contracts and provides continence of solid particles of the chymus. The antral region has different phases of propulsion, grinding and squeezing, and retropulsion. This causes an intensive trituration of the chymus [11] such that more than 90% of the food particles that pass this region are smaller than 0.25 mm. Almost no particles larger than 2 mm are allowed to leave the stomach [12]. Digestible solids and particles smaller than 2 mm are emptied according to zero-order kinetics, and nondigestible solids larger than 7 mm are merely emptied from the stomach in the fed state [13]. These indigestible solids are subsequently swept out through the pylorus by the housekeeper wave during one of the following interdigestive phases. This is why large particles (single units) need phase III activity to be emptied from the stomach. When taken after a heavy breakfast the gastric residence time of an ibuprofen controlled-release formulation was reported to vary on average from 4.3 to 11.0 h [14]. However, when dissolution occurs in the stomach and the dissolved drug is emptied with fluids in consequence, bioavailability need not be affected by a prolongation of gastric residence time [15].

The emptying of liquids from the stomach can be described by zero-order mass versus time equations in the fed state and by exponential terms in the fasted state [8]. For multiple units of small diameter, the pylorus is no obstacle; they are emptied like liquids. The gastric emptying of multiparticulates is much less influenced by concomitant food intake [16]. In the fasted state their emptying occurs exponentially, and when taken with food they exhibit a linear pattern of emptying from the stomach [17]. The active substance cannot be absorbed from enteric-coated dosage forms unless the dosage form has left the strongly acidic surroundings of the stomach and is transported to the intestine. When enteric-coated multiparticulates are administered, this transfer takes place as a continuous process according to the dispersion of beads in the stomach. The drug incorporated in single units stays in the stomach or passes the pylorus as a bolus. In consequence, single units cause a higher intersubject variability in lag time as well as an average prolongation of lag time (Fig. 8). After administration of enteric-coated diclofenac formulations, plasma concentrations above the limit of determination were reached earlier with pellets than with plain tablets.

Moreover, significant differences can be detected in the variability of individual pharmacokinetic parameters (Fig. 9). Referring to Figure 8, fluctuations of the areas under the curves (AUCs), as well as fluctuations of lag times, were significantly lower for the pellet formulation under investigation. The

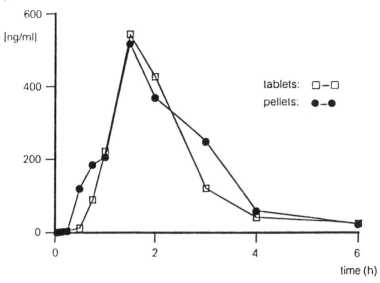

Fig. 8   Plasma concentration–time plots after administration of various enteric-coated diclofenac formulations.

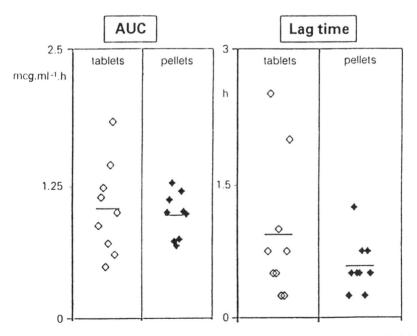

Fig. 9   Individual scatter of the controlled variables AUC and lag time after administration of single-unit and multiparticulate diclofenac enteric-coated formulations.

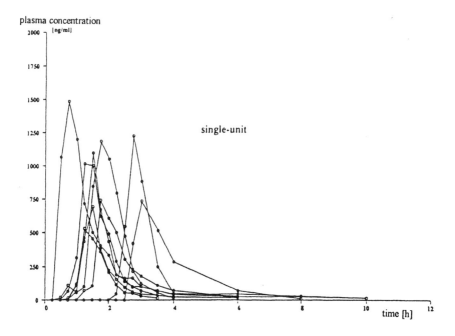

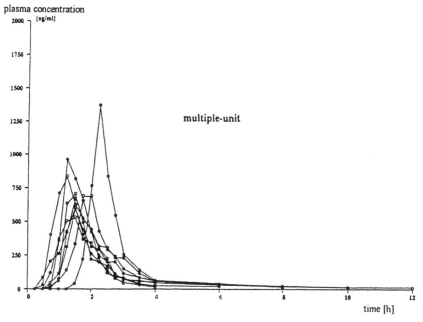

Fig. 10 Individual diclofenac plasma concentration–time plots after administration of controlled-release pellets in comparison with a plain controlled-release tablet.

smaller variability of plasma levels of pellets may be a further advantage of this galenical type (Fig. 10). Accordingly, in comparison with controlled-release single units, a substantially lower variance of all relevant parameters, particularly C_{max} and t_{max}, was documented for pellet formulation, as depicted by the individual plasma profiles in Figure 10.

II. IN VITRO CHARACTERIZATION OF CONTROLLED-RELEASE DOSAGE FORMS

A. Methodology

Dissolution tests can be a powerful tool in investigations of the pharmaceutical quality of any solid oral dosage form in every stage of its development. During galenical development, dissolution tests can be used for optimization of manufacturing processes. Based on the results of dissolution tests, the choice of appropriate samples for further clinical testings is facilitated. It is recommended that dissolution data be provided for at least two batches, both at the time of release and at the end of the shelf life. For medicinal products already on the market, dissolution tests are used to reassure constant manufacturing quality expressed as homogeneity of the lots as well as constant lot-to-lot quality. Most drugs for oral application are organic molecules with the degree of lipophilicity that is necessary for the molecules to penetrate the tight lipophilic membranes of the gastrointestinal tract. This absorption through membranes is normally fast and can often be described according to a first-order time law. Absorption or persorption of undissolved particles occurs much slower and in a negligible amount. Therefore, in the biopharmaceutical fate of a drug, dissolution of drugs in hydrophilic intraluminal fluids of the gastrointestinal tract is the most important step for absorption (Fig. 11). The dissolution rate of most drugs from their dosage form determines primarily the rate and the amount to which drugs are absorbed and appear in the blood or at the site of action, which is called the bioavailability. Numerous drugs that are slightly soluble or poorly wettable can be dissolved within a reasonable time span only by galenical tricks, whereas drugs require an extended or delayed release primarily for pharmacokinetic or pharmacodynamic reasons. Dissolution of solid oral dosage forms can barely be determined in the human body. That is why in vitro experiments are of such great importance in the prediction of biopharmaceutical performance.

B. Suitability of Compendial Dissolution Devices and Methods

Flow-Through Cell

The flow-through cell was entered in the last edition of the *European Pharmacopeia* and the fourth supplement of USP XXII separate from the traditional

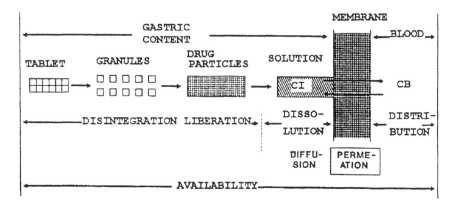

Fig. 11   Scheme of pharmaceutical and biological availability.

basket and paddle models. In addition to these methods for dissolution testing of oral pharmaceuticals, USP XXII describes apparatus <3>, known as the Bio-Dis II method, in its general chapter on dissolution. The traditional beaker models with stirred basket or paddle are closed systems. All drug dissolved during the experiment remains in the system, and the gradient of concentration between the drug in dosage form and the drug in bulk fluid (normally, 900 mL) is gradually getting flatter. In vivo drug dissolved from a dosage form is rapidly absorbed from the site of dissolution in most cases, so that the gradient of concentration always remains steep. This phenomenon cannot be simulated in closed in vitro systems, whereas in open systems, where infinite volumes of dissolution media can be brought into contact with the dosage form, the absorption process concomitantly can be simulated by maintaining a steep gradient of concentration. The flow-through cell is such an open system, with a reservoir of dissolution medium, a precise pump providing flow rates between 5 and 50 mL/min, and a splitting unit for samples. The flow-through cell can also be run in a loop alignment as a closed system (Fig. 12). Moreover, for different dosage forms, appropriate flow-through cells are available. Actually, system suitability tests for the apparatus are worked out by the Federation Internationale Pharmaceutique (FIP) Dissolution Working Group.

Bio-Dis II

The Bio-Dis II apparatus works neither with stirring elements nor with a continuous flow of dissolution medium along the dosage form surfaces. It was developed on the basis of the USP XVII disintegration tester as a dissolution model for coated pellets. Six inner glass cylinders move in parallel with oscillation in outer tubes each of maximum volume of dissolution about 200 mL. The inner glass tubes are locked (Fig. 13) on the bottom and, if necessary, on the top by screens

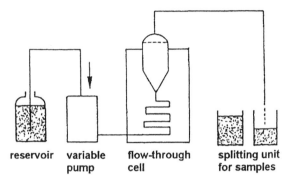

reservoir   variable   flow-through   splitting unit
             pump   cell   for samples

Fig. 12   Flow-through cell depicted schematically in an open-system alignment.

of appropriate mesh size. The screens hold the dosage form in the inner tube and carry it over sequentially to following outer tubes filled with dissolution media of various compositions. The dipping rate can be varied according to the mechanical stress that is suitable for dissolution. With the Bio-Dis II apparatus the products under investigation can be dissolved at one pH level or sequentially in media of up to six different pH stages. The maximum volume of dissolution is about 1200 mL; therefore, the Bio-Dis II is categorized as a closed system.

Suitability of Dissolution Methods

To test the dissolution behavior of multiparticulates, basket and paddle apparatus can be run without modification of the pharmacopeial specifications. For tests of multiparticulates in gelatine capsules, interactions with the gelatine in alkaline media leading to artifacts in the dissolution profiles were described [18]. Normally, this can be avoided either by removing the capsule shell prior to the test, or in the case of enteric-coated gelatine capsules, by performing the dissolution test at higher stirring rates to prevent pellet clugging. During its gastrointestinal passage in the human body, a dosage form meets changing physiological conditions involving changes in the physicochemical surroundings of the dissolution environment. Such changes are simulated using dissolution media of varying compositions. Modeling of pH changes by the half-change procedure can be experimentally difficult for nondisintegrating multiparticulates in stirring devices. However, modeling can be done easily in a flow-through cell. There the dissolution behavior of a drug product brought into contact stepwise with a large number of different dissolution media can be determined. The disadvantage is that the dosage form is isolated from environmental mechanical influences except the very smooth hydrodynamics of solvent flow at the surface of the dosage form fixed in the cell. It should be noted that simulation of the physiological mechanical stress caused by peristaltic movement in the human body cannot be adequately simulated using a flow-through cell device.

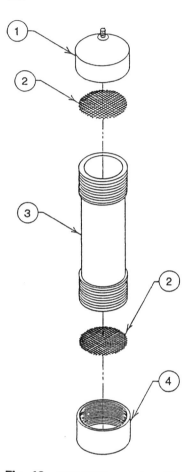

Fig. 13 USP XXII apparatus <3>: inner tube (3) with top cap (1), mesh screens (2), and bottom cap (4).

If the purpose of a dissolution test is to evaluate the ruggedness of a dosage form toward strong mechanical impacts, which may occur especially in the antral region of the stomach, strong agitation in the Bio-Dis II apparatus could be a suitable method of evaluation. Contrary to the basket, paddle, and flow-through cell devices, the Bio-Dis II apparatus is not generally standardized in all details. For example, the mesh size of the screens used to lock the inner cylinders is not delineated. If a screen with a large mesh size is chosen, undissolved particles can settle down in the outer cylinder and are not carried over to the following sequence of dissolution. In the case of disintegrating single units, a dependence of rate and amount of dissolution on screen mesh size was measured (Fig. 14).

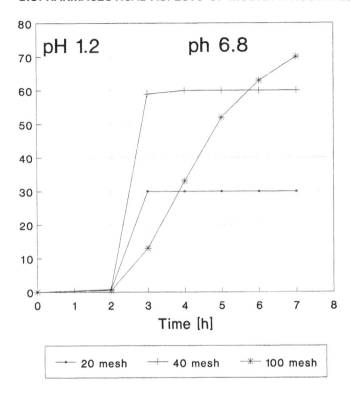

Fig. 14 In vitro dissolution of a controlled-release diclofenac single-unit formulation with USP XXII apparatus <3> at 30 dips/min: influence of bottom screen mesh size on apparent cumulative amount dissolved.

For nondisintegrating multiparticulates the mesh size must be smaller than the beads so that all beads are forwarded to succeeding rows and are completely removed from the final row after the end of the dissolution program. In this way the Bio-Dis II is a suitable instrument for showing the effectiveness of coating films and the dissolution properties of the drug product concomitantly, as shown in Fig. 16 for an enteric-coated diclofenac multiparticulate formulation.

C. Factors Influencing Dissolution of Controlled-Release Dosage Forms

Agitation

To perform dissolution tests of a modified-release dosage form with a chosen apparatus, the test is normally performed at various agitation rates. The "normal" mechanical stress on an oral dosage form on its way through the gastroin-

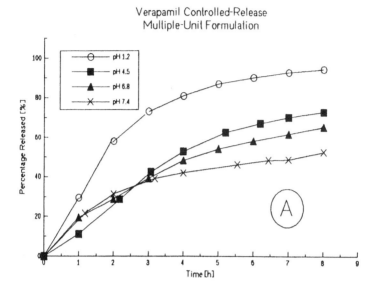

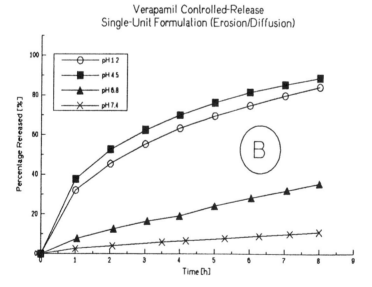

Fig. 15   Influence of pH on in vitro dissolution exemplified for verapamil controlled-release formulations with the paddle apparatus: (A) multiparticulates; (B) and (C) single units.

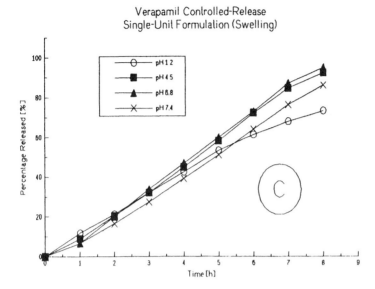

Verapamil Controlled-Release
Single-Unit Formulation (Swelling)

testinal tract is modeled by 100 rpm in the basket apparatus and by 50 or 75 rpm in the paddle apparatus. If monolithic extended-release dosage forms whose dissolution rate is controlled by erosive mechanisms or the propagation velocity of a swelling front within the dosage form has to be tested, moderate agitation should be suitable, to work out possible pH dependencies, but for evaluation of the ruggedness of such dosage forms, stronger agitations have to be installed.

pH Stages

Along its GI transit, any modified-release dosage form contacts body fluids at various pH levels. In the stomach under fasting conditions the pH is strongly acidic. Under the buffering influence of chyme, the pH in the stomach can rise to an average of about pH 5, or even above. In the proximal intestinum under the influence of bile and pancreatic juices, the pH level rises to the physiological neutral point of about pH 7.4, and distally to an even more alkaline pH 8. Therefore, it is recommended that dissolution tests be performed for controlled-release dosage forms at various pH stages within the physiological range. To show up pH dependency, this is performed preferably by parallel dissolution tests (Fig. 15). To evaluate the effectiveness of the enteric coat of delayed-release oral dosage forms by dissolution tests, a sequential design should preferably be chosen (Fig. 16).

Surface-Active Substances and Ionic Strength

To model the influence of surface-active physiological ingredients that naturally occur in the digestive fluids, sodium lauryl sulfate or polysorbate are preferably

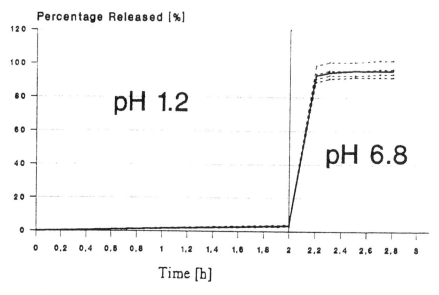

Fig. 16 Influence of pH on in vitro dissolution of enteric-coated diclofenac multiple units in USP XXII apparatus <3>.

added to the dissolution media. This procedure may also be necessary if the drug under investigation is poorly wettable, and consequently, dissolution cannot be performed within an acceptable period of time. Examples of poorly wettable drugs are among others, carbamazepine [19], nifedipine [20], and prednisolone [21]. In order not to influence the overall solubility, surface-active substances should be added in submicellar concentrations. The osmotic pressure of body fluids is usually about 285 mOsm/kg. The concentration of buffering agents, and consequently, the osmotic pressure of dissolution media, was shown to have an impact on the dissolution behavior of verapamil extended-release products of various galenical types (Fig. 17). Therefore, the influence of osmotic pressure on dissolution rate must be checked in vitro. Solubility processes usually depend on the solvent temperature. Consequently, $37 \pm 0.5°C$ is to be maintained throughout the experiments.

III. BIOAVAILABILITY/BIOEQUIVALENCE STUDIES

Testing of bioavailability of drug products is of great importance in the field of generic drugs, where bioequivalence to the innovator, or another reference has to be proven by the mean of comparative clinical studies. Bioequivalence studies serve as necessary evidence for equal biopharmaceutical quality of drugs. To get

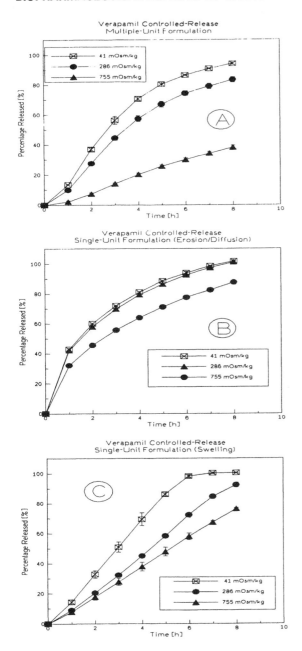

Fig. 17   Influence of ionic strength on in vitro dissolution exemplified for verapamil controlled-release formulations with the paddle apparatus: (A) multiparticulates; (B) and (C) single units.

reliable results in terms of discriminating power, studies need to be standardized as to both study design and evaluation of data.

A. Study Design Exemplified for Food Studies

Study designs of bioavailability studies of controlled-release oral dosage forms that have different objectives generally have a common base. Because the performance of investigations of food influence has turned out to be important, studies on food effects are recommended internationally. We therefore describe proposals for the design of clinical tests of extended-release dosage forms in food studies.

Significant changes in the physiology of the absorption site in the GI tract can be induced by the intake of solid and liquid food. The bioavailability of drug products is influenced by the properties of the drug substance and dosage form. The effects of food on bioavailability with respect to clinical pharmacokinetics and therapeutic implications have been reviewed [22]. Studies of food influence on bioavailability can have various objectives. During the development of a formulation, its biopharmaceutical characterization is often outstanding interest. In addition, constant efficacy in certain populations under special dietary conditions has to be worked out. A variety of study designs have to be considered to work out the defined objective of the prevailing study.

1. Single-dose food effect studies are generally recommended in an early stage of drug development to show up the robustness of the drug product. They can also be helpful to work out the maximum effect of concomitant food intake on absorption. As a reference, the dosage form under investigation should be given after an overnight fast of at least 10 h, together with a defined volume of beverage, and fasting should be continued for another 4 h. Test conditions include application of the dosage form immediately after a heavy, high-fat breakfast of standardized caloric content and a defined volume of beverages. The application should preferably be performed according to a two- or three-way changeover design.
2. In therapeutic practice during long-term treatments, patients generally are given controlled-release pharmaceuticals. The pattern of drug application is also rather constant in terms of the time of drug intake in relation to meals. Multiple-dose studies are suitable to elucidate the effect of changing the dosage regimen relative to the time of food intake, which may occur randomly.

Various study designs for the evaluation of food effects are presented schematically in Fig. 18.

Worldwide consensus has been reached with regard to analytical methods and their performance. Analyses are usually performed according to the general requirements of Good Laboratory Practice. In addition to the determination of

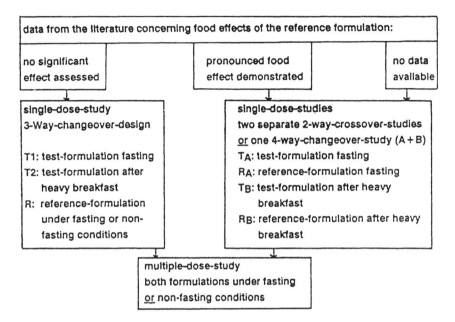

data from the literature concerning food effects of the reference formulation:

| no significant effect assessed | | pronounced food effect demonstrated | | no data available |

single-dose-study
3-Way-changeover-design

T1: test-formulation fasting
T2: test-formulation after
    heavy breakfast
R: reference-formulation
   under fasting or non-
   fasting conditions

single-dose-studies
two separate 2-way-crossover-studies
or one 4-way-changeover-study (A + B)
T_A: test-formulation fasting
R_A: reference-formulation fasting
T_B: test-formulation after heavy
    breakfast
R_B: reference-formulation after heavy
    breakfast

multiple-dose-study
both formulations under fasting
or non-fasting conditions

Fig. 18 Studies necessary to assess bioequivalence of controlled-release dosage forms (modified from Ref. 25).

the active moiety, it might be requested to analyze optical isomers and pharmacodynamically active metabolites at the same time. Drugs with slow elimination kinetics of more than 30 h involve certain problems in practice. For these drugs, too, at least 80% of the individual AUC values have to be determined analytically before extrapolation to infinity is allowed. When there are long washout phases, alternatives such as parallel-group designs may be suitable.

B. Evaluation of Data

AUC values can be calculated either after data fitting to pharmacokinetic models, which are usually compartmental models, or preferably, without model fitting, according to the trapezoidal rule or with the help of deconvolution techniques. The mere comparison of individual AUC values of test and reference as the only criterion for bioequivalence decision is no longer state of the art. AUC reflects only the extent of bioavailability. The rate of bioavailability described by the parameters C_{max} and t_{max} must be evaluated as well. Characterization of controlled-release dosage forms requires additional parameters. After a single-dose application, the half-value duration, the time span during which concentration is above 50% of C_{max}, is often used. When the pharmacodynamic effectivity is taken into consideration, the time span above the minimal effective concentration can be evaluated. In the steady state the quotients of relative minima and relative maxima in the plasma concentration versus time profiles are expressed as peak–trough fluctuations. Also, a comparison of truncated AUC values above and below the average concentration in the steady state, called AUC fluctuation, is used as a very rugged parameter to describe oscillations of the plasma concentration profile. There is at present international consensus as to statistical evaluation that because of the extremity of data spread, Student's t-test is not suitable to prove bioequivalence. Alternatively, the ANOVA, ANOVAlog, and Mann–Whitney statistical procedures are suggested for use, if necessary, in parallel.

C. Bioequivalence Testing of Enteric-Coated Forms

Testing of the bioequivalency of enteric-coated formulations is primarily the same as that for immediate-release dosage forms The rate and extent of bioavailability usually show greater intersubject and intrasubject variability. The variability needs change in the design and evaluation of clinical studies [23]. For AUC values, no modification of the evaluation procedure normally used is necessary. Through direct comparison by means of bioequivalence evaluations, the 90% confidence interval of the test versus the reference ratio has to be within the limits 80% to 125%. This requirement should also be met for multiplicative modeling, as suggested, for example, by CPMP [24], where data have to be transformated logarithmically prior to evaluation.

For C_{max}, bioequivalence evaluation is more complex. The onset of absorption of drug from single units takes place with pH-induced desegregation in the intestine. The entire dose can be absorbed in a rather short time. This can be seen by a steep slope in the plasma concentration profile with a sharply pronounced peak. Multiple units are generally more or less dispersed in the stomach, and absorption of a fractionated dose takes place for each bead with its arrival at the intestine. Depending on the spread and regrouping of dispersed subunits, the rise of plasma concentration and the sharpness of peak can be less pronounced. Taking these facts into account, two possibilities for bioequivalence testing are given.

1. If two formulations of identical galenical type are compared, the usual evaluation should be performed. The 90% confidence interval of the test versus reference ratio has to be within the limits 70% to 143%. This requirement should also be met for logarithmically transformated data. If there is a pharmacokinetic rationale (e.g., highly variable drugs [26]) for some drugs, a wider acceptance range may be necessary.

2. If two formulations of different galencial types of different sizes have to be compared, statistical evaluation according to the procedure described above should not be used. The spread of the subunits in the GI tract as the cause for pronounced differences in C_{max} values suggests descriptive statistics. Further evaluation should be performed taking pharmacodynamic relevance into

account. Similar attempts were undertaken for the evaluation of t_{max} for immediate-release formulations [27].

t_{max} *and* t_{lag}

According to the rational for evaluation of C_{max} for t_{max}, descriptive statistics should also be used. t_{max} ought to be evaluated together with the lagtime, t_{lag}. Differences in both parameters should also be evaluated with respect to their clinical relevance. Performing a lag-time correction prior to evaluation of t_{max} has been discussed in the literature [28]. It should be considered that although the lag-time correction is able to show up the rate and extent of absorption from the dosage form at the site of absorption, it does not reflect the biopharmaceutical capability of the drug product to surmount physiological gates. Food intake may alter the movement of enteric-coated dosage forms in the GI tract. More direct interactions of food with the dosage form can change bioavailability. To check the extent of this effect and the ruggedness of the enteric coat, food studies are desirable.

REFERENCES

1. *Die Rote Liste 1990*, Bundesverband der Pharmazeutischen Industrie, Editio Cantor, Aulendorf, Germany, 1990.
2. H. Möller, *Pharm. Ztg. 125*:1105 (1980).
3. J. Krämer, H. Blume, and M. Siewert, *Pharm. Ztg. 135*:2169, 2209 (1990).
4. G. Parker, C. G. Wilson, and J. G. Hardy, *J. Pharm. Pharmacol. 40*:376 (1988).
5. J. E. Devereux, J. M. Newton, and M. B. Short, *J. Pharm. Pharmacol. 42*:500 (1990).
6. H. Bechgaard, Acta Pharm. Technol. 28:2, 1982, pp. 149–157.
7. A. H. Staib, D. Loew, S. Harder, E. H. Graul, and R. Pfab. *Eur. J. Clin. Pharmacol. 30*:691 (1986).
8. H. Minami and R. W. McCallum, *Gastroenterology 86*:1592 (1984).
9. J. N. Hunt, in *Esophageal and Gastric Emptying* (A. Dubois and D. O. Castell, eds.), CRC Press, Boca Raton, Fla., 1984, pp. 65–71.
10. J. N. Hunt, and D. F. Stubbs, *J. Physiol. (London) 215*:209 (1975).
11. S. Holt, J. Reid, T. V. Taylor, P. Tothill, and R. C. Heading, *Gut 23*:292 (1982).
12. K. A. Kelly, in *Physiology of the Gastrointestinal Tract* (L. R. Johnson, ed.), Raven Press, New York, 1981, pp. 393–410.
13. R. A. Hinder and K. A. Kelly, *Am. J. Physiol. 233*:E335 (1977).
14. M. T. Borin, S. Khare, R. M. Beihn, and M. Jay, *Pharm. Res. 7*:304 (1990).
15. A. F. Parr, R. M. Beihn, R. M. Franz, G. J. Szpunar, and M. Jay, *Pharm. Res. 4*:486 (1987).
16. A. Kyroudis, S. Markantonis, and A. Beckett, *Pharm. Weekbl. [Sci]. 11*(2):44 (1989).
17. S. O'Rheilly, C. G. Wilson, and J. G. Hardy, *Int. J. Pharm. 34*:213 (1987).
18. J. Krämer, Doctoral thesis (in preparation), Heidelberg University.

19. S. Luhtala, *Acta Pharm. Nord.* 4:85 (1992).
20. B. Scheidel, G. Lenhard, M. Siewert, G. Stenzhorn, and H. Blume, *Pharm. Ztg. Wiss.* 2:31 (1989).
21. H. Schott, L. Chong Kwan, and S. Feldman, *J. Pharm. Sci.* 71:1038 (1982).
22. H. Blume, G. Gentschew, J. Krämer, and K. K. Midha, Review, submitted for publication (1993).
23. F. Stanislaus and K. Walter, Evaluation of bioequivalence of enteric-coated products, *Proceedings of Bio-international* (H. Blume and K. K. Midha, eds.), Bad Homburg, Germany, 1992.
24. CPMP Working Party on the Efficacy of Medicinal Products, *Note for Guidance: Investigations on Bioavailability and Bioequivalence*, 1991.
25. V. W. Steinijans and R. Sauter, in *Biointernational* (K. K. Midha and H. Blume, eds.), Medpharm, Stuttgart, 1993, pp. 235–250.
26. H. Blume, M. Siewert, H. Reimann, K. Kübel-Thiel, and E. Mutschler, *Pharm. Ztg.* 132:2025 (1987).
27. H. Blume, A. Blume, M. Siewert, G. Stenzhorn, and W. Stüber, *Pharm. Ztg.* 132:244 (1987).
28. B. Scheidel, in *Bioverfügbarkeit und Bioäquivalenz von Retardarzneimitteln* (H. Blume, ed.), Govi Verlag, Frankfurt, 1990, pp. 67–77.

13

In Vivo Behavior of Multiparticulate Versus Single-Unit Dose Formulations

George A. Digenis

College of Pharmacy
University of Kentucky
Lexington, Kentucky

I. INTRODUCTION

In this chapter we emphasize the great usefulness of gamma scintigraphy and neutron activation techniques in the in vivo assessment of the performance of orally administered formulations. These techniques, in conjunction with blood-level data, have been proven to be quite powerful since their first utilization in our laboratories [1–3].

II. SHORT DESCRIPTION OF GAMMA SCINTIGRAPHY AND NEUTRON ACTIVATION TECHNIQUES

The underlying principles of gamma scintigraphy can be explained briefly as follows. Radiation photons arising from a gamma-emitting radionuclide, incorporated into a formulation, strike the NaI crystal of a gamma-camera detector head. The resultant flash of light is then detected by photomultiplier tubes. The "information" acquired by the tubes is displaced on a cathode-ray oscilloscope and is digitized so that it can be stored on a magnetic disk. Thus quantitative image processing can be performed [4]. Using this technique, the in vivo behavior of two different dosage forms can be monitored simultaneously [4].

Conventional radiolabeling techniques that utilize technetium-99m and/or indium-111 suffer from the following disadvantages: (1) they are applicable to relatively simple dosage forms, and (2) they are limited to labeling small laboratory batches. In contrast, the neutron activation (NA) technique involves the incorporation of a stable isotope into a pharmaceutical formulation prior to manufacturing [3]. Subsequently, the intact dosage form is radiolabeled by subjection to neutron activation utilizing a nuclear reactor.

Utilization of neutron activation requires the appropriate choice of a stable isotope to be incorporated in a pharmaceutical dosage form. The parent stable isotope should exist in a high natural abundance or be available in an enriched form, should possess a high neutron capture cross section, and must be nontoxic and chemically inert. The daughter (radioactive) isotope, which results after neutron bombardment of the parent isotope, must (1) must have a relatively short half-life (2 h to 3 days), (2) should be a gamma emitter with an energy of 100 to 300 keV, and (3) should decay to a stable nontoxic isotope [5].

According to the neutron activation equation [5], the amount of radioactivity that is generated after neutron bombardment is dependent on the number of target atoms, the neutron flux density, the irradiation time (t_i), the decay time, the thermal neutron capture cross section (σ) and the decay constant of the isotope produced (α). The σ and α parameters and the decay time dictate the choice of the isotope to be utilized in radiolabeling a dosage form.

It has been found that erbium-170 (^{170}Er) ($\sigma = 5.8$ barns) and samarium-152 (^{152}Sm) ($\sigma = 208$ barns) best satisfy the foregoing requirements. When neutron activated, they produce ^{171}Er ($t_{1/2} = 7.52$ h, $\lambda = 308$, 296, and 112 keV) and ^{153}Sm ($t_{1/2} = 46.7$ h, $\lambda = 103$ keV). The energies of emission of these radionuclides fall within the required range, 100 to 300 keV, and produce excellent images with a gamma camera. To date, several pharmaceutical preparations have been labeled by NA techniques (15 to 50 s of bombardment) which were proven to exhibit the same dissolution and disintegration profiles as those of nonirradiated dosage forms [3, 6–9].

III GENERAL INFORMATION

A. In Vivo Behavior of Orally Administered Multiparticulates and Tablets

In the early 1970s we were able to attach triethylenetetramine covalently to insoluble, nondegradable polystyrene beads (I), ranging in diameter from 0.15 to 0.42 mm [2]. These beads were found to form a stable complex with the radionuclide ^{99m}Tc in the pH Oral administration to humans of the radiolabeled polystyrene beads with a light breakfast (composed of an oatmeal cereal) showed that after a brief period of stasis (approximately 10 min) the beads emptied from the stomach in a monoexponential fashion (Figs. 1 and 2a) [2]. The slope of the straight line resulting from the plot of natural logarithmic counts in the entire stomach versus time represented the time taken for 50% of the beads to empty from the stomach [2]. The slope of the line has been considered to characterize

(P) = Polystyrene backbone

the gastric emptying time (GET) of a subject. This simple definition of gastric emptying time has been adopted by several laboratories and clinical departments.

Utilizing the aforementioned multiparticulate radiopharmaceutical preparation, the GET for normal subjects was found to range from 45 to 85 min [13]. Gastric emptying times were found to be delayed in patients with diabetes mellitus and were especially long in persons suffering from diabetic gastroenteropathy [13–15].

Oral administration to human subjects of polystyrene–triethylenetetramine–^{99m}Tc-labeled beads (0.15 to 0.42 mm in diameter, with a density of 1.2) in a hard gelatin capsule provided a model for the in vivo behavior of enteric-coated multiparticulate formulations. Indeed, when the labeled beads were administered to volunteers on an empty stomach, release of the beads from the gelatin capsule began at 30 to 40 min after ingestion (Fig. 2b)(1). On a full stomach, the same subjects exhibited much longer times (93 to 120 min) for initial release of the contents [1]. In all cases the capsule appears to remain stationary and the capsule contents disperse immediately to the other regions of the stomach after initial release of the beads. The relatively large (0.15 to 0.42 mm) insoluble polystyrene particles disperse into the stomach only after major collapse of the capsule walls, a rather lengthy process [1].

In contrast to the above, when the gelatin capsule was filled with a water-soluble formulation (principally using lactose as a filler) and the soluble chelating agent, etidronate sodium labeled with ^{99m}Tc, was administered on an empty stomach, initial release of the radioactivity occurred rapidly (6 min)(Fig. 3) [12]. Considerable swelling of the capsule was also observed before the release [2,12]. The decrease in radioactivity from the capsule region indicated a gradual loss of its contents (Fig 3). At this point, a sharp break in the curve, representing counts in the capsule region, indicated virtually complete disappearance of radioactivity from the capsule. The relatively rapid release of the water-soluble formulation occurs because the gastric juices gain access to the interior of the capsule via a small orifice and/or diffusion through its walls. These findings are consistent with the observation reported previously that increasing the water solubility of a

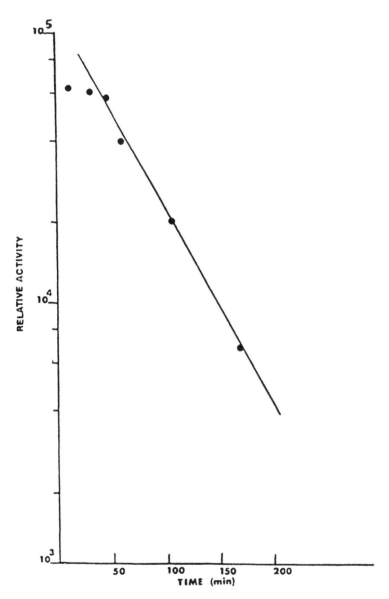

Fig. 1 Semilogarithmic plot illustrating the rate of clearance of radioactive beads (polystyrene–triethylenetetramine–^{99m}Tc from the stomach of a normal human male subject (GE$t_{1/2}$ = 45 min). The beads (0.15 to 0.42 mm in diameter and d = 1.2) were administered dispersed in an oatmeal cereal breakfast. (From Ref. 1.)

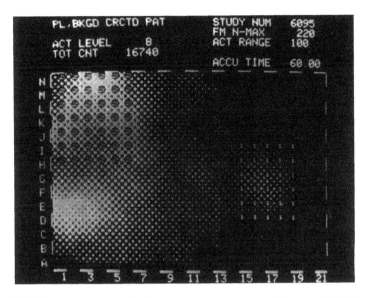

Fig. 2a Scintiphoto showing two computer areas of interest within the stomach of a human subject. Right side, area of interest over a hard gelatin capsule; left side, area of interest over the pyloric region of the stomach, after dispersal of polystyrene beads. (From Ref. 1.)

formulation will increase its rate of release from a gelatin capsule [16,17]. It is also known that particle size and type of diluent or filler used in various formulations can drastically affect the rate of their release from gelatin capsules [18].

Recent studies in our laboratories concentrated on further exploration of the in vivo behavior of commercially available multiparticulate formulations. The effect of density on the rate of the downward migration of beads in the gastrointestinal (GI) tract of human volunteers was studied. Two batches of beads (0.6 to 0.7 mm) covered with an insoluble water-permeable polymer ("onionskin" type) were prepared in an identical manner but with densities of 1.0 and 1.3. The two batches were radiolabeled with two different radionuclides (one with ^{99m}Tc and the other with ^{111}In). When the two types of beads were administered simultaneously to each of the 10 fasted subjects they exhibited identical residence times in the various segments of the GI tract (Fig. 4). Additionally, the time of the arrival at the ileocecal junction was identical for both types of beads (270 to 320 min) (Fig. 5).

The results from several laboratories that followed this initial study can be summarized at this point. Multiparticulates of particle size from 0.1 to 3 mm with densities of 1 to 3 appear to exhibit identical gastric emptying rates under

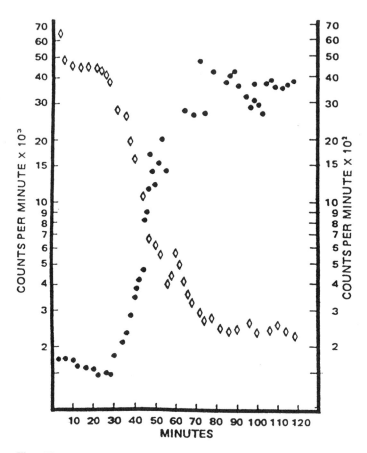

Fig. 2b   Graphic display illustrating release of radioactivity from a hard gelatin capsule, filled with insoluble radiolabeled polystyrene beads (0.15 to 0.42 mm in diameter and $d = 1.2$), in the human stomach (fasted conditions). $\Diamond$, disappearance of radioactivity from the capsule (right side, Fig. 2a); $\bullet$, concomitant appearance of radioactivity in the pyloric region of the stomach (left side, Fig. 2a). (From Ref. 1.)

fasted conditions. Bigger particles (3 to 5 mm) with densities above 3 are *suspected* to empty at a slower rate from the stomach under fed conditions [19]. Such studies are currently in progress in our laboratories. It is our opinion that although such claims have been made in the literature, no definitive studies have been conducted to fully prove the foregoing hypothesis. The small number of subjects involved and the method of labeling of multiparticulates, which may not be considered relevant to actual pharmaceutical preparations, casts doubts on the findings of these studies [20].

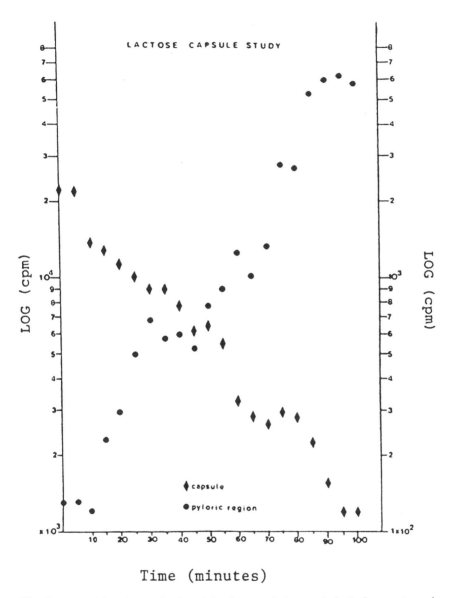

Fig. 3 Plot of the release of radioactivity from a gelatin capsule in the human stomach fasted using a labeled water-soluble formulation. ◆, disappearance of radioactivity from the capsule region; ●, concomitant appearance of radioactivity in the pyloric region of the stomach. (From Ref. 12.)

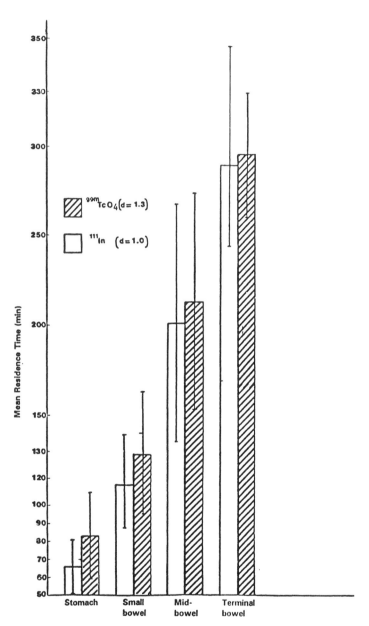

Fig. 4   Mean residence time of water-insoluble beads (0.6 to 0.7 mm in diameter) in selected regions of the gastrointestinal tract in humans (n = 10) as a function of specific gravity. [Heavier beads (d = 1.3) were labeled with ^{99m}Tc and lighter (d = 1.0) with ^{111}In.]

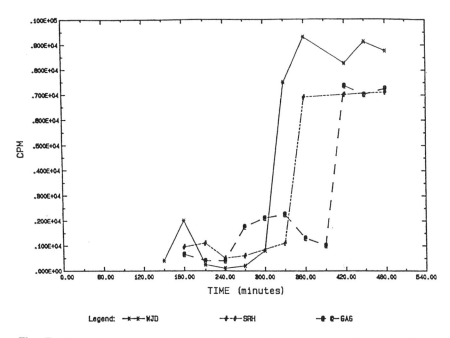

Fig. 5 Plot of counts per minute at the ileocecal junction region of human subjects versus time [99mTc]-radiolabeled beads (density = 1.3).

It is our experience that on an *empty* stomach, no differences between heavy and light pellets can be observed [21]. Claims have been made, however, that light (density of 0.86) particles (10 × 6 mm in size) exhibit delayed gastric emptying under fed conditions. Similar claims have been made for osmotic pump devices (Osmets) of size 25 × 7 mm. In a study involving only four subjects it was shown that gastric retention of these pumps appears to be prolonged by food intake [19].

In general, it can be said that multiparticulates empty gradually and to some extent predictably from the stomach [22,23]. Up to about 2 to 3 mm they appear to be small enough to pass through the contracted pylorus. In this connection it can be said that the critical size of 2 mm claimed is *not* absolute and needs further investigation in humans. Larger units (12 mm) stay in the stomach from 50 min to 10 h after a light meal, and are less predictable with regard to their gastric retention times under fed conditions [24].

The influence of food on larger particles (tablets 12 to 14 mm) has recently been studied in our department [25]. It was found that in five out of eight fed subjects a radiolabeled sustained-release tablet of ibuprofen (800 mg) remained in the stomach and eroded slowly over 7 to 12 h resulting in gradual increases in small bowel radioactivity. In three out of the total eight fed subjects, however, the intact tablet was ejected from the stomach in about 4 h [25]. This is in marked contrast to a similar study involving fasted volunteers in which gastric retention times of the sustained-release tablet ranged from 10 to 60 min [26]. Interestingly, in the first five subjects, with prolonged gastric retention, the bioavailability of ibuprofen from the tablets above was approximately equal to that obtained with the three subjects who exhibited rapid gastric emptying times (Fig. 6).

In two independent studies it has been shown by scintigraphy techniques that the in vivo disintegration of a tablet is linearly related to the time of the first detectable level of the drug in blood [27,28] (Fig. 7). This correlation has been shown to be true with p-acetaminophenol (acetaminophen) [27] and 5-amino-salicylic acid [28]. Both of these drugs are believed to be absorbed throughout the GI tract. This relationship has not yet been shown to be true for drugs exhibiting a narrow "window" of absorption. The data obtained so far empha-size that the in vivo disintegration of a novel dosage form may dramatically influence the blood concentration profile of a drug. This is a good illustration of how gamma-scintigraphic techniques can help in the development of a novel and efficacious oral drug delivery system.

Recent data from our laboratories show that the complete in vitro disin-tegration of controlled-release tablets is usually faster than their in vivo disper-sion by approximately a factor of 4 (E. Sandefer and G. A. Digenis, unpublished data) (Fig. 8). This is not always true and is dependent on the condition and physiology of the patient's stomach at the time of ingestion of the tablet. Of importance is our observation that the slower-releasing tablets usually survive

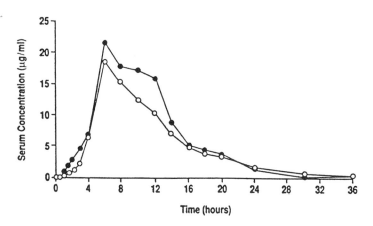

Fig. 6 Average ibuprofen serum concentrations (from a tablet) versus time plots for five subjects with prolonged gastric emptying (●) and three subjects with rapid gastric emptying (○). (From Ref. 25.)

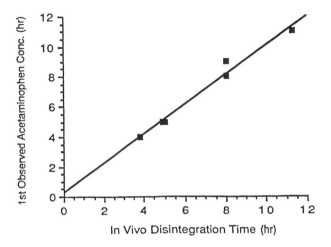

Fig. 7 Correlation of observed time of in vivo disintegration of a 325-mg *p*-acetaminophenol (APAP) tablet to the time of the first detectable level of the drug in blood. (From Ref. 27.)

longer as a single unit in the fed stomach and thus tend to exhibit a longer residence time in the stomach.

B. Inherent Properties of Enteric-Coated Versus Noncoated Formulations

We must pay special attention to the residence time of a dosage form in the stomach. This is, of course, of importance for drugs that are acid sensitive and more significantly for drugs that are absorbed from a limited segment of the upper small intestine. Toward this end, we felt that gamma scintigraphy provides an opportunity to examine the in vivo performance of several polymers used as enteric-coating agents. Our findings showed that with some polymeric enteric-coating materials, the longer the gastric residence time of the tablet, the longer was its in vivo dissolution time. Alternatively, the opposite could be observed with other polymeric enteric-coating agents. For example, with these novel materials the longer the mean residence in the stomach of the enteric-coated tablet, the faster is its in vivo mean dissolution time.

In the case of an enterically coated erythromycin multiparticulate formulation, the shorter the mean residence time in the stomach of the erythromycin-containing beads, the higher the bioavailability of the antibiotic. A possible explanation of this observation is given later. It is important to mention at this point the fact that in the case of fed studies, a single-unit dosage form may stay in the stomach for as long as 12 h. In such a case the chemical effects of the enteric

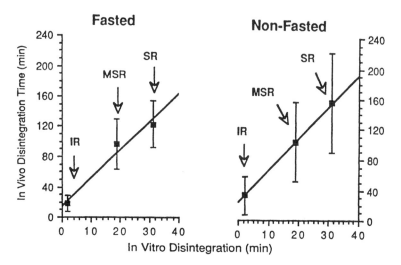

Fig. 8 In vitro and in vivo correlation of 100% disintegration of a controlled-release tablet. IR, immediate release; MSR, moderately slow release; SR, slow release.

coating may be so pronounced that bioavailability may be impaired. Thus the prudent formulator must conduct in vitro studies in which the dosage form is immersed for at least 12 h in a milieu of pH 1.2. Subsequent to this time, the dosage form must be washed thoroughly, and then studies may begin on drug release in a neutral environment (pH 7.2).

The following statements summarize some important observations that have been made by external scintigraphic techniques.

1.   Nondisintegrating particles (0.6 to 3 mm) empty monoexponentially from the stomach under fasted conditions. Under these conditions the beads are less dispersed in the stomach and exhibit a gastric emptying time (GET $t_{1/2}$) of 45 to 80 min in humans. Under fed conditions multiparticulates spread extensively in the stomach and exhibit longer gastric emptying times (up to 180 min with a light meal and even longer with a heavier meal).

2.   Administration of food prior or simultaneously with a dose of multiparticulates causes more spreading of the pellets than under fasted conditions.

3.   Ingestion of food 30 min before administration of a dosage form causes an *increase* in the residence time of multiparticulates or a tablet in the stomach [1,25,44]. In contrast, food administration 30 min *after* oral dosing results in more uniform gastric emptying times of multiparticulates and tablets and decreases their transit time through the small intestine.

4.   Prolonged gastric retention of enteric-coated dosage forms often leads to physical and/or chemical alteration of their polymeric coatings. This results in faster or slower release of the drug in the upper intestine (pH > 5.5) and

consequently affects its bioavailability. This, coupled with variable gastric emptying, is a major reason for the inconsistent in vivo behavior of enteric-coated formulations and the frequent lack of in vitro-to-in vivo correlation with enteric-coated formulations. As mentioned previously, because multiparticulates (up to 3 mm) can exit from the contracted pylorus (due to the presence of food in the stomach), they will exhibit a relatively shorter gastric residence time. Consequently, under fed conditions, less variable blood levels are obtained, with multiparticulates than with single-unit dosage forms. To benefit from this beneficial property of multiparticulates, they must be formulated in a rapidly disintegrating hard gelatin capsule. This is because the human stomach recognizes a gelatin capsule as a single-unit dosage form. Consequently, rapid dispersion of the beads from the gelatin capsule is an important requirement, particularly under fasted conditions.

IV. SPECIFIC EXAMPLES

A. Plasma Levels of Erythromycin from Enteric-Coated Multiparticulates and Tablets in Humans

Erythromycin pellets (1.5- to 2-mm-diameter pellets of Eryc II) were prepared by the manufacturer following procedures identical to those for the commercial product but with the addition to the formulation of 12.8% w/w of naturally occurring samarium oxide (26.7% ^{153}Sm) [9] or 0.67% w/w of erbium oxide (^{170}Er) [8].

^{152}Sm-containing pellets were irradiated in a nuclear reactor, with neutrons, for 33 s to produce radioactive ^{153}Sm pellets (103 keV) ($t_{1/2}$ = 46.7 h) with 10 μCi per pellet 24 h after irradiation. Dissolution tests were performed to verify that the erythromycin content of the irradiated pellets was the same as in nonirradiated virgin pellets. Erythromycin extracted from the irradiated pellets was analyzed microbiologically and found to be unaffected by neutron activation treatment [8].

Five of approximately 320 unlabeled erythromycin enteric-coated pellets in a capsule were replaced with five radiolabeled pellets. Thus each capsule contained 250 mg of erythromycin base, including five ^{155}Sm = labeled pellets with a total radioactivity dose of 50 μCi. By following the live radiolabeled beads, the site and time of disintegration within the GI tract could be determined [9].

Seven healthy volunteers were fasted for a minimum of 9 h. A single 250-mg erythromycin capsule was administered with 200 mL of H_2O. Food was restricted: 100 mL of H_2O was permitted at 2 h postdose; ad libitum H_2O intake was allowed at 4 h postdosing; a standard meal was served 8 h postdosing [9].

The same procedure was followed in the fed mode with the exception that a standardized breakfast was ingested 30 min after administration of the capsule/

tablet. Supine subjects were then monitored (anterior dynamic) by a gamma scintillation camera.

Subjects were scanned continuously during the first 60 min, then for 10 min at 30-min intervals up to 6 h, and after 6 h post dose for 10 min at 1-h intervals to 12 h. This imaging procedure allowed accurate determination of anatomical position of the beads in the GI tract from stomach to colon. Blood samples were withdrawn and the plasma analyzed microbiologically.

The ^{155}Sm-labeled 250-mg erythromycin gelatin capsule formulation described above was administered orally to each of seven healthy volunteers under fasted and nonfasted conditions. When the anatomical position of the beads in the GI tract was correlated to the resulting blood levels of the antibiotic, the following observations could be deduced:

1. At the end of the second hour after ingestion of the beads, 30 of 35 pellets persisted under fasted conditions (Fig. 9), while only 15 of 35 beads could be detected at that time under nonfasted conditions (Fig. 10). (It should be noted that five radiolabeled beads were introduced in each capsule. A total of 35 beads were therefore administered to 7 subjects.) The results showed clearly that under fasted conditions, these enteric-coated beads persisted for longer than under nonfasted conditions (Fig. 9 versus Fig. 10).

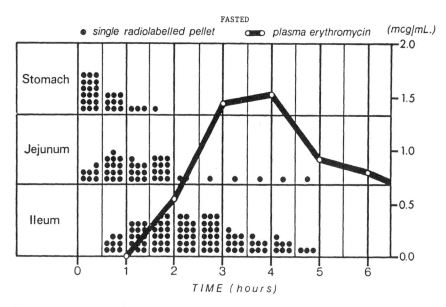

Fig. 9 Fasting subjects ($n = 7$): mean number of radiolabeled pellets per anatomical region, correlated with plasma erythromycin concentrations following single 250-mg doses of erythromycin capsules, as a function of time. (From Ref. 9.)

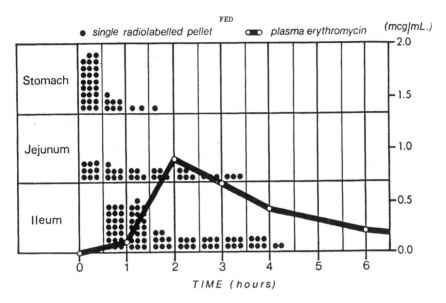

Fig. 10   Nonfasting subjects ($n = 7$): mean number of radiolabeled pellets per anatomical region, correlated with plasma erythromycin concentrations following single 250-mg doses of erythromycin as a function of time. (From Ref. 9.)

2. All pellets exited from the stomach intact under fasted or nonfasted conditions. This showed that the enteric coating of the beads was appropriately formulated.

3. The time to reach maximum (t_{max}) concentrations of erythromycin was 3 h under fasted conditions (Fig. 9), while under nonfasted conditions $t_{max} = 2$ h (Fig. 10).

4. The bioavailability of the erythromycin from the pellets and its maximum plasma concentrations (C_{max} values were considerably higher under fasted conditions) (Fig. 11).

5. The first detectable plasma levels of erythromycin were registered at 1.4 h after administration of the capsule under fasted conditions. Earlier times (0.8 h) of first detection were achieved under nonfasted conditions.

6. Interestingly, the majority of the beads were located at the jejunum and proximal portion of the ileum at the time (1.4 h) of the first detection of erythromycin. This indicated that the site of absorption of erythromycin from this formulation is the jejunum and the upper portion of the small intestines.

7. The pronounced shift toward later times in the bioavailability curve of erythromycin and its greater magnitude under fasted conditions (Fig. 11) are

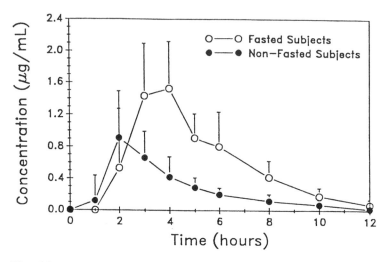

Fig. 11   Mean plasma concentrations (μg/mL) multiparticulate formulations in a hard gelatin capsule of erythromycin following single 250-mg oral doses under fasted and nonfasted conditions, as a function of time. (From Ref. 9.)

in direct contrast with observations quoted in the literature, where "dumping" of the drug (theophylline) from a controlled-released multiparticulate was noticed. The later phenomenon was attributed to the possible interaction of digestive enzymes, bile secretions, and pancreatic fluid with the formulation. This appears to have resulted in enhanced bioavailability of theophylline following a heavy meal.

B.   Comparison of GI Tract Behavior of Enteric-Coated Formulation with an Enteric-Coated Tablet After Simultaneous Administration

Plasma concentrations of erythromycin arising from the enteric-coated multiparticulate formulation showed that in each subject studied, the erythromycin plasma levels were considerably higher than the minimum inhibitory concentration (MIC) for the antibiotic under fasted and nonfasted conditions (Fig. 12). In contrast, however, when an equivalent dose of erythromycin (250 mg) was administered in an enteric-coated form, in 50% of subjects the antibiotic failed to achieve the MIC value under nonfasted conditions (Fig. 13). In those patients who exhibited poor plasma levels of the antibiotic, rapid transit of the tablet through their upper GI tract was observed.

For example, in one volunteer the tablet reached the cecum in 45 min before it disintegrated. Those tablets that reached the cecum intact exhibited long

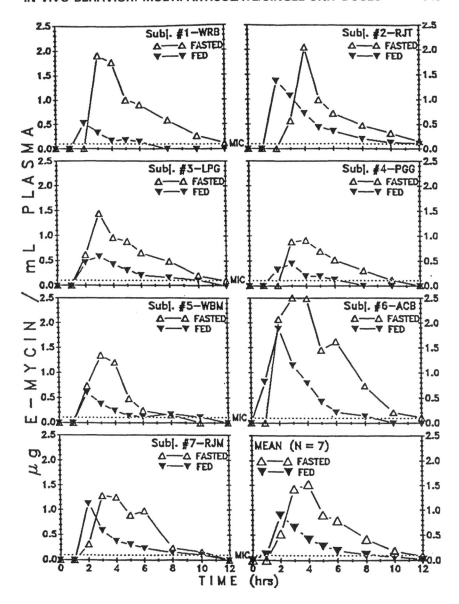

Fig. 12   Individual and mean erythromycin concentration–time curves following a 250-mg oral dose (free base) from enteric-coated multiparticulate formulation [9] in a hard gelatin capsule.

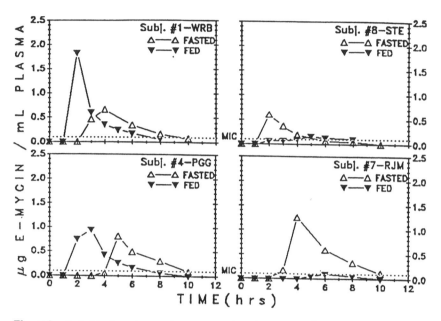

Fig. 13 Individual erythromycin concentration–time curves following a 250-mg (free base) oral dose from an enteric-coated tablet.

disintegration times in that region. Blood-level profiles of erythromycin resulting from the administration of an enteric-coated tablet under fed and fasted conditions seemed to exhibit the same differences as in the case with the enteric-coated multiparticulate formulation (Fig. 14). Here again the C_{max} value appeared earlier in the case of the fed mode (Fig. 14). The bioavailability, however, of the antibiotic appeared to be almost equal under fasted or fed conditions. In a separate study an erythromycin-containing enteric-coated tablet (250 mg) was coadministered with a hard gelatin capsule containing 320 enteric-coated beads. The latter contained [152]Sm but no erythromycin and were identical in composition with the multiparticulate formulation described above. The table was labeled with indium-111. Gamma scintigraphy confirmed that the multiparticulate formulation (beads) emptied from the stomach gradually and predictably within 2 h. However, the tablet exhibited large intrapatient variation with regard to its gastric emptying time. Erythromycin blood levels resulting from the administration of the tablet were compared to those obtained from the study with the beads containing erythromycin. Figure 15 shows that the erythromycin plasma levels resulting from the multiparticulate formulation were considerably higher under fasted conditions than those obtained from the tablet.

In the fed mode, however, the bioavailability of erythromycin from the multiparticulate was not appreciably different than from the tablet. However,

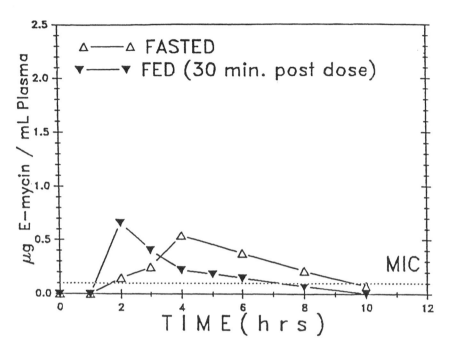

Fig. 14 Erythromycin mean concentration–time curves (fasted versus fed) following a 250-mg (free base) oral dose of an enteric-coated tablet.

when the individual plasma levels of the antibiotic are compared (Fig. 12 versus Fig. 13) it can be seen that the plasma levels of erythromycin from the tablet are highly variable under fed conditions, and in fact, in some cases failed to reach the minimum inhibitory concentration. This was attributed to the rapid transit of the tablet through the upper GI tract.

C. Additional Examples of In Vivo Pellet Behavior Versus Tablet

Feely and Davis [24] have published the results of a study that compared the in vitro and in vivo release of phenylpropanolamine from 20 pellets (3.1 mm in diameter) and one 12.5-mm-diameter hydroxypropyl cellulose (HPMC) single-unit matrix. It was found that the in vitro release of the drug was identical from the two formulations at pH 7.0 (Fig. 16). In contrast, at pH 1.0 the release of phenylpropanolamine from the small pellets (3.1 mm) was more rapid than from the single-unit matrix (Fig. 16). When the above-mentioned two formulations were administered orally to eight volunteers 30 min after a light breakfast, the following observations were made:

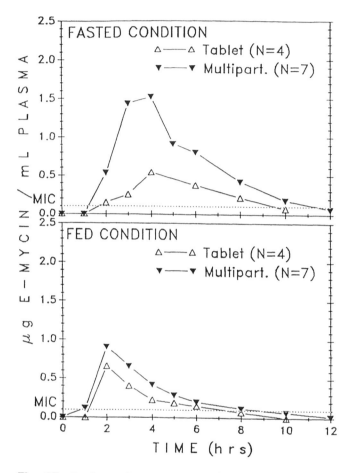

Fig. 15 Erythromycin mean concentration–time curves (fasted and fed) following a 250-mg (free base) oral dose of an enteric-coated multiparticulate formulation in a hard gelatin capsule versus a single enteric-coated tablet (250 mg, free base).

1. The multiple units (pellets) emptied from the stomach gradually over a period of 180 min. The single-unit dosage formulations exhibited extremely variable gastric emptying times (GETs). These varied from 60 to 570 min [24].
2. The intestinal transit times of both formulations were found to be similar.
3. Despite the GET differences in the two formulations, the bioavailability of phenylpropanolamine from these dosage forms was *not* significantly different (Fig. 17).

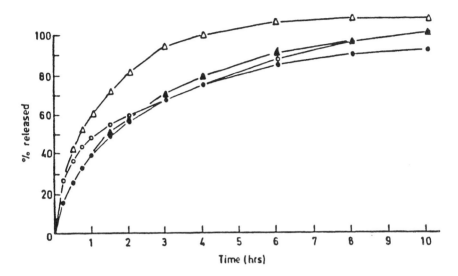

Fig. 16 Release of phenylpropanolamine from 3.1- and 12.5-mm-diameter HPMC matrices. Key: open circle, 3.1 mm, pH 7; closed circle, 12.5 mm, pH 7; open triangle, 3.1 mm, pH 1; closed triangle, 12.5 mm pH 1. (From Ref. 24.)

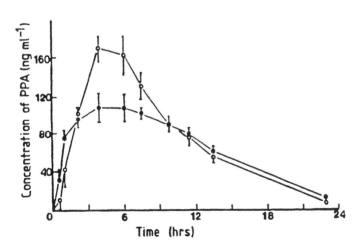

Fig. 17 Mean serum phenylpropanolamine concentrations plotted against time for two HPMC matrix formulations. ○, 3.1-mm matrices; ●, 12.5-mm matrices. Mean ± SE; $N = 6$. (From Ref. 24.)

4. The differences in the blood-level profiles was attributed to differences in the in vitro rates of release of the drug at a low pH value, *not* to their substantially different GET. The fact that the bioavailability of the drug was similar from the two formulations was attributed to the fact that this molecule is absorbed throughout the major portion of the GI tract.

V. CONCLUDING REMARKS

In conclusion, the choice of formulating a drug in a multiparticulate or single-unit dosage form must be based on:

1. The site of drug absorption in the GI tract ("narrow window" or not)
2. The susceptibility of the drug to gastric juice (enteric coating)
3. The intended site of drug delivery in the GI tract
4. The nature of polymeric coating on the dosage form

REFERENCES

1. D. L. Casey, R. M. Beihn, G. A. Digenis, and M. S. Shambhu, *J. Pharm. Sci.* *64*:1412 (1976).
2. G. A. Digenis, R. M. Beihn, M. G. Theodorakis, and M. S. Shambhu, *J. Pharm. Sci.* *66*:442 (1977).
3. A. Parr, M. Jay, G. A. Digenis, and R. M. Beihn, *J. Pharm. Sci.* *74*:590 (1985).
4. J. T. Fell and G. A. Digenis, *Int. J. Pharm.* *22*:1 (1984).
5. G. A. Digenis, A. F. Parr, and M. Jay, in *Drug Delivery to the Gastrointestinal Tract* (J. G. Hardy, S. S. Davis, and C. G. Wilson, eds.), Ellis Horwood, Chichester, West Sussex, England, 1989, p. 111.
6. A. Parr, R. M. Beihn, and M. Jay, *Int. J. Pharm.* *32*:251 (1986).
7. A. F. Parr, R. M. Beihn, R. M. Franz, G. J. Szpunar, and M. Jay, *Pharm. Res.* *4*:486 (1987).
8. A. F. Parr, G. A. Digenis, E. P. Sandefer, I. Ghebre-Sellassie, U. Iyer, R. U. Nesbitt, and B. M. Scheinthal, *Pharm. Res.* *7*:264 (1990).
9. G. A. Digenis, E. P. Sandefer, A. F. Parr, R. M. Beihn, C. McClain, B. M. Scheinthal, I. Ghebre-Sellassie, U. Iyer, R. U. Nesbitt, and E. Randinitis, *J. Clin. Pharmacol.* *30*:621 (1990).
10. M. C. Theodorakis, G. A. Digenis, R. M. Beihn, M. B. Shambhu, and F. H. Deland, *J. Pharm. Sci.* *69*:568 (1980).
11. M. C. Theodorakis, R. M. Beihn, and G. A. Digenis, *J. Nucl. Med. Biol.* *7*:374 (1980).
12. G. A. Digenis, in *Radionuclide Imaging in Drug Research* (C. G. Wilson and J. G. Hardy, eds.), Croom Helm, London, (1982), p. 103.
13. P. A. Domstad, E. E. Kim, J. J. Coupal, R. M. Beihn, et al., *J. Nucl. Med.* *21*:1098 (1980).
14. W. J. Shih, L. Humphries, G. A. Digenis, F. X. Castellanos, P. A. Domstad, and F. H. DeLand, *Eur. J. Nucl. Med.* *13*:192 (1987).

15. P. A. Domstad, W. J. Shih, L. Humphries, F. DeLand, and G. A. Digenis, *J. Nucl. Med.* 28:816 (1987).

16. L. L. Augsburger, in *Sprowls' American Pharmacy,* 7th ed. (L. W. Dittert, ed.), J. B. Lippincott, Philadelphia, 1974, Chapter 10.

17. J. M. Neuton, G. Rowley, and J. F. V. Thornblom, *J. Pharm. Pharmacol.* 23:452 (1971).

18. J. G. Wagner, *Biopharmaceutics and Relevant Pharmacokinetics,* Hamilton Press, Hamilton, Ill. 1971, p. 116.

19. S. S. Davis, J. G. Hardy, M. J. Taylor, D. R. Whallay, and C. G. Wilson, *Int. J. Pharm.* 35:331 (1984).

20. S. S. Davis, J. G. Hardy, M. J. Taylor, D. R. Whalley, and C. G. Wilson, *Int. J. Pharm.* 21:167 (1984).

21. S. Sangekar, W. A. Vadino, I. chaudry, A. Parr, R. Beihn, and G. A. Digenis, *Int. J. Pharm.* 35:187 (1987).

22. H. Bechgaard, *Acta Pharm. Technol.* 28:149 (1982).

23. H. Bechgaard and F. N. Christensen, *Pharm. J.* 229:373 (1982).

24. L. C. Feely and S. S. Davis, *Pharm. Res.* 6:274 (1989).

25. M. T. Borin, S. Khare, R. M. Beihn, and M. Jay, *Pharm. Res.* 7:304 (1990).

26. A. F. Parr, R. M. Beihn, R. M. Franz, G. J. Szpunar, and M. Jay, *Pharm. Res.* 4:486 (1987).

27. J. G. Hardy, J. N. C. Healey, and I. R. Reynolds, *Aliment. Pharmacol. Ther.* 1:273 (1987).

28. E. Sandefer, R. M. Beihn, J. Tossounian, K. Igbal, A. W. Malik, and G. A. Digenis, *Pharm. Res.* 5:S106 (1988).

14

Capsule Shell Composition and Manufacturing

Ronnie Millender

Capsugel, Division of Warner-Lambert Company
Greenwood, South Carolina

I. INTRODUCTION

The unique characteristics of the hard gelatin capsule offer some distinct advantages when used as a carrier for multiunit oral dosages. Because the capsule is an open container, the formulator has the flexibility to experiment easily with different proportions of particles to obtain different dosage characteristics. This flexibility is enhanced by use of different-sized capsules to yield different potency strengths. The capsule is usually gravity-fed and thus no appreciable pressure is applied to the multiparticulates. As a result, rate-controlling film coatings on the particles as well as matrix-controlled multiparticulates will remain intact during the encapsulation process. Finally, the hard gelatin capsule offers unique identification characteristics by being able not only to vary the size of the capsule but also to vary the colors that can be used. In this chapter we describe the composition and manufacture of the hard gelatin capsule and how some of the unique features are realized.

II. HISTORY

The history of the pharmaceutical industry teaches us that capsules as a dosage form have been around a very long time. Indeed, capsules were mentioned in an ancient Egyptian papyrus dating back to about 1500 B.C. But it was not until the middle of the nineteenth century that they were found to be of commercial interest (Table 1).

As their name implies, hard gelatin capsules are largely composed of gelatin. The *U.S. Pharmacopoeia* [1] defines capsules as

> solid dosage forms in which the drug is enclosed in either a hard or a soft, soluble container or "shell" of a suitable form of gelatin. . . . Hard gelatin capsules are made from special blends of bone and pork skin gelatins of relatively high gel strength. The bone gelatin gives a tough, firm fiber, while pork skin gelatin contributes plasticity and clarity to the blend. Hard gelatin capsules may also contain FD&C colorants, opaquing agents such as titanium dioxide, dispensing agents, hardening agents such as sucrose, and preservatives. They normally contain between 10% and 15% of water.

Hard gelatin capsules have been available commercially (in one form or another) for well over 100 years. The same process is used today by all modern manufacturers of capsules and based on machinery patented by Arthur Colton for Parke-Davis in 1931 (U.S. Patent 1,787,777) and still in use today. The basic flow chart is shown in Fig. 1; see also Table 2. Capsules are normally filled with a powder, either compressed to a slug or loose. However, with different modifications to the filling machine, other formulations can be put in the capsule. Formulations such as pellets, pastes, tablets, caplets, and even other capsules can be added to the capsules as an individual ingredient or in combination with the other formulations. The degree to which this occurs is limited somewhat by the

Table 1 Significant Events in Capsule Discovery

Year	Event	Inventor
1730	Oblate containers to mask the taste of turpentine	dePauli
1834	Patent issued for immersing filled leather pouch in gelatin	Mothes and Dublanc
1838	Gelatin-coated pills	Garot
1845	Process patent for making capsules	Steinbreener
1877	First machine capable of producing and filling two-piece capsules	Limousin
1895	Developed first modern capsule-producing machine	Colton

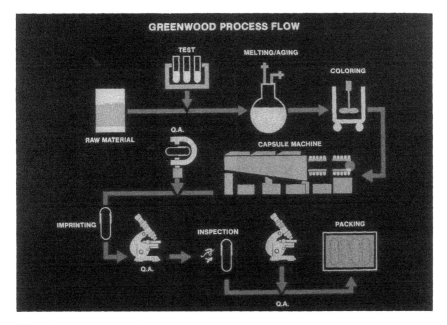

Fig. 1 Process flow chart.

size of the capsule (i.e., diameter of the capsule parts and volume of the closed capsule). The dimensions of the standard sizes are given in Table 3.

III. GELATIN

Gelatin is made of purified collagen. Collagen, the principal protein of animal connective tissue, is derived mainly by processing hides and bones. Whole hides are normally processed by tanneries, where the splits (the inner nongrained hide surfaces) and other trimmings, fleshings, and so on, provide the raw material for gelatin manufacture. Pigskin provides a considerable source of gelatin and is received from meatpacking houses in frozen condition. Pigskin is normally processed under acid conditions. Bone may be either fresh material that has undergone a degreasing process to reduce the fat content to a maximum of 1.0% or sun-dried material from India and Pakistan, which has been naturally weathered and degreased by the action of monsoon rains and hot sunshine. Bone is normally processed under alkali conditions.

The commercial production of gelatin, by hydrolysis of collagen, may be carried out by two main processes: the acid process and the alkali process. For purposes of defining raw gelatin in the manufacture of the hard gelatin capsule, it

Table 2 Process Flow

Step	Process
Raw material	Gelatin
	Water
	Dyes
Test	Chemical analysis
	Microbial analysis
Melting/aging	Gelatin melted in hot water
	Vacuum pulled to remove bubbles
	Aging to complete hydrolysis
Coloring	Gelatin solution put in smaller tanks
	Additives introduced to gelatin
	Viscosity adjusted
Capsule machine	Dipping capsule
	Drying capsule
	Stripping capsule
	Cutting capsule
	Joining capsule
Quality assurance	Dimensional checks
	Physical defect checks
Imprinting	Offset engravure system
Quality assurance	Print defect check
Inspection	Manual sort
	Electronic sort
	Mechanical sort

Table 3 Dimensions of Standard Capsule Sizes

Size	Individual length (mm)		Individual diameter (mm)		Closed length (mm)	Final volume (mm^3)
	Body	Cap	Body	Cap		
000	22.20	12.95	9.55	9.91	26.14	1.37
00	20.22	11.74	8.18	8.53	23.30	0.95
0	18.44	10.72	7.34	7.65	21.20	0.68
1	16.61	9.78	6.63	6.91	19.00	0.50
2	15.27	8.94	6.07	6.35	17.50	0.37
3	13.59	8.08	5.56	5.82	15.50	0.30
4	12.19	7.21	5.05	5.31	13.90	0.21
5	9.32	5.54	4.64	4.83	11.00	0.13

is classified by type (porkskin, bone, or calfskin) and Bloom jelly strength (Fig. 2). *Bloom jelly strength* is a standard but arbitrary measurement. It is defined as the weight in grams required to produce, by means of a plunger 12.7 mm in diameter, a depression of 4 mm in the surface of a gel of concentration 6.67 wt %, which has been matured at $10 \pm 0.1°C$ for a period of 16 to 18 h.

Gelatin is produced from biological materials, which by their nature are variable. Consequently, each production run of gelatin yields, in effect, a distinct entity of varying molecular size and form, even under rigidly controlled and standardized manufacturing conditions. Therefore, different production runs of gelatin may produce correspondingly different process characteristics when making capsules. This is one of the major challenges in the manufacture of capsules.

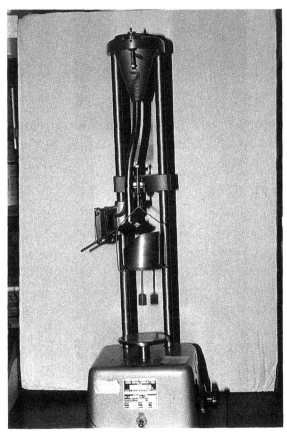

Fig. 2 Bloom gelometer.

IV. FORMING THE CAPSULE

A. Preparation of the Gelatin Solution

Melting of Raw Gelatin Solution

Gelatin is supplied to the capsule manufacturer as a relatively dry (10% w/w), granular structure (Fig. 3). Gelatin is insoluble in cold water, but when immersed

Fig. 3 Dry gelatin.

it absorbs a considerable amount of the water. In so doing, it swells and becomes soft. If the gelatin is then removed from any unabsorbed water and heated, it dissolves in the absorbed water to form a gelatin solution. This heating process is known as melting (Figs. 4 and 5). This was one of the first methods used as a melting process and may still be used today.

A second method and probably the most widely used method today is to omit the cold-water soak stage. Instead, the dry granular gelatin is gradually added to hot water ($\approx$82°C) in a mixing tank with continuous high-speed stirring. High-speed stirring is essential to "wet down" the granules and prevent them from forming clumps or agglomerates, which are difficult to dissolve [2]. The completed melting stage is dependent on the process used and the concentration used. The complete melting process takes about 8 h by the cold-water soak process and about 1½ h in the hot-water melt process.

The gelatin solution is held a certain period of time to allow for the complete hydrolysis of the gelatin. The viscosity of the solution is at its highest point just after the melt process begins. This viscosity drops rapidly until the hydrolysis is complete. Then the gel enters a fairly steady state (approximately 0.4%/h viscosity degradation) [3] until it again drops off rapidly due to degradation of the gelatin (Fig. 6).

Fig. 4 Open melt tank.

Fig. 5 Closed melt tank.

Gel Tank Preparation

The gelatin solution is now a clear liquid solution of approximately 25 to 45% concentration, available to make the hard gelatin capsule shell. After the aging process is complete, the gelatin solution is drawn into smaller individual tanks or pots, where the gelatin solution will be properly prepared to make the *correct* gelatin capsule (Fig. 7). This preparation requires addition of additives for enhancement of chemical properties, preservative properties, and color; temperature adjustment of the gelatin solution; and viscosity adjustment of the gelatin solution.

Additives.   Substances that are present in the finished capsule must be added to the liquid gelatin solution during the preparation stage. These additives normally used are listed below.

Dyes and Other Colorants.   Dyes are used to give the capsule shells distinctive colors. These individual colorants must be approved for compatibility with the fill medicaments and the particular application of the final product. In addition to soluble dyes, other colorants may be used, such as iron oxides. These are used to obtain a certain color and to satisfy local regulatory requirements. Figure 8 is an example of the acceptability status of certain colorants in different

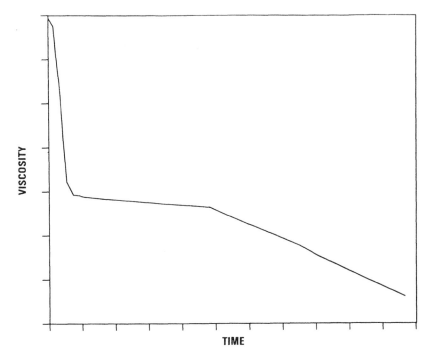

Fig. 6   Viscosity versus time chart.

countries. The need for different colors of capsules is to differentiate one product visually from another.

One of the major advantages of the gelatin capsule is this ability to distinguish products by the distinct color of the capsule. One other very interesting aspect of color is the effect on the patient. This is explained in the paper "Influence of Capsule Color Schemes on Anxiety States" by Max Luscher [4], a well-known European color psychologist. In his publication "Luscher Palette," prepared for Capsugel, he identifies certain color combinations as the ideal for certain types of medicaments. As an example:

1. *Laxatives*: olive and light brown
2. *Antiobesity agents*: yellow–dark blue
3. *Cardiovascular drugs* (stimulating): aubergine and orange
4. *Appetite stimulants*: green and orange

Opacity.   Titanium dioxide (TiO_2) is used to add opacity to the gelatin film. The amount of opacity varies to obtain the desired color shade and in some cases to conceal the medicament completely inside the capsule, as in the case of a placebo. The amount of TiO_2 is varied to arrive at the desired opacity.

Fig. 7 Liquid gelatin filling small tanks.

Coloring matters and pigments used for coloring Capsugel hard gelatin capsules																			
Coloring agent	EC No.	Origin	EC	B	D	DK	GB	SF	F	NL	I	N	A	P	S	CH	E	USA
Tartrazine	E 102	Synth.	+	+¹	+¹	+	+¹	—	+¹	+¹	+¹	—	—	+	—	+	+¹	+¹
Quinoline yellow	E 104	Synth.	+	+	+	+	+	+	+	+	+ⁿ	+	+	—	+	+	+	+ⁿ
Yellow orange S	E 110	Synth.	+	+	+	+	+	—	+	+	+	—	+	+	—	+	+	+ⁿ
Azorubin	E 122	Synth.	+	+	+	+	+	—	+	+	+ⁿ	—	—	+	—	+	+ⁿ	—
Amaranth	E 123	Synth.	+	+	+	·+	+	—	+	+ⁿ	+ⁿ	—	—	+	—	+	+	—
Ponceau 4R	E 124	Synth.	+	+	+	+	+	—	+	+	+ⁿ	—	+	+	—	+	+	—
Erythrosin	E 127	Synth.	+	+	+	+	+	+	+	+	+	+	+	+	+	+	+	+
Patent blue V	E 131	Synth.	+	+	+	+	+	+	+	+	+ⁿ	+	+ⁿ	+	+	+	+	—
Indigotin	E 132	Synth.	+	+	+	+	+	+	+	+	+	+	+	+	+	+	+	+
Brilliant black BN	E 151	Synth.	+	+	+	+	+	—	+	+	+	—	+	—	—	+	+	—
Titanium dioxide	E 171	Natural	+	+	+	+	+	+	+	+	+	+	+	+	+	+	+	+
Iron oxide, yellow	E 172	Natural	+	+	+	+	+	+	+	+	+	+	+	—	+	+	+	+*
Iron oxide, black	E 172	Natural	+	+	+	+	+	+	+	+	+	+	+	—	+	+	+	+*
Iron oxide, red	E 172	Natural	+	+	+	+	+	+	+	+	+	+	+	—	+	+	+	+*

+ authorized   — non-authorized   +ⁿ restricted authorization   Status: 31.01.1986
+¹ product must specifically declare presence of Tartazine   (without guarantee for the accuracy of the data)
* In USA, iron intake is limited to 5 mg elementary iron per day

Fig. 8 Example of colorant acceptability. (From *The Capsugel List of Colorants for Oral Drugs.*)

Preservatives. The gelatin solution is a good media for bacteria growth, especially in the temperature range (40 to 50°C) and moisture (60 to 80% water) that is used in making hard gelatin capsules. Under these conditions no approved preservative will, by itself, prohibit bacteria growth. However, several preservatives have shown some success as long as the bioburden is low. The effectiveness of preservatives is dependent on the bioburden and the preparation of the preservative. The following is a list of the preservatives known to have been used:

1. Sodium bisulfite
2. Methyl/propyl parabens
3. Potassium sorbate
4. Sodium propionate

These preservatives may be used in the manufacture of gelatin capsules.

A more ideal solution is to use no preservatives at all. This requires an extremely clean process with special attention to microcontamination issues.

Temperature Adjustment of the Gelatin Solution. The temperature of the gelatin is a very important parameter not only for its effect on the pickup of gelatin on the mold pin during the dipping process but also its effect on the physical and microbial properties of the gelatin solution. Higher temperature will speed up the gelatin degradation process. Each manufacturer weighs the advantages and disadvantages of this factor to determine the temperature for his process. The temperature of the gelatin solution in the small tanks is controlled by the means of a water jacket around the tank. The controller maintains the temperature by means of either electrical heaters or steam.

Viscosity Adjustment of the Gelatin Solution. The apparent viscosity of gelatin at a given concentration varies significantly with changes in temperature. The temperature dependency of the viscosity of gelatin solution has been shown by Croome and Clegg [5] to be represented by an equation of the type

$$\text{kinematic viscosity} = A e^{K/T}$$

where A is a function of the concentration of the gelatin, K is a constant, and T is the absolute temperature of determination.

Control of the viscosity of the gelatin is the most important parameter to be controlled. It is the value of the apparent viscosity or kinematic viscosity that controls the deposition of gelatin on the mold pin. Since the same solution (gelatin) is being measured and will be measured at a constant reference temperature, the viscosity to be controlled will be the absolute viscosity in centipoise. All viscosities are expressed in terms of the viscosity of pure water at 20°C [1]. This is approximately 1 cP. A material that would be 100 times as viscous as water at this temperature has a viscosity of 100 cP. Gelatin is usually used at a viscosity of ≈500 to 1000 cP.

Several methods are available to measure the viscosity of the gelatin solution in the gelatin feed tank:

1. *Capillary tube viscometer*. This type determines the time required for a given volume of liquid to flow through a capillary. Many capillary tube viscometers have been devised. The Ostwald and Ubbelohde viscometers are among the most frequently used [1].
2. *Rotational viscometer*. This measures the resistance to movement of a bob or spindle immersed in the sample. Several suppliers are Brookfield, Routouisco, and MacMichael [1]. The target viscosity will depend on parameters at the machine, such as dip cam profile, gel solution temperature, and pin temperature. However, the target viscosity should remain constant as long as the other parameters stay constant.

B. Dipping Process

The gelatin solution is fed from the small gelatin tanks through a tube into the dipping dish. There is one dish for the body half of the capsule and one for the cap. Each dish holds about 15 to 20 L of gelatin solution. The gelatin solution in the dipping dish is about 80% w/w water. It has a temperature of between 40 and 50°C, depending on capsule size and process. The viscosity as measured at 52°C is between 500 and 1000 cP. The temperature and viscosity of the gelatin in each hard capsule dipping dish is very closely monitored and controlled.

The dish is also arranged such that the gelatin is continuously circulated in the dish to ensure uniformity of the gelatin solution. At the temperature and concentration of the gelatin solution on the machine, there will always be some evaporation of water. To compensate for this and to control the viscosity of the gelatin precisely, the viscosity of the gelatin being fed into the dish from the gelatin feed tank must be lower than that in the dish. Some manufacturers also monitor the viscosity in the dish and automatically adjust the viscosity by adding water directly to the gelatin in the dish [6,7]. The deposition of a gelatin film on the mold pin is one of the most important properties that each manufacturer must control. The factors that influence this deposition are:

1. Temperature of the mold pin
2. Temperature of the gelatin solution
3. Viscosity of the gelatin solution
4. Rate (velocity) that the pins are withdrawn from the gelatin solution

Most of the specific information concerning this process is proprietary to each manufacturer. Several technical facts are well known. By definition, viscosity is defined as the shear stress/shear rate ratio. The amount of gelatin that is picked up by the mold pin is a product of viscosity of the gelatin solution times pin velocity. Except near the gel point ($\approx 45°C$), gelatin solutions behave as a Newtonian fluid rather than as a non-Newtonian fluid [3].

Fig. 9 Pins over dish before dipping.

The greater the velocity of the mold pin withdrawal from the gelatin, the greater will be the gel deposition on the mold pin. The dipping process starts when a group of mold bars (pin end down) is moved over the top of the dipping dish (Fig. 9). This group of mold pins is lowered into the gelatin solution in the dipping dish. As the mold pins emerge from the gelatin solution, the combination of lower temperatures/increased concentration/zero shear rate all combine to increase the effective viscosity from ≈600 to 900 cP to ≈100 times that value. This is unique to a gelling liquid like gelatin and causes the gel to "set." Before the gelatin can completely set, the mold bars are allowed to "hang" to allow some of the gelatin to "run" to the end of the pin (Fig. 10). This is done to create a thicker film at the shoulder end of the capsule. A thicker portion at this point is needed to give the capsule the strength it will need to withstand the forces that will exerted on it later in the capsule-filling machine. In between the process of dipping and the capsules being transferred to the top deck, a stream of cold air is blown over the gelatin film to set the gel further so that the film now holds its shape [8].

C. Drying Process

There is now a gelatin film on each mold pin. These pins have been transferred to the top deck of the hard capsule machine and the drying process begins. The pins

Fig. 10 Pins over dish after dipping.

travel under several drying kilns where large volumes of air are blown across the pins to dry the film. Each kiln covers a particular drying zone where the airflow, relative humidity, and temperature of the air is controlled precisely. The problem of drying wet solids has received considerable attention in the technical and scientific literature.

Highly efficient industrial dryers with sophisticated controls are common practice in food processing, textiles, and paper industries; however, technical data as they relate to drying gelatin capsules are limited. The same general concepts apply, and through theory and experimentation, the drying process can be defined. When a solid dries, two fundamental processes are involved:

1. The transfer of heat to evaporate the liquid
2. The transfer of mass as vapor and internal liquid

The equilibrium between the heat and mass transfer rates can be expressed mathematically as [9]

$$\frac{dw}{d\Theta} = \frac{h_t A t}{H} = kAP$$

or

rate of drying = rate of heat transfer = rate of mass transfer

where

$\dfrac{dw}{d\Theta}$ = drying rate, pounds of water/hour

h_t = total heat transfer coefficient

A = area of heat transfer and evaporation

H = enthalpy of evaporation at t_s

$t = t_a - t_s$ = temperature difference between air and surface of evaporation

t_a = air temperature

t_s = temperature of surface evaporation

k = mass transfer coefficient

$P = P_s - P_a$ = vapor pressure difference

P_s = vapor pressure of water at t_s

P_a = partial pressure of water vapor in air

The heat and mass transfer process occur simultaneously and the drying rate is controlled by the slower of the two.

The factors governing the rate of drying are either internal or external. For the gelatin capsule, the internal factor is the water diffusion through the film of the capsule to the outside surface. The external factors are the dry-bulb temperature of the air, the relative humidity of the air, and the air quantity and flow pattern. Figure 11 shows a typical drying-time curve for hard gelatin capsules (or any wet solid for that matter).

While Fig. 11 shows that the moisture content varies continuously with time, a more precise illustration of the nature of this variation can be obtained by differentiating the curve ($dw/d\Theta$) and plotting the drying rate versus percent moisture of the capsule. Figure 12 shows that the drying process is not a smooth, continuous one; it exhibits three distinct drying rates. The a-b portion is a horizontal straight line indicating a constant rate of drying. During this constant-rate period, the drying mechanism is that of vaporization from a free liquid or a very wet solid. Since the surface of the film remains saturated with water during this phase, this process is similar to the evaporation of an open pan of water exposed to air having less than 100% RH. Streeter [10] describes the process as follows: "Liquids evaporate because of molecules escaping from the liquid surface. The vapor molecules exert a partial pressure in the space known as vapor pressure. If the space above the liquid is confined, after a sufficient time, the number of vapor molecules striking the liquid surface and condensing are just equal to the number escaping in any interval of time and equilibrium exists." That is why it is necessary continuously to provide a fresh supply of dry air (less

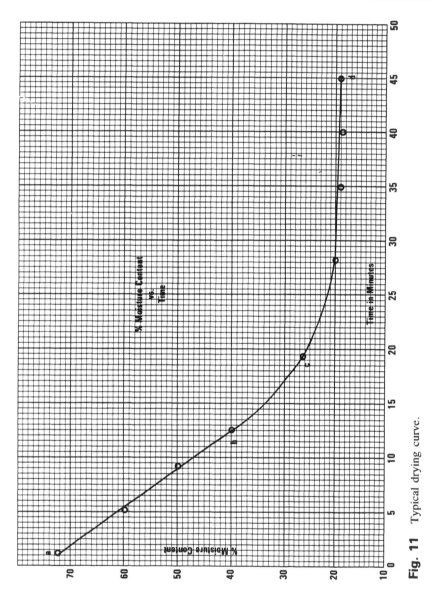

Fig. 11   Typical drying curve.

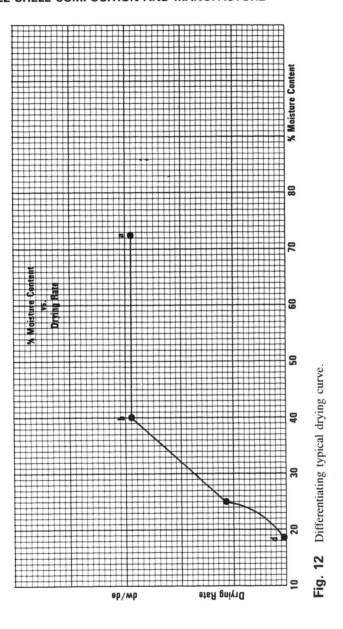

Fig. 12 Differentiating typical drying curve.

than 100% RH) around the capsule. In this way, the vapor pressure of water at the film surface is greater than the partial pressure of water vapor in the air, and the water tends to evaporate and migrate as water vapor into the air immediately around the capsule. This migration will be restricted if the air is stagnant around the film. High velocities of air flowing past the film tend to reduce the resistance to vapor diffusion; therefore, the rate of vaporization increases with air velocity. W. H. Carrrier demonstrated that the rate of drying during this constant rate phase can be expressed as [9]

$$\frac{dw}{d\Theta} = (a + bv)(e^{l} - e)$$

where a is the rate of drying in still air, b the rate of increase with velocity, v the velocity of air, e^{l} the vapor pressure of the liquid, and e the vapor pressure in the atmosphere.

The vapor pressure of the liquid (water at the film surface) depends on the temperature of the film. Two factors will affect this temperature. First, during the constant-rate phase, the gelatin film will actually decrease in temperature because as the liquid evaporates from its surface, the heat absorbed in the evaporation process will cool the film, called evaporative film cooling. The second factor is the temperature of the air. In the constant-rate phase, this is the one factor that will raise the temperature of the gelatin film. The last part of the equation is e, the vapor pressure in the atmosphere. This is usually expressed as relative humidity.

By definition, the drying rate $(dw/d\Theta)$ in the constant rate phase can be increased by:

1. Increase in air velocity (bv)
2. Increase in air dry-bulb temperature (effect on e^{l})
3. Decrease in the relative humidity of the air (effect on e)

The second phase of drying is the falling-rate phase. In Fig. 12 this is at point b. At this point, the film of the gelatin is no longer like a very wet solid. Point b is termed the critical point. It is at this point that the drying rate is influenced by other factors. This phase from b-c is called the phase of unsaturated surface drying. It becomes more difficult for the moisture to diffuse through the gelatin film. The diffusion rate then becomes the controlling factor, which is dependent on the following factors:

1. Nature of the gelatin
2. Temperature of the gelatin
3. Thickness of the gelatin film

The last phase, c-d, is a continuation of the falling-rate phase but is now near a point where the moisture content of the gelatin approaches equilibrium with the air around the capsule. Figure 13 shows the relationship between the relative

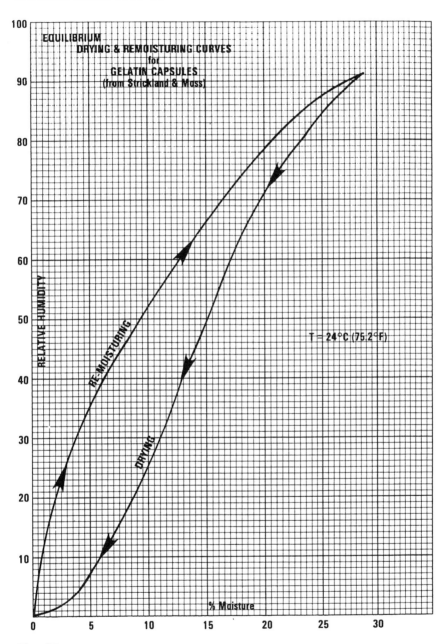

Fig. 13 Equilibrium moisture curve.

humidity of the air and the equilibrium moisture content of gelatin. This shows that at $\approx 50\%$ RH, the capsule will settle out at about 15% moisture content.

There are several important considerations to be taken into account for the drying process. First, the production rate of the machine dictates that the drying process be accomplished in a certain amount of time. Second, the critical point between the first and second phases must be understood and determined. At this point it is very critical that the external drying forces *do not* overcome the internal drying forces. If they do, the outside film of the capsule will overdry, actually creating a barrier for the migration of moisture through the film and will retard the drying rate. Finally, there are five parameters that are carefully controlled and manipulated to control the overall drying process:

1. Machine speed or production rate
2. Amount of conditioned air at the machine
3. Airflow across the capsule during each phase of drying
4. Air temperature across the capsule during each phase of drying
5. Size of capsule to be dried or total drying load

D. Automatic Section

The capsule has gone from an 80% w/w water film on a mold pin to an 18% w/w water capsule. The cap is still on its mold pin and the body is on its pin. The

Fig. 14 Bars feeding into automatic section.

gelatin film formed on the pin is longer than required. This allows the capsule to be stripped off the pin and trimmed so that a good, clean edge will exist on the open end of the capsule.

At the table section, a pair of mold pin bars are flipped over to 90° angle and fed into the automatic section (Fig. 14). A stripper bar with 22 or 30 metal, clothespin-shaped strippers open and go behind the capsule film (Fig. 15). The stripper bar then closes around the pin and pulls the capsule film off the pin and into a multifluted collet (Fig. 16). The collet then closes around the capsule parts and begins to rotate. As this collet, with capsule parts, begins to rotate, a razor sharp and hardened knife is brought into contact with the capsule, trimming the capsule part to the required length (Fig. 17). This leaves the open end of the capsule with a good clean edge. Trimmings are taken away, usually to be recycled.

The collets with the trimmed capsule part carries the capsule parts to a joiner block. This is where the two capsule parts, body and cap, now come together with the open cap-half sliding over the open body half to just *the* correct prejoined length (Fig. 18). The joined capsule is then transferred away from the hard capsule machine for further processing.

To complete the cycle for the mold pin bars, they are passed through the next section of the machine where the pins are cleaned with felt pads (Fig. 19).

Fig. 15 Strippers behind capsule on pin.

Fig. 16 Stripper pulling capsule into collet.

Fig. 17 Knives cutting capsule to proper length.

Fig. 18 Two halves of capsule at joiner block.

Fig. 19 Felt pads to clean pins.

At the same time, a small amount of release agent is applied to the mold pin to prevent the gelatin film from adhering too strongly to the mold pin and to allow the gelatin film to slide easily over the surface of the metal mold pin. The capsules are now taken away by conveyor from the hard capsule manufacturing machine, put into containers, and then processed for the next stage of the process.

E. Process Control

In the early days of hard capsule manufacturing, process control usually meant compliance with certain regulatory requirements. It was a matter left up to the quality control department. In today's environment, however, that system cannot survive. With the introduction of high-speed filling machines and the never-ending need for capsule manufacturers and their customers to become more efficient, it requires a higher quality and more consistently performing capsule. This requires that process control be practiced from the time the raw gelatin is supplied to the capsule manufacturer until the finished capsule arrives at the customer. Table 4 shows most of the traditional steps in this process control.

Gelatin is derived from a natural product and thus does not always exhibit the same characteristics from lot to lot. Even when the gelatin passes required tests, it is not always a good indicator of the actual gelatin solution behavior in areas such as setting and drying characteristics. With assurances that the raw gelatin will perform satisfactorily, the gelatin will be melted next. At this stage the variables monitored and controlled are:

1. Gelatin temperature during melt
2. Vacuum during melt
3. Gelatin solution concentration

These processes must be controlled very precisely in order to end up with a gelatin solution with a consistent concentration and have that gelatin with the same viscosity each time. The use of microprocessors is now common in the industry to ensure this consistency. The next stage is the preparation of the gelatin solution for the body and cap feed tanks or pots. There are two primary concerns here: viscosity and color. Again, viscosity is the major factor in controlling gelatin deposition in the dipping process. Two methods have been used to control this. The first method was to measure the viscosity indirectly by measuring the density of the solution with a Baumé hydrometer [11]. The other method is by the use of a standard glass U-tube or by many different types of electromechanical rotational viscometers. This last method has gained popularity in recent years because of its ability to feed information to a microprocessor. The operator no longer has to do a manual interpretation of the data. Instead, the computer will interpret the data and make the necessary adjustments automat-

Table 4 Steps in Capsule Manufacture and Quality Controls

Production step	Corresponding quality control
Preparation of raw materials:	
Gelatin	Appearance
	Odor
	Color
	Grain size
	Solubility
	Gelling (bloom)
	Viscosity
	pH value
	Isoelectric point
	Bacteriology
	Chemical purity
Water	Electrolyte content
	pH value
	Bacteriology
Coloring matters and	Identity
pigments	Solubility
	Bacteriology
	Chemical purity
Gelatin solution	Viscosity
	Temperature
	Color shade
	Color composition
Production machine	Temperature
Dipping	Relative humidity
Rotation	Viscosity
Gelling	Dimensions
Drying	Color shade
Stripping	
Cutting	
Joining	
Ejection	
Inspection	Optical check
	Defects
Imprinting	Laboratory checks on ink
	Statistical control
Counting, packaging	Statistical end control (AQL)
	Laboratory check

ically. Another advantage of this is the ability to gather data automatically and analyze it for further process improvement.

The other important parameter to control in this gelatin feed tank is the color. One of the distinct advantages of the capsule is the ability to differentiate a product by its own distinct color. Because of this desire to be somewhat different, it is not uncommon to have over 1000 different colors of capsules. Control of the colorant strengths being added to the gelatin feed tanks must be controlled by either a spectrophotometer or other appropriate means.

A final color check is made of the gelatin solution itself. A manual method is to make a dry film of gelatin and compare it visually to a standard dry film of gelatin. The solution to maintaining color shade consistency is to establish a method to measure color automatically and convert that color into numbers by use of a colorimeter (Fig. 20). Once a numerical standard is established, comparisons against it can be made without fear of drift. The use of numerical control of color in textiles, paints, plastics, and dyes enables many firms throughout the world to assure uniformity of color in production. The use of this type of instrumentation is an essential process control technique for good color control.

It is in the next stages of the operation, the capsule-forming machine, that the more dramatic changes in process control have taken place. The age-old

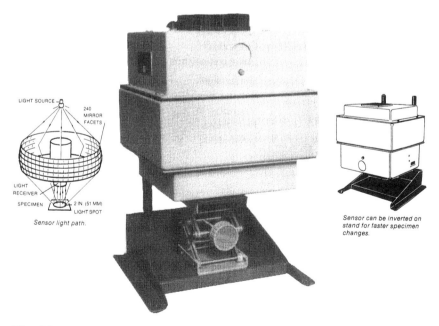

Fig. 20   Hunter colorimeter.

method of quality control inspecting finished capsules alone to determine if the process is operating successful will no longer product the quality of capsule needed in today's market. Previously, the details of how to make capsules were discussed along with a discussion of the important parameters that affect the quality of the capsule. It is important, then, that all these parameters be monitored and controlled:

1. Gelatin temperature and viscosity at the point where the capsule is formed
2. Amount and temperature of air at different points in the drying cycle
3. Relative humidity of the drying air
4. Mechanical action of the dipping process itself

These process conditions are not only monitored but also plotted on a control chart (manually or automatically) so that opportunities for making an improvement in the process can be recognized. As the joined capsule comes off the machine, the physical dimensions of the capsule itself are measured and plotted both to assure that the process is in control and to look for opportunities for improvement. The dimensions most commonly measured are:

1. Capsule parts dome thickness.
2. Capsule wall thickness, either as the double-wall thickness at the cut edge of the parts or the single-wall thickness at a standard distance from the cut edge of the capsule.
3. Cut length of both body and cap parts.

A statistical sampling system of the finished capsule is used to determine the quality characteristics. The classification and definition of common visual defects are listed below along with a description of these defects.

F. Capsule Defects [12]

Capsule defects are generally defined into critical, major, and minor classifications. *Critical defects* are defined as those that cause loss of container properties of the capsule, or cause a low dosage weight of the capsule, or cause a major filling-machine stoppage or production operation delay. *Major defects* are defined as those that *may* cause problems in the filling operation, such as nonopening, failure to rectify, or incorrect closure. *Minor defects* are defined as those defects that do not affect the filling operation but detract from the visual or cosmetic appearance of the product. A description of these defects is shown in Fig. 21.

V. PRINTING

Hard gelatin capsules can be printed with a variety of logos. Information such as product identification code number, company logo, dosage size, or even direc-

CAPSUGEL

Classification of visual defects

	Defect	Reason for classification
Critical defects	Oil hole	Powder leaker
	Scrape hole	Powder leaker
	Uncut	Low dosage risk
	Cracked	Powder leaker
	Short body*	Low dosage
	Double dip	Bushing blockage
	Telescoped	Rectification channel blockage and bushing blockage
	Mashed	Rectification channel blockage
	Trims	Rectification channel and bushing blockage
	Splits*	Powder leaker
Major defects	Loose pieces	Non-rectification
	Closed capsule	Non-separation
	Bad join	Non-separation
	Double caps	Non-rectification
	Rough cuts (major)*	Non-separation
	Collet pinches (major)*	Non-separation
	Punched ends (major)*	Non-separation/non-rectification
	Thin spots	Potential powder leaker
	Short cap*	Potential risk in joining
	Long body*	Punched end during closure
	Long cap*	Punched end during closure
Minor defects	Splits (minor)*	Visual appearance defects with no risk of defects during filling operation
	Rough cuts (minor)*	
	Collet pinches (minor)*	
	Punched ends (minor)*	
	Bubbles*	
	Wrinkles*	
	Star ends*	
	Specks*	
	Scrapes	
	Dirt marks*	
	Strings*	

* detailed criterion specified separately

Fig. 21   Capsule defect definition.

Defect	Dimensional limit	Criteria for classification
Short body		Length of body is 2 mm shorter than specified length. (For body the SNAP-FIT™ ring is absent).
Split		A split in the cap reaching the SNAP-FIT™ groove is defined as a critical defect. Smaller splits are defined as minor defects. Splits on capsule body greater than 1 mm are critical defects for all sizes.
Rough cut		Irregular cuts which result in abnormal separation resistance are defined as major defects. Rough cuts above 1 mm and not causing separation problems are classified as minor defects.
Collet pinches		Major collet pinches are classified based on abnormal separation resistance. Longitudinal pinch marks greater than 4.0 mm in length for sizes 00, 0, 1 and greater than 3.0 mm in length for sizes 2, 3, 4 are classified as minor defects.
Punched ends		Major punched ends are classified based on abnormal separation resistance or difficult rectification. Indentations greater than 3.0 mm diameter for sizes 000, 00, 0 and greater than 2.0 mm for sizes 1, 2, 3, 4 are defined as minor defects.
Short cap		Length of cap is 1 mm shorter than specified length.
Long body or cap		Length of cap or body is 1 mm greater than specified length.

Defect	Dimensional limit	Criteria for classification
Wrinkles		Longitudinal wrinkles greater than 3 x 2 mm for sizes 000-4 are classified as minor defects.
Star ends		Top stellated wrinkles greater than 4.0 mm diameter for sizes 000, 00, 0, 1 and 3.0 mm for sizes 2, 3, 4 are classified as minor defects.
Specks		Specks of 1 mm or more in size are judged as minor defects.
Dirt		Marks which can be recognized at a distance of 30 cm under natural light with an area of 3.0 mm² for sizes 000-4 are judged as minor defects.
Strings		Strings at the cutting edge of 3 mm or more for sizes 000, 00, 0, 1 and 2 mm or more for sizes 2, 3, 4 are judged as minor defects. Large strings which will block the rectification channels are judged as critical defects.
Bubbles		Bubbles greater than 1.0 mm are defined as minor defects. Three or more bubbles in a capsule greater than 0.5 mm are also defined as minor defects.

tions on how to use the product can be printed on the capsule. The capsule may be printed along the length of the capsule (axially) or around the capsule (radially) (Fig. 22). There is also the possibility to print a different color on the body part and on the cap part. With all these possibilities of print along with all the different options available in capsule color combinations, any capsule product can readily be distinguished from another capsule product.

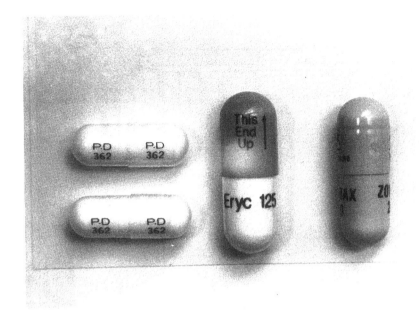

Fig. 22 Types of printed logo.

A. Machines

The rotary offset gravure method is the method of choice to print capsules (Fig. 23). The desired log is "etched" at set intervals around the circumference of a metal gravure cylinder or roll. This roll is usually made of stainless steel or of brass that has been chromium plated. The cylinder rotates in an ink dish or pan and picks up ink in the etched cavities as it rotates. A metal doctor blade scrapes the excess ink from the cylinder face, leaving ink only in the etched cavities. As the cylinder continues to rotate, the gravure roll comes into contact with a rubber roll called a transfer roll (Fig. 24). The ink is transferred from the etched cavities on the gravure roll to the surface of the transfer roll. As the capsules pass under the transfer roll, the ink is transferred to the capsule in the form of the logo.

This printing process is usually performed by the capsule manufacturer, although a few companies do print their own capsules. The majority of the commercially available print machines are made by three companies in the United States: Ackley, Hartnett, and Markem. Although the offset gravure roll process is used by all three manufacturers, the method of handling the capsules is somewhat different for each.

Hartnett produces three basic models: the Delta, the Model A&B Axial, and the Model-C radial printer. The Delta (Fig. 25) uses buckets or inserts

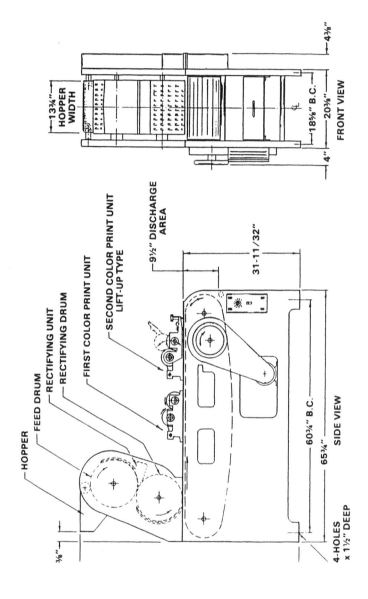

CAPSULE PRINTER
RADIAL AND/OR LINEAR TWO
COLOR PRINTER

Fig. 23 Typical print machine.

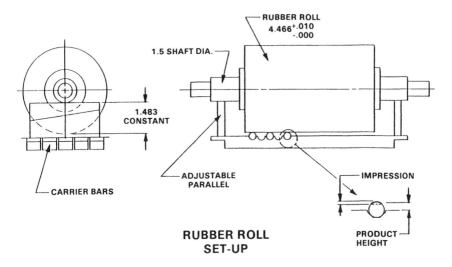

Fig. 24   Transfer roll setup.

attached to a chain drive to carry the capsules to the printing head. It can print on both sides of the capsule (body and cap) with different colored ink on each side, if desired. The capsules can be axially or radially printed; however, the capsule cannot be rectified. This eliminates the possibility of having the logo always placed on a set part, cap or body. The production rate is 75,000 to 250,000 capsules per hour.

The Model-A and Model-B (Fig. 26) print axially and have higher production rates than the Delta. The Model-A and Model-B feed capsules through a hopper into a rotating drum. This drum then transfers the capsules to a moving belt that has pockets (inserts) for each capsule. These capsules are then transported to the print head, where the capsules are printed (Fig. 27). The only difference between the Model-A and Model-B is the number of rows. The Model-A will print about 200,000 per hour and the Model-B will print about 400,000 per hour.

The Model-C is used for radial printing and it can orient the capsule prior to the printing head. A bulk hopper feeds the capsules to a feed drum with two directional pockets, which along with the guides, gaging block, air jets, and vacuum slots cause the capsules to be oriented with the cap end always to the right. The capsules are then transferred to the conveyor carrier bars, which take the capsules to the printing head. The logo is transferred around the capsule by the rubber roll, spinning the capsule in the insert as it goes by. To print two colors, one color for body and one for the cap part, a duo-tone print unit can be

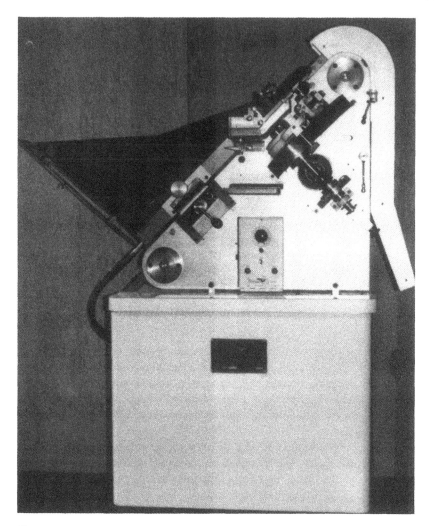

Fig. 25 Hartnett Delta model.

added to the print machine. The machine prints 10 rows (capsules) simultaneously with a capacity of 320,000 per hour.

Ackley makes three basic machines. The multitrack axial machine is very similar in principle to the Model-A described above, the major obvious difference being that it prints more rows—capsules—simultaneously (up to 18) with a capacity of up to 1,000,000 capsules per hour. Ackley also makes a 10-track rectified radial printer (Fig. 28). This unit uses a feed hopper to load the capsules into a feed drum. The capsules are then fed through the rectifying unit as they are

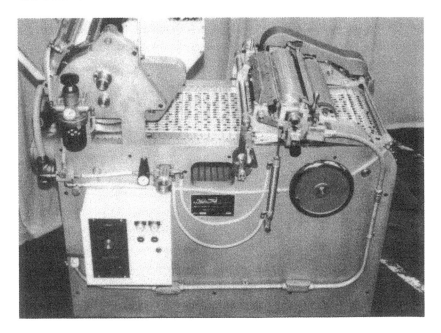

Fig. 26 Hartnett Model-B print machine.

Fig. 27 Print head.

Fig. 28   Ackley two-color spin print machine.

being transferred to the rectifying drum. The capsules are then fed to specially designed carrier bar inserts body part first and are turned in these inserts as the capsules settle down. The capsules are then transferred to the print head. Two color printing is again possible by installing a second print head on the print machine.

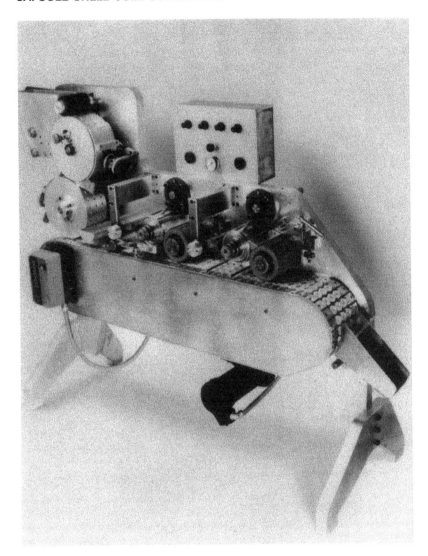

Fig. 29 Ackley three-row print machine.

Ackley also produces a three-row print unit that operates similar to its radial printer (Fig. 29). It can be used to print either axial or rectified radial print. Its capacity goes from 120,000 per hour for rectified radial printing up to 300,000 per hour for axial printing.

Markem makes three print machines. The 156AMII is a back-fed unit using a metal disk to carry the capsules to the print head (Fig. 30). It is designed

Fig. 30 Markem 156AMII print machine.

Fig. 31 Markem 156AMIII print machine.

394

to print either one or both sides of the capsule at the same time. It is a single-track machine with a capacity of 60,000 to 100,000 capsules per hour. The 156AMK III is also a compact benchtop bowl-feed unit (Fig. 31). The major operational difference is that this unit uses two tracks, which extend the capacity of 75,000 to 200,000 capsules per hour.

The third Markem machine, the 280A (Fig. 32), is designed for higher speed, from 200,000 to 350,000 per hour. It uses a revolving cylinder with slots cut to accommodate the capsule. The capsules are fed in the middle of this cylinder, and by centrifugal force the capsules fill the slots. The capsules are then transferred to the print head, where they are printed on either one or both sides at the same time.

B. Printing Process Control

The quality of the printed capsule is very dependent on two factors: (1) the type and color of printing ink and (2) the logo design.

Printing Ink

All printing inks have three basic ingredients: (1) colorant, (2) film formers, and (3) solvents.

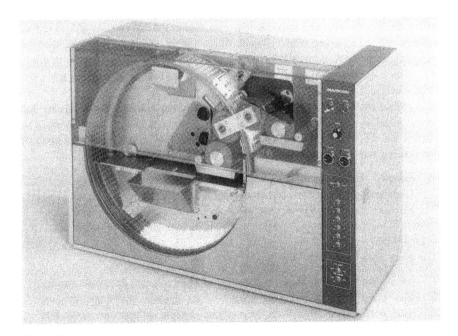

Fig. 32 Markem 280A print machine.

Colorant.   There are two types of colorants:

1. *Pigments:* insoluble substances such as carbon, iron oxides, and titanium dioxide
2. *Lake dyes:* dyes that have been absorbed to a substrate of alumina or titanium dioxide

These colorants may be used alone or in combination and make it possible to manufacture practically any color of ink. However, since the logo is small in comparison to the capsule itself, the *capsule color* is varied to catch the eye of the consumer, and the printing ink stays with the basic colors, the predominant ink color being black.

Film Formers.   The most commonly used film formers are shellac and modified shellac. Shellac is a resinous substance produced by a scale insect, *Laccifer lacca* Keri (Coccidae).

Solvents.   The choice of solvent is dependent on the particular viscosity and drying characteristics required of the ink. The volatility or drying rate of the ink is most important. This is because the ink must dry on the capsule before the capsule is ejected from the machine. If it does not, some of the ink can be transferred to other capsules and cause a defect called ink specks. The actual time for this to occur is usually less than 2 s.

On the other hand, if the ink dries too fast, it will dry on either the gravure roll or the transfer roll and will not be transferred to the capsule, creating a defect called print interruption. The viscosity of the ink is important to ensure that the etchings on the engravure roll are filled and to ensure that when the solvent dries, the correct amount of colorant and shellac will be left on the capsule to leave the impression desired.

Since the ink is kept in an open reservoir, there is a loss of the solvent due to evaporation. For this reason, an addition of solvent to the dish is needed to correct for this evaporation. The viscosity/drying rate relationship is controlled by varying the amounts of solvent and type of solvent:

1. *Ethanol:* fastest-drying alcohol
2. *Isopropyl alcohol:* fast-drying alcohol
3. *Butanol alcohol:* slow-drying alcohol

Logo Design

The print styles available are:

1. *Axial print.* Print is down the length of the capsule (42°-wide surface).
2. *Opti print.* Print is down the length of the capsule (84°-wide surface).
3. *Spin print.* Print around the axis of the capsule or approximately 360°.

Axial print (Fig. 33) is the most popular style. The capsules are not rectified prior to printing and are randomly oriented in the print machine. If the logo is to be

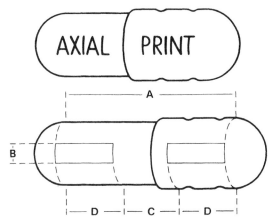

A = Overall length of printable area
B = Maximum height of imprint
C = Distance between imprints on cap and body
D = Maximum length of imprint on cap and body

Capsule size	A mm	B mm	C mm	D mm
000	17.40	3.40	5.10	5.70
00	16.40	3.00	4.60	5.90
0 el*	18.00	2.80	4.60	6.70
0	15.50	2.80	4.60	5.50
1	13.80	2.50	4.60	4.60
2	12.70	2.30	4.40	4.10
3	11.50	2.10	4.00	3.70
4	10.40	1.90	4.00	3.20
5	7.60	1.70	1.90	2.50

*0 el = 0 elongated: size 0 with elongated body

Fig. 33 Axial print parameters.

seen on both the body and cap, the color of the print ink must be chosen so that it will be distinctive from the capsule color. Opti print (Fig. 34) is the same as the axial print except that a different transfer roll is used. This roll produces the 84°-wide imprint. Spin print (radial print) in theory allows the printing to go around the capsule 360° (Fig. 35). The capsule can also be rectified such that a distinct logo can be printed on the cap and another distinct logo on the body. The two-color print machines allow for one color ink to be used for the cap and a different color for the body.

With the information above, the decision can now be made to the logo. Once the type of logo has been decided upon, artwork can be developed. Acid is

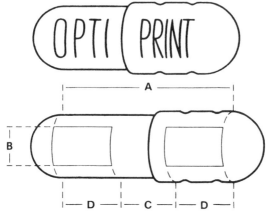

A = Overall length of printable area
B = Maximum height of imprint
C = Distance between imprints on cap and body
D = Maximum length of imprint on cap and body

Capsule size	A mm	B mm	C mm	D mm
000	17.40	6.80	5.10	5.70
00	16.40	5.90	4.60	5.90
0 el*	18.00	5.30	4.60	6.70
0	15.50	5.30	4.60	5.50
1	13.80	4.80	4.60	4.60
2	12.70	4.40	4.40	4.15
3	11.50	4.00	4.00	3.70
4	10.40	3.60	4.00	3.20
5	7.60	3.30	1.90	2.50

*0 el = 0 elongated: size 0 with elongated body

Fig. 34   Opti print parameters.

then used to chemically etch the logo on the engravure roll. For larger lines on these logos, a screened etch is used. This reduces the amount of ink picked up by the engravure roll and reduces the chance of ink specks and smears. All these factors must be considered from the time the logo is developed until the capsule is printed.

C. Print Defects

Common print defects are described in Fig. 36.

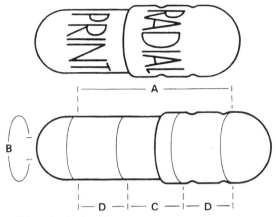

A = Overall length of printable area
B = Maximum height of imprint
C = Distance between imprints on cap and body
D = Maximum length of imprint on cap and body

Capsule size	A mm	B mm	C mm	D mm
000	16.00	21.00	4.00	6.00
00	14.00	19.00	4.00	5.00
0 el*	14.00	17.00	5.00	4.50
0	12.00	17.00	4.00	4.00
1	10.00	15.50	4.00	3.00
2	9.30	14.20	4.00	2.70
3	8.50	13.00	4.00	2.30
4	8.00	11.80	4.00	2.00
5	7.00	9.00	3.00	2.00

*0 el = 0 elongated: size 0 with elongated body

Fig. 35 Radial print parameters.

VI. MOLD PINS

The mold pin is the single-dimensional constant around which all the other physical attributes of the capsule can be determined. The design of this stainless steel mold pin must take into account the following:

1. The mold pins are slightly tapered from the open end of the capsule toward the closed end of the capsule. The larger diameter is at the base of the pin, which corresponds to the open end of the capsule. This taper is part of the patent specification for each manufacturer and is generally on the order of 0.1 to 0.3

Classification of print defects

Defect	Defect example		Criteria for classification
Unprinted			**Major defect:** capsules with complete absence of the logo.
Multiple print	Major	Major 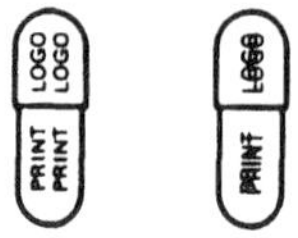	**Major defect:** capsules with print in multiple positions, resulting in illegibility of the print logo.
Print interruption	Major	Minor 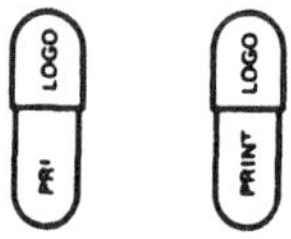	**Major defect:** capsules with at least half of the lettering missing, resulting in non-identification. **Minor defect:** capsules with at least $\frac{1}{2}$ of one letter missing but where the logo remains legible.
Print smeared	Major	Minor 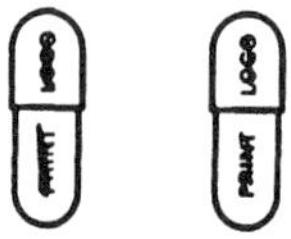	**Major defect:** capsules with extensive smearing, resulting in illegibility of the logo. **Minor defect:** capsules with smearing, resulting in partial illegibility of the logo.
Print off-register	Major	Minor 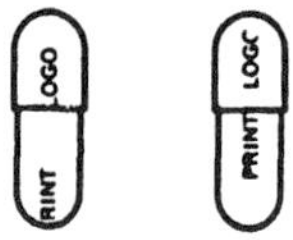	**Major defect:** capsules with letters missing, resulting in a non-identification of the logo. **Minor defect:** capsules with at least $\frac{1}{2}$ of one letter missing but where the legibility of the logo remains intact.
Print off-set	Major	Minor 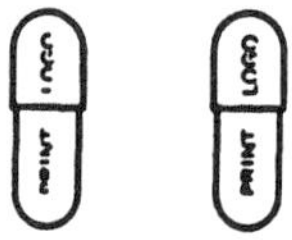	**Major defect:** capsules with at least $\frac{1}{2}$ the logo letter height missing due to logo misplacement. **Minor defect:** capsules with an incomplete logo letter height but where the logo remains legible.
Ink specks	Major	Minor 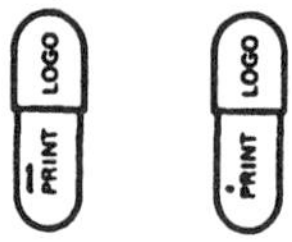	**Major defect:** capsules with ink specks 5.0 mm or greater for sizes 00–2 and 4.0 mm or greater for sizes 3 and 4. **Minor defect:** capsules with ink specks 1.0 mm or greater for sizes 00–4.
Print line		Minor 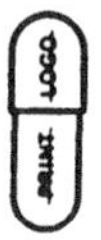	**Minor defect:** capsules with line through the printing logo. Legibility is not affected.

Fig. 36 Definition of print defects. (From Ref. 13.)

400

mm/cm length of pin. This taper is to ensure that the dry capsule halves can be stripped from the mold pin. If there were no taper, a vacuum would be created when the capsule is pulled off the pin which could cause the capsule wall to collapse.

2.   The diameter of the mold pins must be designed such that when a certain thickness of gelatin film is applied to the mold pin, the resulting diameters of the body and cap halves of the capsule will allow the open end of the cap to slide over the open end of the body. This clearance is generally in the range of 0.05 to 0.08 mm.

3.   The diameter of the mold pin will correspond to standard capsule sizes 000 through 5, the size 000 being the largest capsule and size 5 the smallest. The dimensions of the capsules are essentially the same for each size regardless of manufacturer, the most critical being that the volume of the capsule match a standard. Table 5 shows the standard volume plus the effect of powder density on the final capsule weight capacity. These figures apply to all major suppliers of capsules. The dimensions must be controlled so that the capsules will perform on a capsule-filling machine without adjustments necessary for each supplier's capsule. The dimensions that are most critical are body and cap diameters, body and cap individual lengths, and the overall filled joined lengths. These dimensions vary only slightly among suppliers. Dimensions of body and cap for the three suppliers are given in Table 6.

Table 5   Average Capsule Volume and Filled-Capsule Weight Capacity for Typical Powder Dose Densities[a]

Capsule size	Capsule volume (mL of water)	Capsule weight capacity (mg) for powder dose density:			
		$0.6 \ g/cm^3$	$0.8 \ g/cm^3$	$1.0 \ g/cm^3$	$1.2 \ g/cm^3$
000	1.37	822	1096	1370	1644
00	0.95	570	760	950	1140
0 el	0.78	468	624	780	936
0	0.68	408	544	680	816
1	0.50	300	400	500	600
2	0.37	222	296	370	444
3	0.30	180	240	300	360
4 el	0.25	153	204	255	306
4	0.21	126	168	210	252
5	0.13	78	104	130	156

[a]Data are to be considered as approximations since filling product variables and filling machine type can significantly affect the capsule weight capacity.

Table 6 Capsule Body and Cap Dimensions

a. Capsugel

Size	Individual diameter (mm)		Individual length (mm)		Filled joined length (mm)
	Body	Cap	Body	Cap	
000	9.60	10.00	22.20	12.95	25.75
00	8.23	8.60	20.22	11.74	23.30
0 el	7.38	7.70	20.19	11.68	22.70
0	7.38	7.70	18.44	10.72	21.20
1	6.68	6.98	16.61	9.78	19.00
2	6.13	6.41	15.27	8.94	17.50
3	5.61	5.88	13.59	8.08	15.50
4 el	5.11	5.38	13.69	7.84	15.46
4	5.11	5.38	12.19	7.21	13.90
5	4.70	4.89	9.32	5.54	11.00

b. Pharmaphil

Size	Individual diameter (mm)		Individual length (mm)		Filled joined length (mm)
	Body	Cap	Body	Cap	
00	8.16	8.51	20.19	12.00	23.70
0 el	7.34	7.64	20.09	11.73	23.50
0	7.34	7.64	18.69	11.04	22.10
1	6.65	6.91	16.55	9.82	19.60
2	6.09	6.35	15.29	9.03	18.00
3	5.58	5.84	13.66	8.12	16.20
4	5.08	5.33	12.39	7.36	14.70

c. Elanco

Size	Individual diameter (mm)		Individual length (mm)		Filled joined length (mm)
	Body	Cap	Body	Cap	
00	8.16	8.51	20.20	11.40	22.90
0	7.33	8.63	18.50	10.90	21.80
1	6.62	6.90	16.50	9.90	19.50
2	6.07	6.35	15.10	8.90	17.80
3	5.56	5.82	13.50	7.90	15.90
4	5.06	5.32	12.30	7.20	14.50

REFERENCES

1. Capsules, p. 699; Gelatin, p. 217; Viscosity, pp. 667–668; in *United States Pharmacopeia XIX*, United States Pharmacopeial Convention, Rockville, Md., 1975.
2. H. J. Oglevee, *Hard Gelatin Capsules Manufacturing Manual (II)*, 1976, p. 3.
3. A. W. Rahn, *Viscosity Characterization of Gelatin Solutions*, 1972, pp. 1, 5.
4. M. Luscher, *Influence of Capsule Color Schemes on Anxiety States, Luscher Palette*, 1984, pp. 3–5.
5. R. J. Croome and F. G. Clegg, The physical properties of gelatin, *Photogr. Gelatin*, p. 35 (1965).
6. H. J. Oglevee and B. R. Clement (to Parke Davis & Company), U.S. Patent 3,632,700 (1972).
7. G. W. Martyn, Jr., The people–computer interface in a capsule molding operation, *Drug Dev. Commun.*, p. 39 (1974–75).
8. L. L. Haring, Hard shell capsule manufacture, *Capsule Manufacturing Technology Meeting*, December 1985.
9. M. A. Elshazky, *Drying of Gelatin Capsules*, 1970. p. 4.
10. F. V. Streeter, Fluid properties and definitions, in *Fluid Mechanics*, 5th ed., 1971, p. 16.
11. W. G. Norris, *Manuf. Chem. 32*:249 1961.
12. *Capsugel Hard Gelatin Capsules, Quality Assurance System—Visual Defects*, BAS-1141-E, May 1985.
13. *Capsugel Hard Gelatin Capsules, Quality Assurance System—Print Defects*, BAS-1142-E, May 1984.

15

Packaging of Multiparticulate Dosage Forms
Materials and Equipment

K. S. Murthy

Parke-Davis Pharmaceutical Research
Warner-Lambert Company
Morris Plains, New Jersey

Franz Reiterer and Jürgen Wendt

Teich AG
Obergrafendorf-Mühlhofen
Austria

I. INTRODUCTION AND SCOPE

Multiparticulate dosage forms such as pellets, beads, and coated or uncoated granules are popular within the pharmaceutical industry in the preparation of sustained-release (SR), controlled-release (CR), or extended-release (ER) and in some instances, immediate-release solid oral dosage forms. Spherical particles have the lowest surface-to-volume ratio and hence require the lowest quantity of coating material to achieve a target release profile. A low effective surface area is also exposed during the dissolution process. Pellets and beads are thus well suited to provide the slow release desired for oral SR or ER products since different types of beads with different release rates can be combined into a single dosage unit to obtain a desired release profile [1,2]. Examples of multiparticulate dosage forms currently marketed in the United States are listed in Table 1. These products are typically composed of pellets filled into hard shell gelatin capsules or compressed into tablets or less commonly as sachets. Sachet as a dosage form is not common in the United States but is more widely available in Europe for administering pharmaceutical powders or granules [4,5]. By far the most common delivery system for presenting pellets and spherical granules is the hard

Table 1 Examples of Pellet Products Currently Marketed in the United States

Product	Active ingredient	Marketing company
Capsules		
Actifed 12-hour	Triprolidine HCl, pseudoephedrine HCl	Burroughs Wellcome
Cotazym-S	Pancrealipase, USP	Organon
Compazine Spansule	Prochlorperazine	SmithKline Beecham
Cardizem CD	Diltiazem	Marion Merrell Dow
Contact Allergy Timed Release	Phenylpropanolamine HCl, chlorpheneramine maleate	SmithKline Beecham
Dexedrine Spansule	Dextroamphetamine sulfate	SmithKline Beecham
Dilacor-XR	Diltiazem	Rhône-Poulenc Rorer
Doryx	Doxycyline hyclate	Parke-Davis
Dilatrate-SR	Isosorbide dinitrate	Reed & Carnrick
Entolase HP	Pancrealipase, USP	A.H. Robbins
Erythromycin Delayed Release	Erythromycin	Abbott
Eryc	Erythromycin	Parke-Davis
Indocin SR	Indomethacin	Merck
Levsinex TimeCaps	Hyoscyamine sulfate	Schwarz Pharma
Fedahist Gyrocaps	Pseudoephedrine HCl, chlorpheniramine maleate	Schwarz Pharma
Micro-K Extencaps	Potassium chloride	A.H. Robins
Micro-K 10 Extencaps	Potassium chloride	A.H. Robins
Norpace Controlled Release	Disopyramide phosphate	Searle
Nicobid	Nicotinic acid	Rhône-Poulenc
Novafed Controlled Release	Pseudoephedrine HCl	Marion Merrill Dow
Novafed A Controlled Release	Pseuodephedrine HCl	Marion Merrill Dow
Prelu-2 Timed Release	Phendimetrazine tartrate	Boehringer Ingelheim
Sudafed 12-hour	Pseudoephedrine HCl	Burroughs Wellcome
Theoclear	Theophylline	Central
Slo-bid Gyrocaps	Theophylline	Rhône-Poulenc Rorer
Theo-24	Theopohylline	Searle
Ornade Spansule	Phenylpropanolamine HCl, chlorpheneramine maleate	SmithKline Beecham
Verelan	Verapamil HCl	Lederle
RU-Tuss II	Antihistamine/ decongestant	Boots

Table 1 Continued

Product	Active ingredient	Marketing company
Theo-Dur Sprinkle	Theophylline	Key
Valrelease	Diazepam	Hoffmann–La Roche
Tablets		
PCE	Erythromycin	Abbott
Theo-Dur	Theophylline	Key

Source: Ref. 3.

gelatin capsule. During dissolution, the shell dissolves rapidly, exposing the contents to the dissolution fluids as individual units. Since there is a limit to the amount of pellets that can be filled into a capsule, to develop high-dose products it sometimes becomes necessary to compress the pellets into tablets. Compressing multiparticulates into tablets entails additional formulation efforts to ensure that the individual units are regenerated in the body fluids upon ingestion or administration. Packaging of multiparticulate products involves many of the same considerations that go into the packaging of any dry oral dosage form.

A marketed product package includes active drug, pharmaceutical adjuvants, packaging, package insert, and educational materials [6]. Package development is an integral part of product development since packaging is one of the major factors that determines the stability and, in some instances, even the therapeutic performance of the product [7]. Packaging provides the necessary elegance and convenience for administering the dosage form and is a critical parameter that affects the integrity of the finished product during storage, transportation, and distribution.

Selection of packaging materials for a specific formulation is influenced by a number of considerations, including some of the recent developments that have affected the American pharmaceutical industry. These are an extension of Current Good Manufacturing Practices (CGMPs) to the preparation of investigational drug substances as well as drug products, stricter enforcement of CGMPs in the manufacture of commercial and investigational products, globalization of manufacturing, research and development (R&D) activities, environmental and safety considerations in the production and packaging operations of pharmaceuticals, and disposal of pharmaceutical waste [8–11]. Regulatory acceptance is an indispensable element in the selection of packaging materials [12]. Internationalization or globalization of manufacturing and R&D activities by multinational corporations has brought to the fore the need to become familiar and compliant with the regulatory practices in foreign countries; harmonization of specifications and test procedures for components, including packaging materials; and qualification and validation of these materials in actual use [13,14]. For over-the-counter (OTC) products, the use of tamper-evident or tamper-resistant packaging is mandated in the United States by regulations issued in 1982 (CFR) in the aftermath of the Tylenol tampering incident that occurred in 1981 [15]. Mounting environmental concerns dictate the need to implement procedures in the handling and disposal of packaging waste both to protect the environment and to comply with the applicable local regulations [16–18]. Thus, in addition to product protection and cost, a number of other considerations (e.g., regulatory, global, and environmental concerns) play important roles in packaging selection for a product.

A pharmaceutical product is subjected to a number of hazards, both physical and mechanical, during its storage, handling, shipping, and distribution. The physical or climatic hazards involve temperature and humidity fluctuations, permeation of atmospheric gases and moisture as well as deleterious effects due to light. Finished products experience a number of mechanical hazards from the time of manufacture through movement and storage in the distribution channels until the product reaches the consumer. During storage and transportation, drug products are subjected to shock drops, impact, compressions due to stacking of one carton over the other, abrasion, and vibration. Protection from physical hazards, especially those stemming from exposure to humidity or the ingress of gases is accomplished mainly through primary packaging, that is, packaging that comes into direct contact with the finished product. For solid oral dosage forms, adverse effects due to mechanical hazards can be countered through appropriate secondary packaging and by eliminating the head space inside the bottle by stuffing with a filler such as cotton. Protection from light is provided by both primary packaging (e.g., use of amber containers) and by secondary packaging (e.g., carton and fiberboard combination) [19]. It should be noted that while packaging protects the product from the deleterious effects of humidity and light, it cannot mitigate the effects of heat.

In this chapter we deal with primary packaging materials used in presenting pellets and granules filled into hard shell gelatin capsules or compressed into tablets. We provide information pertaining to the composition and properties of different materials used in unit dose and multidose packaging of these products, quality control, and regulatory considerations in the selection and use of packaging materials. We also address environmental concerns associated with the disposal of some of the plastic and polymeric materials. The information focuses on solid oral products, tablets and capsules, since these constitute the most common dosage forms into which the pellets, beads, and coated granules are presented to the consumer. It is to be expected that new materials and packaging innovations, changes in societal mores, product technology, pharmacy practice, governmental regulations, prescription to OTC switches, advertising methods, and related factors will result in the redesign of pharmaceutical packaging [20].

II. CLASSIFICATION OF PACKAGING SYSTEMS FOR MULTIPARTICULATE PRODUCTS

Pellet products, tablets, or capsules are packaged either in multi-dose containers or in unit-dose packages.

A. Multi-dose Bulk Containers

Multi-dose containers are by far the most popular in the United States and account for nearly 90% of solid oral drug products sold in the country. In retail pharmacies, prescription drug products are transferred from the manufacturer's bulk container to an amber plastic, usually polypropylene (PP), vial by the pharmacist and dispensed to the customer. In some European countries, notably Germany and France, solid oral products are predominantly sold directly to the consumer in blister or foil packages, which account for nearly 80% of sales of these products [21]. In the United States, unit-dose packages are employed primarily in hospitals as physician samples or sold retail as special convenient packages (e.g., oral contraceptives or methyl prednisolone tablets, marketed as Medrol Dosepak), where the purpose of blister packaging is to serve as an educational tool or to assist in patient compliance [6].

For multi-dose products, the container, closure, liner, innerseal, stuffing, and desiccant (if needed) constitute the components of the container/closure system. Containers commonly used for storing solid oral products are made up of plastic or glass [22]. Plastic containers are far more common than glass for use with solid oral products because they are not fragile and do not shatter or cause spillage when dropped, and their storage is easier in a stack. They are lightweight and require less protective secondary packaging. They also have a lower unit cost. The principal disadvantage of plastic containers is that they are not totally impermeable to atmospheric gases and water vapor and may allow ingress of moisture and permit odors and solvents to escape into the environment. The plastic containers widely used to store and dispense dry pharmaceutical dosage forms are made from high-density polyethylene (HDPE) or polypropylene (PP) resins. Vinyl polymers such as poly(vinyl chloride) (PVC) and polyvinylidene chloride (PVDC) and, less commonly, polystyrene (PS) and polyethylene terephthalate (PET) are used as blister-packaging materials. Bulk plastic containers are rigid and are usually injection blow molded. The basic properties of some of the bulk container materials are discussed below.

HDPE.   High-density polyethylene is the most widely used plastic resin for packaging and storing dry pharmaceutical products. It is crystalline and has a density between 0.941 and 0.965 g/mL. Polyethylene (PE) grades with densities below 0.940 g/mL are classified as low-density polyethylene (LDPE) [23].

HDPE consists of essentially all linear PE chains with no branching or com-onomer units and hence are capable of packing into crystallites, unlike LDPE and linear LDPE, which have occasional chain branching or comonomer units dis-tributed along the PE chains leading to a lower degree of crystalline packing and hence lower densities. HDPE is a highly impact-resistant thermoplastic material. Thermoplastic materials have thermal and physical stability at processing tem-peratures and can be melt fabricated into articles of different shapes and sizes and remelted for reuse. HDPE is synthesized by the polymerization of ethylene monomer under controlled conditions of heat and pressure with the aid of catalysts and may contain in addition to the base resin, stabilizers, plasticizers, and processing aids. The tightness and rigidity of the molecular packing of PE chains contributes to the moisture barrier properties of the material. Many of the HDPE resins are listed in the Drug Master Files (DMFs) of the Food and Drug Administration (FDA) and meet the regulatory criteria for use as primary phar-maceutical packaging materials.

Polypropylene. PP consists of long polymer chains composed of pro-pylene monomer units ranging from 1,000 to 30,000 or more. It is crystalline and is widely used in the packaging industry due to high moisture barrier properties and resistance to heat and solvents. It has low barrier properties toward gases [24]. However, its high water vapor resistance helps in protecting the product from the deleterious effects of moisture during storage by the consumer. PP is used widely as the base resin in prescription vials and in blister packages in Europe [25]. Table 2 describes some of the properties of the various plastics used widely to package tablets and capsules. Table 3 describes the chemical structures of the polymeric resins used in pharmaceutical packaging.

Glass. In contrast to plastics, which are basically organic compounds, glass contains mostly inorganic materials, such as silica, sodium and calcium carbonate, and oxides of sodium and calcium. The primary advantages of glass over plastic bottles is its impermeability to gases and vapors and its inertness. Glass also provides light resistance if needed (amber bottles) and the ability to view products stored within the bottle (clear flint glass). It is nonreactive with practically all chemicals. The disadvantages of glass are its high density and fragility. The type of glass used in pharmaceutical industry is borosilicate or neutral glass where the alkaline content is reduced by replacing some of the sodium, potassium, or calcium oxide with acidic components such as boric oxide. Bottles for storing capsule and tablet products are of white flint (colorless) or amber glass, listed in USP XXII as type NP (non parenteral) or type III [27].

Sealing Systems

The cap, liner, and innerseal constitute the components of a sealing system. The primary function of the sealing system is to serve as protective containment

Table 2 Properties of Plastics Used in the Packaging of Solid Oral Products

	Material			
Property	HDPE	PP	PS	PVC
Density (g/mL)	0.94–0.96	0.88–0.90	1.04–1.07	1.3–1.5
Melt temperature (°C)	130–137	167	100–105	75–105
Water absorption (g/24 h)	<0.01	0.01	0.01	0.04
Maximum use temperature (°C)	79–121	107–149	66–77	66–79
Barrier to gases (oxygen, carbon dioxide)	Poor	Poor	Poor	Variable
Barrier to water vapor	Excellent	Excellent	Poor	Variable

Source: Ref. 26.

against loss of contents or entry of contaminants and to provide access and resealability based on convenience and need. An essential requirement for any sealing system is that there be no physical or chemical interaction with the product. Furthermore, all the components of a sealing system must be child-resistant if dispensed in the same containers as used for storing. This implies that bottle packages with 100 dosage units or less marketed directly to the consumer are also required to be child resistant [28]. In addition, products offered for OTC sales must be packaged in tamper-evident sealing systems [29]. The closure selection during packaging development is thus based on containment, control, convenience, and cost.

Closures for bulk containers used to package solid oral products are made from thermoplastic materials or metal. Commonly used plastic caps are made from PP, PE, or PS and are threaded to attain a seal by attaching to the corresponding threads molded into the finish of the container neck. Metal caps used are usually made up of tinplate or aluminum and are coated on the inside with enamel or lacquers for resistance against corrosion.

A closure is designated by a series of numbers or letters (e.g., 48–400). The first number refers to the inside diameter of the closure stated in millimeters. The second number refers to the "finish serial number" based on thread design, size, pitch, profile, length, and thickness of the engagement. For example, "400" refers to a shallow continuous thread and "410" refers to a medium continuous thread [30].

Table 3 Chemical Structures of Polymeric Materials Commonly Used in Pharmaceutical Packaging

Polymer	Chemical structure
Poly(vinyl chloride)	$\left[CH_2-CH \atop \quad\quad Cl \right]_n$
Poly(vinylidene chloride)	$\left[CH_2=C{<}^{Cl}_{Cl} \right]_n$
Polypropylene	$\begin{array}{c} CH_3 \\ \mid \\ (CHCH_2)_n \end{array}$
Polyethylene	$(CH_2-CH_2)_n$
Polystyrene	$\left[CH_2-CH \atop \quad\quad C_6H_5 \right]_n$
Poly(ethylene terepthalate)	$H \left[OCH_2CH_2O\overset{O}{\overset{\|}{C}} - C_6H_4 - \overset{O}{\overset{\|}{C}} \right]_n OH$
Nylon (polyamides)	$H_2N-R-\overset{O}{\overset{\|}{C}} \left[\overset{}{N}_{H}-R-\overset{O}{\overset{\|}{C}} \right]_n N_H-R-\overset{O}{\overset{\|}{C}}OH$

Liners are inserted into the cap to effect a seal between the cap and the container and are required because screwcaps and bottles usually have irregularities that prevent proper seal. Liners are commonly made from paper, plastic film such as PP, PVDC, or PET, or coated paper. A typical liner is an extruded polyethylene coated on paper with or without a liner wax. They may also be laminations of pulp/white paper/coated PE/white pigmented vinyl chloride copolymer wax.

The innerseals used in packaging solid oral products are made up of glassine and wax laminated to the liner. They allow visual evidence of tampering. Innerseals are pressure-sensitive foamed PS or wax-laminated glassine using FDA-approved wax materials.

Desiccant

The inclusion of a desiccant in a package is dictated by the moisture sensitivity of the product. Although the barrier properties of the container are the principal protective elements of the packaging, desiccant provides an additional layer of protection against humidity. A desiccant absorbs water vapor from the atmosphere inside the packaged container, thus lowering the relative humidity to the point that damage to the product is avoided [31–33]. It relies on physical adsorption involving van der Waals forces and electrostatic adsorption between the water molecules and the surface of the desiccant. Adsorption is a surface phenomenon and increases with increase in surface area. Typically, desiccants have a large surface area relative to the mass. When a desiccant is included in a bottle containing capsule products, it is important to be aware of the potential absorption of moisture from the shell, which can render the capsules brittle.

The most common desiccants employed within the pharmaceutical industry are silica gel and calcium sulfate. Silica gel adsorbs water up to 40% of its weight at 25°C. The amount of desiccant to be included in the package is governed by the rate and the amount of moisture that the desiccant is capable of adsorbing and the needs of the package protection based on the properties of the drug product and the desired self life [34].

Figure 1 illustrates the influence of desiccant on the dissolution properties of an experimental SR-coated tablet (Parke-Davis files, unpublished data). The product, when stored for 1 month at 45°C in tightly closed HDPE bottles, without a desiccant, showed a distinctly slower in vitro release profile than that of the freshly prepared product. The inclusion of a desiccant improved the dissolution stability of the product markedly, although there were still some differences in the dissolution patterns of the product before and after storage with the desiccant, presumably due to the effects of heat.

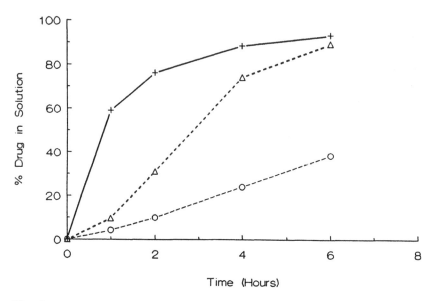

Fig. 1 Effect of desiccant on the in vitro release profiles of tablets stored at 45°C for 1 month. +, Initial; ○, without desiccant; △, with desiccant.

Cushioning Materials

Cushioning or stuffing materials are placed on top of the tablets or capsules to fill the headspace inside a bottle. The purpose of including a cushioning agent is to protect the product from shock, vibration, or rattle during transportation and distribution. Such a stuffing is generally inert and is included in the form of an unwinding coil inside the container. It may consist of cotton, rayon, or polyester. Of these, polyester is the most popular material because of its negligible moisture content and inability to retain moisture.

B. Unit-Dose Packaging

Unit-dose packaging is a package that holds a single dose of the medication in a ready-to-use form. There are two types of unit-dose packages, strip packaging and blister packaging. In strip packaging, the dosage form is packaged in individual pouches that are attached to each other and separated by means of a perforation in a foil strip. Today, strip packaging has been replaced, for the most part, by the blister package, which involves the formation of a plastic transparent bubble. The transparent bubble is made up of thermoformable polymers, that is, capable of being shaped through the application of heat and pressure. The forming film is heat sealed to the lidding materials to form the complete blister pack as shown in Fig. 2.

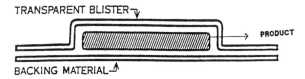

Fig. 2 Basic configuration of a blister pack.

Blister Packaging

Blister packaging offers a number of advantages over multidose packaging. These are [35–38]:

1. Protection of product against microbiological organisms and dust particles
2. Protection against moisture and atmospheric gases providing long shelf life, especially if used with high-barrier films and laminates
3. Provides convenience to the patient
4. Facilitates compliance
5. Easy-to-recognize tampering
6. Can be fabricated child resistant
7. Minimizes hazards associated with transportation relative to multicomponent packaging

The materials for blister packaging consist of two parts, the forming film and the lidding or backing material. The lidding material can be either a push-through or peelable type. In push-through package, the plastic blister is affixed to a rigid or semi-rigid backing material, through which an individual dosage unit is expelled. In the peelable package, the lidding material is peeled off to remove the dosage unit. There are different types of forming and lidding films. An essential requirement is that the lidding material must precisely match the forming film for a well-designed functional blister package.

The forming film is the packaging component that receives the product in deep-drawn pockets. It may consist of a single uncoated polymer or be coated with an agent to decrease the water vapor transmission rate (WVTR). If the product is very moisture sensitive and not protected by coated plastic, a laminate containing aluminum foil (at least 45 μm thickness) can be used as a forming film. Laminates containing aluminum are cold-formed. The plastic forming films are usually colorless and transparent. However, if the product needs protection from light, an amber-colored plastic may be used. After the dosage form, tablet or capsule, has been fed into the deep-drawn pockets of the forming film, the thermoformable plastic is sealed directly against the lidding material.

Forming films used in blister packaging are primarily (1) rigid PVC or PP, (2) PVDC-coated PVC, marketed commercially as Saran (Dow Chemical), and (3) laminations of PVC with PVDC or a fluoropolymer film such as Aclar (Allied

Signal) [39]. The properties of the films, in relation to their application in blister packaging, are discussed below.

PVC.　This is the most widely used blister material. It can be thermoformed over a temperature range of 120 to 140°C, making it very practical for all thermoforming machines. PVC films used for blister packaging are rigid materials, free from plasticizers, and range in thickness from 200 to 300 μm (8 to 12 mils). Unplasticized PVC has good thermoforming properties but may not provide good moisture protection for some products [40]. After the forming process on the packaging machinery, a 250-μm film will have a final thickness of 50 to 100 μm in some deep-drawn pockets. The reduction in thickness may increase WVT.

PP.　PP is increasingly being substituted for PVC in Europe as the base material for blister-packaging machines. Typical film thickness range from 250 to 300 μm (10 to 12 mils). PP is chlorine free, and unlike plain PVC and PVDC-coated PVC, does not suffer from the liability of discharging hydrochloric acid into the environment during incineration. It may eventually replace PVC totally because of this advantage. Its WVT rates are even lower than those of PVDC-coated PVC.

The major disadvantage with PP is its narrow thermoforming temperatures of 140 to 150°C. This means that the film heating system must be powerful and accurate. Temperature controls must be precise. In addition, unlike PVC films, PP films need a cooling step following thermoforming to dissipate the heat content of the film and to reduce shrinkage. Shrinkage of PP web due to temperature drop can lead to misalignment of the blisters during perforation. Since PP is softer than PVC, warping of PP blisters can occur occasionally, making it difficult to package them into cartons. Not withstanding these problems, PP is becoming increasingly popular as the blister-forming material.

PVDC-Coated PVC.　The object of coating PVC films with PVDC is to reduce the WVT rates. The thickness of the PVDC coat is 25 to 50 μm (1 to 2 mils) while the thickness of the PVC film is 200 to 250 μm (8 to 10 mils) and thickness in the drawn pocket area is 50 to 100 μm. The PVDC coat is applied on the side facing the dosage form, to obtain the best water-vapor barrier effect.

Poly(chlorotrifluoroethylene)-coated PVC (PCTFE-Coated PVC).　Some of the common types of marketed Aclar films are 0.75-mil type 88A, 1.5-mil type 22A, 0.75-mil type 33C, and 2.0-mil type 33C. The films are usually laminated to either 7.5- or 10.0-mil PVC bonded to 2.0-mil PE film. Aclar films have the lowest water vapor permeation compared to all other plastic films used in blister packaging and have thermoforming properties similar to plain PVC. During heat sealing, the uncoated PVC side faces the product. One major disadvantage of the PCTFE-coated PVC is that it is more expensive compared to PVC- or PVDC-coated PVC films.

Nylon (Oriented Polyamide)–Aluminum–PVC Laminate (OPA/Al/PVC). This laminate, which consists of 25-μm OPA/45-μm aluminum/60-μm PVC (1-mil OPA/1.8-mil Al/2.4-mil PVC), serves as an excellent moisture barrier. Care has to be taken to ensure that the aluminum foil is pinhole free. Aluminum laminates are cold-formed and require proper adhesives and precise composition to obtain the desired properties. In general, cold forming requires more packaging material than thermoformed films for the same number of dosage forms. The blister lines for this laminate must be specially built. The cost of the laminate is comparable to that of PVDC-coated PVC.

Other less commonly used blister-packaging films are PET and PS. PET has excellent thermoforming properties; but poor WVT compared to plain PVC. Also, PET exists in different crystalline modifications and only the amorphous form has good heat-sealing abilities. PS has good thermoforming properties, but because of the high WVT rate, it is not very popular as a blister material. The relative barrier properties of the polymers are presented in Table 4.

Lidding Materials

The lidding or the backing material may be made of soft or hard aluminum, aluminum/paper laminate, or aluminum/PET/paper laminate. The aluminum is coated on the underside with a vinyl lacquer and sealed to the forming film. The weight of the sealing agent is in the range of 6 to 10 g/m^2, constituting about 10 to 15% of the lidding weight. The sealing agent, usually a heat-sealing lacquer, must be compatible with the forming film. It should have the required strength of the finished package, capable of being run on the blister-packaging equipment, and meet all the regulatory requirements.

The outside of the lidding is normally coated with a print primer, which must withstand the sealing temperatures without showing discoloration or tackiness. It must serve as a substrate to which printing inks can adhere strongly

Table 4 Relative Permeabilities of Some Commonly Used Packaging Materials

Plastic	WVTR[a]	O_2[b]	CO_2
HDPE	0.3	185	585
LDPE	1.0 – 1.5	500	
PP	0.7	150 – 240	500 – 800
PS	7 – 10	250 – 350	900
PVDC	0.2 – 0.6	0.8 – 6.9	3.8 – 44
PVC	2.0 – 5.0	10.0 – 20.0	25.0

Source: Ref. 41.
[a]Units: g/100 in^2/mil/24 h at 37.8°C.
[b]Units: cc/100 in^2/mil/24 h/atm at 25°C.

enough to withstand the peeling force of adhesive tapes and must have sufficient resistance to abrasion.

Hard aluminum is used as a push-through lidding material. The thickness of the foil for these applications ranges from 15 to 25 μm (0.6 to 1 mil). A thickness of 25 μm assures that the film is pinhole free, although some manufacturers guarantee films with thicknesses of 20 μm to be free of pores. Even 15-μm foils are suitable for blister packaging in combination with the forming-film PVC, PVDC-coated PVC, PP, or PET. The barrier properties depend mainly on the thermoformed plastic components.

The outer aluminum side is lacquered with a heat-resistant primer which provides the necessary adhesion for printing inks. Sometimes, if the ink coverage is very high or the printed colors are very dark, an overcoat of lacquer will be necessary to give more resistance to abrasion. The aluminum side facing the product is first coated with a heat-sealing lacquer for sealing to the forming film. For example, PVC, PVDC-coated PVC, PCTFE-coated PVC, OPA aluminum-PVC, and PS can all be sealed with one type of heat-sealing lacquer. The heat-sealing lacquer is applied in two layers. The first layer is a primer lacquered on the plain foil. It is PVC based and designed to ensure optimum adhesion to the aluminum. The second layer is a sealing lacquer and is PVC/acrylic based and is designed to provide optimum sealing strength to PVC. The heat-sealing lacquer for PP forming films usually has a distinct composition; it has to be sealed over a narrow temperature range, 185 to 210°C, in contrast to PVC films, which are sealed over the broader temperature range 150 to 250°C. At temperatures much below 185°C, the blister package will not be tight. If the temperature is more than 210°C, the blisters will be too wrinkled and cause problems during secondary packaging inside a carton.

Soft Aluminum.   This is a material in annealed condition. It is produced by heating the metal and slowly cooling it for a predetermined period of time. Soft aluminum is frequently used for child-proof push-through foils with a foil thickness of 25 μm. The softness and thickness of the film help prevent children from pushing the tablets or capsules through it. Except for this feature, the basic structure of the lidding material is similar to that for hard aluminum.

Paper–Aluminum.   For a combination of paper and aluminum, paper with a weight of 40 to 70 g/m^2 is used. In Europe, a thinner aluminum foil, usually about 7 to 12 μm, is used in child-proof push-through packages. In the United States, a somewhat thicker foil, 15 to 25 μm is popular as a peel-off foil. To be able to push the tablets or capsules through the lidding, the foil must be relatively thin and the paper weight must not exceed 50 g/m^2. For effective peeling, the heat sealing coat must have peeling properties. The heat-sealing lacquer must be adjusted to give a lower sealing strength for peeling than for push-through blisters. At present, only heat-sealing coats for PVC, PVDC-coated PVC, PCTFE-coated PVC, OPA–aluminum–PVC laminates, and PS forming films can be made softer.

Paper–PET–Aluminum.   Lidding material made up of paper–polyester–aluminum laminates are often called peel-off-push-through foil. Such a lidding is more common in the United States than in Europe. The idea is to peel off the paper–PET laminate from the aluminum and then the tablets are pushed through the aluminum. The design requires adhesion between the PET and the foil to be weaker than the bonding between the foil and the film.

Figure 3 illustrates schematically the structures of some of the commonly used lidding materials in pharmaceutical blister packages.

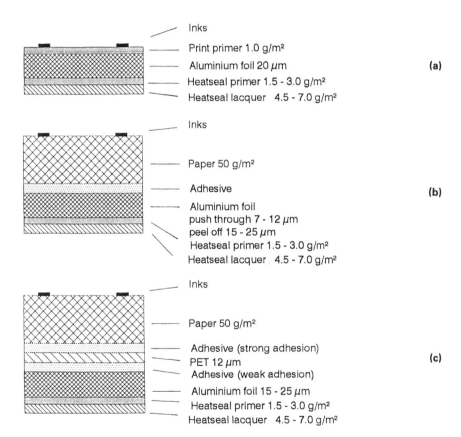

Fig. 3   Schematic of some of the lidding material structures commonly used in the blister packaging of oral solid dosage forms: (a) hard aluminum lidding; (b) paper–aluminum lidding; (c) paper–PET–aluminum lidding.

Strip Packaging

Strip packaging is the forerunner of the blister package for packaging tablet products. In strip packaging, the product is enclosed between two webs of either the same or different materials, depending on whether the dosage units are required to be visible or covered completely in opaque material [42–44]. Like blister packaging, strip packaging can also be designed to offer a high degree of protection. Strip packs can be made from much cheaper materials and are better suited than blister packs for larger tablets. On the other hand, blister packages occupy much less space than strip packs, resulting in savings in shelf or storage space.

The packaging material for strip packaging consists of a bottom laminate and a lidding laminate. Both laminates usually have similar specifications. Previously, during strip packaging, the bottom and the lidding laminate were not preformed before the tablets were fed into the packaging material. The packets were formed by wrapping the laminate around the tablet or capsule while the bottom laminate and lidding laminate were sealed together. Thus a two-dimensional packaging material was folded around a three-dimensional product, causing, occasionally, wrinkles and creases to occur, resulting in loose packaging.

Usually, the basic material of packaging is aluminum foil 20–40 μm thick with the outer side of the foil printed after priming and sealing agent applied on the side facing the drug product. The sealing agent is usually a heat-sealing coat and may be a heat-sealing film of polyethylene or an ethylene ionomer. Ionomers have ionic groups, in addition to the crystalline and amorphous phases, as part of their molecular structure. They are used for their excellent sealability, optical clarity, and ease of thermoforming. The amount of heat-sealing coat applied is in the range of 6 to 10 g/m^2. The plastic films have a thickness of 20–40 μm. They can be added to the foil by extrusion coating or laminated by adhesives.

The newer version of strip packaging is based on blister packaging principles: first the bottom laminate is preformed by either thermoforming or cold forming. Then the product is fed into the preformed bottom film, and finally, the lidding material is sealed to the bottom material. The basic material of preformed strip packaging is aluminum with a minimum thickness of 40–50 μm. The outer side is print-primed and printed or laminated with an oriented film (OPP/OPA/OPET). The foil side that faces the drug product is coated with low density PE or an ethylene-based ionomer. For drug products that are not moisture sensitive, a plastic film such as PVC can be employed as the base material of the preformed strip packaging.

The plastic films can be added to the foil by extrusion coating or laminated by adhesives. The basic material of preformed strip packages is aluminum with a minimum thickness of 40 to 50 μm. The outer side can be print primed and printed or laminated with an oriented film (OPP/OPA/OPET). The foil side that faces the drug product is coated with low-density polyethylene or an ethylene-based ionomer. For moisture-insensitive drug products, a plastic film such as PVC can be employed as the base material of the preformed strip packaging.

The aluminum foil used in blister packaging is a rolled product, rectangular in cross section and used in blister packages in the thickness range 7.0 to 20 μm. For this purpose the important properties to be considered are:

1. Mechanical (tensile strength, elongation, burst strength)
2. Surface appearance (oil free, surface reactivity)
3. Porosity (both pore size and size distribution)
4. Flatness and uniformity of thickness
5. Suitability for use in blister-forming machines

The most important property that controls the WVT rates in aluminum film is its porosity [45]. In foil rolling, pinholes can result from flaws that may occur uniformly or in lines across the foil surface. Foil porosity may be caused by inhomogeneities in the film because of localized deformation that occurs during the rolling process. The larger holes result from damage due to hard particles [e.g., aluminum oxide or titanium diboride (TiB_2)]. Frequent roll changing helps in the production of high-quality foil.

During storage, moisture or gas can breach the barrier through heat seal fins and then through the walls. The WVT rate through the walls of the package is influenced by pinholes. The heat seal and protective lacquer act as additional barriers to transmission. Using Fick's law, it can be shown that in 24 h, with 90% RH on the outside of the package and an initial humidity of 0% inside, 0.2 mg of water will penetrate the foil through a pinhole with a radius of 10 μm. Tests performed on plain aluminum foils with thickness ranging from 7 to 20 μm showed that annealing reduces the permeability of gas and water vapor.

The data presented in Table 5 indicate that the influence of pinholes on the barrier properties of aluminum foil is negligible when the latter is thicker than 12 μm and is coated with lacquer. The barrier properties of a lacquered 20-μm foil is several orders of magnitude superior to that of film and transparent laminates, including those that incorporate PVDC.

Table 5 Moisture Permeability of Aluminum Foils

Material	Thickness (μm)	Permeability ($g/m^2/24$ h)
Aluminum foil	7	0.053
	9	0.019
	12	<0.005
Annealed aluminum foil	7	0.029
	9	<0.005
	12	<0.005

Source: F. Reiterer, date on file with Teich AG.

III. BLISTER-PACKAGING MACHINERY

Blister packaging of pharmaceutical products involves web feeding, blister forming, product loading, printing and registration, sealing, slitting or perforation, and cross cutting [46]. All these operations have to be carried out under CGMP conditions of cleanliness and hygiene, absence of product mix-up or cross contamination, maintenance of product stability, and impeccable documentation. Blister packages for pharmaceutical products are of several different types (e.g., push-through, peel-off, child-proof, calendar packs, etc.). The machines are therefore of varying complexity and design. The simplest equipment uses blisters that are preformed on a separate machine. The product is inserted into the blisters by hand and sealed. Thermoform–fill–seal machines can form blisters, load product into the blister, seal, and cut them apart. Some machines can also automatically pack the finished packages into cartons.

Packaging lines in use for form–fill–seal machines can be either intermittently operating or continuous motion units. In the intermittent operating packaging line, the form web stops during each machine operation. In the continuously working machine, the forming elements, sealing units, and related parts travel with the web as it moves through the machine. The sealing temperatures of 140 to 200°C are much lower in the intermittently operating machines compared to the continuous units, which require temperatures on the order of 200 to 300°C. Thus, while the latter have the advantage of higher output per hour, they incur the disadvantage of exposing the product to higher temperatures.

A. Intermittently Operating Machines

The basic operations of an intermittently operating blister packaging machine in sequence are: web feeding, web heating, formation of blister, dosage delivery into the blister, sealing of lidding material fed from the reel to the tops of the filled blister material, perforation, punching into strips, and finally, feeding the blister strips into a cartoning machine [46,47]. The operations that take place at the various stations are discussed below.

1. *Web feeding stand.* Forming films and lidding materials are fed into the machine. The desired rate of web supply is provided with the help of dancing rolls.

2. *Heating station.* The web passes through the preheating oven that raises its temperature to the forming level. Forming films containing PVC are heated up to 120 to 140°C for periods varying from 0.1 to 0.5 s. The temperature range is narrower (140 to 150°C) for PP forming films. Aluminum-containing film formers are not heated before the forming process.

3. *Forming stations.* Plastic films are formed when the preheated web is drawn by air pressure or for deeper depth, formed by a mechanical plug assist plate. The temperature of the web as it leaves is affected by the temperature of

the heater and the length of time it is at the heating station. Therefore, both temperature and exposure time must be closely controlled.

4.   *Cooling stations.* After the forming process is completed, cooling is necessary with PP forming films. There is no need to cool laminates containing PVC.

5.   *Product feeding.* There are different types of feeding systems. Orienting and feeding the tablets or capsules is done by either vibratory feeders or sometimes through a belt feeding arrangement. Flood feeders are very common. Manual feeding has been replaced by mechanical feeding systems where the tablets or capsules to be packaged are swept into the pouch.

6.   *Filling control.* Misfilled packages or flaws are detected by optical, electromechanical, or video control systems. The misfilled packages are rejected and scrapped.

7.   *Sealing station.* At the platen-type sealing station, the lidding material is heat sealed to the forming film, which now contains the product. The sealing temperatures could range from 145 to 200°C and the exposure times are on the order of 0.1 to 0.5 s.

8.   *Cooling station.* The films are cooled following the sealing process. The cooling times are longer for PP forming films.

9.   *Registration.* When a pattern or printed web is used, a registration system is needed to control the movement of the web and to ensure that the printed web material is properly positioned on each finished package. A pattern printed on flexible material may not be exactly the same size each time it is printed and there may be variations in the dimensions of the web.

10.   *Coding station.* At this station, the packages are marked with the information displayed on the package (e.g., batch number, production date and expiration date).

11.   *Perforating station.* To produce child-proof blister packaging, a cross-shaped perforation, of varying depth, is made along the sealing seams at this station.

12.   *Punching or slitting station.* The strip is punched to individual blisters by using single-, double-, or multistroke punches.

13.   *Cartoning station.* The blister is put into the surrounding carton or secondary package along with a package insert.

Figure 4 is a schematic describing the various operations that take place in blister-packaging machine. Photos of the actual machines used in blister packaging of pharmaceutical dosage forms are displayed in Fig. 5 and 6.

B.   Continuously Operating Machines

Continuous-motion thermoform–fill–seal machines are generally used for high-volume products and on lines that do not require frequent changeovers. In these

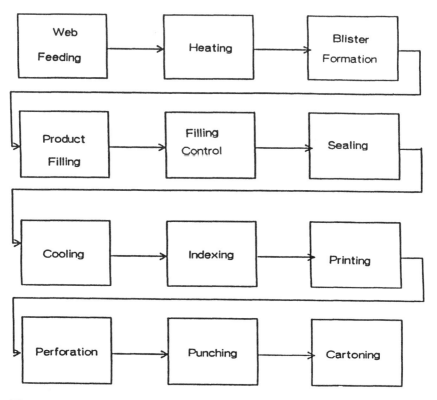

Fig. 4 Schematic of various operations involved in a typical blister-packaging machine.

Fig. 5 Typical blister-packaging machine. (Courtesy of Uhlmann Pac-Systems.)

(a)

(b)

Fig. 6 Stations of a blister-packaging machine. (Courtesy of Uhlmann Pac-Systems.) (a) Unwind stand, (b) heating station, (c) forming station, (d) product feeders, (e) fill control system, (f) sealing station, (g) index station, (h) coding station, (i) perforating station, (j) punching station.

(c)

(d)

Fig. 6 Continued

machines, the package and the product move forward continuously during forming, loading, and sealing operations, and higher sealing temperatures ranging from 200°C–300°C are used due to the short contact time, 0.1 to 0.3 s, of the sealing roller and the lidding material. In a continuously operating machine, a web can be stretched but not shrunk, and therefore the size and the position of the pattern on the package are controlled by printing slightly undersize and stretching the web to make the pattern fit the package. Consequently, the periodicity of

(e)

(f)

(g)

(h)

Fig. 6 Continued

printing is slightly less than 100% (between 99.9 and 99.0%). Normally, the lidding material is held fixed over a distance of 30 cm by grippers. Stretching is accomplished with the aid of bolts. On continuously operated machines, the printing distance has to be smaller than the nominal eye-mark distance. On the other hand, the cooling station, perforating station, punching station, and the cartoning station are similar to those for the intermittently operating line. A cooling step after sealing is necessary for both intermittent and continuously operating machine.

A number of compact blister-packaging machines are available in the market for use in R&D laboratories, clinical trials, and small-volume production runs. They operate on the same general principles as intermittent operating

(i)

(j)

machines used for manufacturing operations. The dosage units are loaded either manually or automatically before passing through a sealing and coding station. The output of these units is on the order of 100 to 2000 units per minute. Some of the newer machines have been adopted for using PP as the base web of the blister pack. Table 6 lists the names, addresses, and telephone numbers of some suppliers of packaging materials and machinery.

Table 6   Some Vendors/Suppliers/Manufacturers of Materials and Equipment
for Packaging Solid Oral Dosage Forms

Containers for bulk packaging (plastic and glass)

Wheaton Glass Company, 1501 North Tenth Street, Millville, NJ 08332. Phone:
(609) 825-1400.

Cole-Parmer, 7425 North Oak Park Avenue, Chicago. IL 60648. Phone: (708)
647-7600.

Nalge Co., Subsidiary of Sybron Corp., Box 20365, Rochester, NY 14602.
Phone: (716) 586-8800 (plastic containers only).

Kerr, Plastic Products Division, 500 New Holland Avenue, Lancaster, PA 17602.
Phone: (717) 299-6551.

West Co., 1041 West Bridge Street, Phoenixville, PA 19460. Phone: (215)
935-4500.

Munnerstadter Glaswarenfabrik GmbH, Schindbergstasse 27, D-8732 Munnerstadt/
Ufr., Germany. Phone: 09733/790.

Heinz Plastics GmbH, Glashuttenplatz 1, D-8648 Kleintettau/Ofr., Germany.
Phone: 09269/770.

Alpla Werke, Allmendstrasse, A-6971 Hard/Voralberg, Austria. Phone:
05574/316310.

Blister-packaging materials

Alusisse Flexible Packaging Inc., 6700 Midland Industrial Drive, Shelbyville, KY
40065. Phone: (502) 633-6800.

American National Can Co., 125 King Street, Greenwich, CT 06836. Phone:
(203) 863-8084.

American Mirrex Corp., 1389 Schoolhouse Road, Box 728, New Castle, DE
19720. Phone: (800) 233-2400.

Klockner-Pentaplast of America, Box 500, Klockner Road, Gordonsville, VA
22942. Phone: (703) 832-3600.

Mobile Chemical Co., Films Division, 1150 Pittsford-Victor Road, Pittsford, NY
14534. Phone: (800) 828-6381.

DRG Medical Packaging Inc., Box 7730, Madison, WI 53707. Phone: (608)
249-0404.

Allied Signal Inc., Engineered Plastics, Box 2332, Morristown, NJ 07962. Phone:
(201) 455-5010.

3M Packaging Systems Division, 3M Center, Building 220-8W, St. Paul, MN
55144. Phone: (612) 733-0634.

Plus Packaging Inc., P.O. Box 12, Madison, NJ 07940. Phone: (201) 538-2216.

Haendler & Natermann Flexpack GMBH, Natermann-Platz 1, D 3510 Hann,
Munden 1. Phone (05541) 7040.

Teich AG, A-3200 Obergrafendorf-Mühlhofen, Austria. Phone: 02747/8484-0.

Hoechst AG-Kalle, Rheingaustrasse 190, D-6200 Wiesbaden-Biebrich, Germany.
Phone: 06121/681-0.

VKW GmbH, D-7813 Staufen, Germany. Phone: 07633/8110.

Wasdell Packaging Machines Ltd., Upper Mills Estate, Bristol Road, Stonehouse,
Gloucester, United Kingdom GL10 2AT.

Table 6 Continued

Packaging machines

Iman Pack Inc., 527 Atlantic Avenue, Freeport, NY 11520. Phone: (516) 868-7888.

Klockner Borsch Sales & Service, 2111 Sunnydale Boulevard, Sarasota, FL 34243. Phone: (813) 359-4000.

Packaging Dynamics Inc., Box 5332, Walnut Creek, CA 94596. Phone: (415) 938-2711.

Automated Packaging Systems, 8400 Darrow Road, Twinsburg, OH 44087. Phone: (815) 338-9500.

Supermatic Packaging Machinery, 7 Spielman Road, Fairfield, NJ 07006. Phone: (201) 575-6350.

Efficient Automated Machine Corp., 39-13, 23rd Street, Long Island City, NY 11101. Phone: (718) 937-9393.

Raymond Automation Co. Inc., Box 5838, Norwalk, CT 06856. Phone: (203) 845-8900.

Robert Bosch GmbH, Produktbereich Hofliger & Karg, Stuttgarter Strasse 130, D-7050 Waiblingen, Germany. Phone: 07151/142264.

Uhlmann Pac-Systeme, D-7958 Laupheim 1, Germany. Phone: 07392/702-0.

Contract packaging

Abbott Laboratories/Contract Mfg. Services, D-45F/1401 Sheridan Road, North Chicago, IL 60064. Phone: (708) 937-2459.

Applied Analytical Industries, 1206 North 23rd Street, Wilmington, NC 28405. Phone: (919) 763-4536.

Clinical Packaging Services Inc., 61 Mattatuck Heights, Waterbury, CT 06705. Phone: (203) 597-9353.

Paco Pharmaceutical Services Inc., 1200 Paco Way, Lakewood, NJ 07024. Phone: (908) 367-9000.

Robert Bosch Corp., Packaging Machinery Division, 121 Corporate Boulevard, South Plainfield, NJ 07080. Phone: (908) 753-3700.

Schering-Plough Corp. Third Party Business, Box 526, Kenilworth, NJ 07033. Phone: (908) 709-2600.

Unipak Ltd. Oakhill Trading Estate, Worsley Road North, Walkden, Greater Manchester M28 5PT, United Kingdom. Phone: 0204 792747/700900.

Pharmaserve Limited, Clifton Technology Park, Wynne Avenue, Swinton, Manchester, United Kingdom M27 2HB.

Klocke Verpackungsservice GmbH, Randolf Diesel Strasse, D-7504 Weingarten, Germany. Phone: 07244/610.

SK Pak, Thorslundswej 7, DK-5000 Odense, Denmark. Phone: 09/142766.

Lamp S. Prospero s.p.a., Via Bella Pace 25/A, I-41030 S. Prospero Sulla Secchia, Italy. Phone: 059/908961.

IV. QUALITY CONTROL OF PACKAGING MATERIALS

A. Bulk Containers

The bulk containers used in packaging virtually all solid oral dosage forms are composed of HDPE or glass. From the standpoint of CGMPs, all packaging components such as bottles, closures, and sealing materials are classified as raw materials. Therefore, the usual quality control measures applied to other raw materials are also followed for packaging components. For example, all packaging materials should have specifications including characterization and identification tests (e.g., infrared or ultraviolet absorbance) and be subjected to physical evaluation, including dimensional and functional analysis. It is extremely important to provide complete identification and specifications for the packaging materials used with any given product. Bulk containers, plastic or glass, are approved for use based on appropriate testing to assure that they meet the requirements of physicochemical tests for plastics listed in USP XXII [48,49].

Plastic Containers

The following tests are included in the official compendium as part of quality control for HDPE and LDPE containers:

1. *Multiple internal reflectance.* This is a qualitative test that involves a comparison of infrared (IR) spectrum of the test sample with that of a reference standard using an IR spectrophotometer equipped with a multiple internal reflectance accessory. The IR spectrum from 3500 to 600 cm^{-1} of the sample exhibits the same major absorption bands at the same wavelength as the spectrum of the standard preparation.

2. *Thermal analysis.* Thermal analysis, either differential scanning calorimetry (DSC) or differential thermal analysis (DTA), is used widely to characterize polymeric materials. They measure the heat loss (exotherm) or gain (endotherm) resulting from physical or chemical changes within a sample as a function of temperature. The test, as applied to the characterization of plastic containers, compares changes in the thermal properties of test sample with that of the standard as the samples are heated at a given heating rate under an inert atmosphere maintained in the heating chamber. The compendial standards require that the thermogram of the test specimen should be similar to the thermogram of the HDPE or LDPE reference sample, and the temperatures of the endotherms and exotherms in the thermogram of the test specimen should not differ from those of the standard by more than 6.0°C for HDPE and 8.0°C for LDPE.

3. *Light transmission.* Since a plastic container is intended to provide protection of the contents from light, it should meet the requirements specified for light transmission. The transmittance of the test sample, mounted so that the light beam is normal to the surface of the sample and reflection losses are at a

minimum, is recorded continuously on a recording spectrophotometer in the region of 290 to 450 nm. The observed light transmission of the test plastic container intended for packaging solid oral products does not exceed 10% of the limits given in the USP, at any wavelength between 290 and 450 nm.

4. *Water vapor permeation (WVP)*. The rate of moisture permeation is determined by the weight gain of the container/closure system containing a known quantity of anhydrous calcium chloride as the desiccant, after storing for a period of 14 days in an atmosphere of 75% RH ($\pm$ 3%) and 23°C ($\pm$ 2°C). The weight gain in the systems containing the desiccant is compared to the weight gain observed in control systems not containing desiccant. Under these storage conditions, it is assumed that the weight gain is due solely to the absorption of water vapor by the container/closure system and not by its transmission to the product.

5. *Nonvolatile residues, heavy metals, and residue on ignition*. The tests are based on the extraction of a designated amount of plastic material of a specified surface area using a specified extracting medium at a given temperature. For nonvolatile residues the limits are specified using water, alcohol, or hexane as the extracting medium. The difference in the residue between the sample and the blank should not exceed the specified limit. For heavy metals the limits for lead are set at 1 ppm using purified water as the extracting medium. For residue on ignition the limits are set for the water extract. The difference in the residue between the sample and the blank should not exceed the specified limit.

A typical packaging component specification for an HDPE bulk container would include the following:

1. A description and/or mechanical specifications for the container: size, style, shape, finish, color, wall thickness, description of the resin used, weight, overflow volume, diameter, height, drawing(s), and the name and address of the supplier

2. Physicochemical specifications for the container as outlined in USP XXII and described earlier

3. A description and/or mechanical specifications of the closure system: size, style, description of the material used, color, description of the liner, innerseal, and name and address of the supplier(s)

Glass Containers

Glass containers are used primarily for bulk packaging of dry products that are moisture sensitive or photosensitive drugs. For such applications, USP type NP (nonparenteral) glass is used. The USP specifications for glass containers are:

1. *Mechanical*. Style, material, finish, overflow capacity (volume) of the container, weight, and conformance to drawn specifications.

2. *Physicochemical*. The tests specified in the USP under this category are aimed at determining the chemical resistance of the glass, measured as the

amount of alkali extracted with water under specified conditions. The tests include a powdered glass test conducted on crushed sample of the glass and a water attack test conducted on the whole container.

B. Blister-Packaging Materials

The properties of blister-packaging materials that define the quality are:

　　1. *Thickness or gauge.* The thickness of the films is measured with the aid of a micrometer according to the test procedures of ASTM (American Society for Testing and Materials). The thickness is usually expressed in mils (1/1000 of an inch).

　　2. *Tensile strength or elongation.* The tensile strength of a film is the amount of force necessary to break a sample strip in tension. Experimentally, the property is measured by taking sample strips of known dimensions gripped at each end of a clamp (jaws) and moving the jaws apart at a controlled speed until the sample breaks. The instrument records the change in the length of the sample, at various applied forces, due to elongation.

　　3. *Moisture permeation characteristics.* The moisture permeation characteristics are determined by the test procedure outlined in the USP XXII. In method 1, 10-unit-dose containers each containing desiccant pellets are sealed. A second set of 10-unit-dose containers is sealed without desiccant pellets to serve as controls. Both are stored at 75% ± 3% RH and 23° ± 2°C. Any moisture passing through the blister is absorbed by the desiccant. At the end of the specified storage period, the weight gained by the test and control samples is noted. The rate of moisture permeation is calculated from the difference in the weights of the test and control samples. The unit-dose package classification is based on the amount of moisture absorbed. The different types of unit-dose packages are described in Table 7.

　　4. *Adhesion.* The adhesion test measures the force necessary to separate two sheets of nonporous materials: for example, glassine laminated with wax. The sealed sheets are pulled apart at a specified rate and the force required to separate the sheets is recorded.

　　5. *Print tolerance.* Since packaging materials invariably carry a printed message, they must be capable of being printed. The printing surface of a lamination should also be thermally stable since it comes into contact with the heating elements. It is important to test the material for compatibility of the particular printing process and the ability to produce the required print quality.

　　Typical specifications for a forming film include [50]:

1. Fracture load
2. Elongation
3. Bond strength
4. Water vapor, oxygen, and carbon dioxide transmission

Table 7 USP/NF Classification of Permeability
of Unit-Dose Packages

Category	Moisture permeation
Class A	Not more than 1 of 10 containers exceeds 0.5 mg/day and none exceeds 1 mg/day.
Class B	Not more than 1 of 10 containers exceeds 5 mg/day and none exceeds 10 mg/day.
Class C	Not more than 1 of 10 containers exceeds 20 mg/day and none exceeds 40 mg/day.
Class D	Containers meet none of the foregoing requirements.

5. Thickness of the deep-drawn pockets
6. Amount of heat-seal coating (e.g., g/m^2)
7. Sealing capability measured at the temperature, pressure, width of the film, appropriate sealing partner, and peeling angle as used in packaging the commercial product

C. Packaging Validation

Pharmaceutical manufacturers have a moral and legal obligation to the customers to provide high-quality products that meet all the standards set forth in the approved New Drug Application (NDA) or Abbreviated New Drug Application (ANDA) for the product. It is the responsibility of the manufacturer to ensure that the product is manufactured according to CGMPs and that it is labeled properly and packaged correctly before release to the market. To accomplish these objectives, the various processes and operations that encompass the manufacturing and control have to be validated. Validation may be defined as a program aimed to provide documented evidence that a process has done, is doing, and/or will do what it purports to do, reliably and repeatedly [51,52]. Validation may also be described as a process of establishing documentary evidence that any product, process, activity, procedure, system, equipment, or software used in the control and manufacture of a drug product consistently performs or meets its predetermined specifications and quality attributes [53,54]. The concept of validation is applied within the pharmaceutical industry to a number of processes, operations, and systems (e.g., computer validation, process validation, analytical methods validation, packaging validation, etc.) as part of a CGMP mission to ensure consistent production of a quality product [55]. Thus validation is an integral and essential part of quality assurance and CGMP compliance.

There are several tangible benefits to packaging validation. These include increased productivity with greater throughput and capacity, reduced variability, improved quality, and reduced complaints and rejects. Validation also enhances understanding of how the process, equipment, or system works [56].

The need for validation stems from the fact that quality cannot be inspected or tested into the finished product. Quality, safety, and effectiveness must be designed and built into the product. Deficiencies in product design or manufacturing cannot be corrected by exhaustive finished-product testing [57]. It is impossible accurately to detect and segregate all defective units in any given batch. Each step in a manufacturing process must be validated to be assured that the total manufacturing operation is under control.

A general discussion of the concepts involved in a validation program can be found in published literature on the topic [52–56]. Validation data may be generated according to a preplanned protocol (prospective validation) or during actual running of the process (concurrent validation) or based on a review and analysis of historic information (retrospective validation).

Validation Program

There are four major components in a packaging validation program:

1. Raw materials control
2. Equipment qualification
3. Packaging process validation and maintenance of the validated state
4. Finished product testing

Raw Materials Control.   An important element of quality assurance is the control of raw materials during storage and dispensing [57,58]. High-quality raw materials are a necessary, albeit not a sufficient condition for the production of high-quality finished products [59]. Berry [58] has outlined several steps in raw material validation in the manufacture and processing of a pharmaceutical product. The following directions serve as a general guide for the validation of packaging raw materials:

1. List all the raw materials required in the packaging of the product, including containers, caps, liners, innerseal, cartons, labels, and so on.
2. Identify at least two suppliers for each raw material. It is important to have a primary and one or more alternative suppliers. If technical or business-related problems were to develop with one supplier, the packaging operations may be affected adversely unless more than one reliable and qualified supplier is on hand.
3. Obtain samples and certificates of analysis (C of As) from the supplier(s) to determine the extent of variation between different lots of material from the same supplier as well as variations between different suppliers for the same material.

4. Establish specifications for each raw material. Such specifications may be compendial, if the material is listed in USP/NF, or may be based on prior experience with the material in identifying properties that are critical to the performance of the product. Ensure that the test procedures and standards are incorporated in a written document.

5. Establish sampling procedures. A standard procedure for sampling raw materials, including the number of containers and closures, method of sampling, and other relevant details should be established and implemented. Any deviations from the established sampling procedures should be documented and explained.

6. Establish and document storage conditions, including location and temperature and humidity conditions in the storage area.

7. Establish expiration dating where appropriate, and retest procedures.

8. Write and implement standard operating procedures (SOPs) for the storage and control of all raw materials.

9. Challenge the raw materials by testing the packaged product for stability, compatibility, and absence of interaction with the product.

Complete testing of individual lots of raw materials may be avoided if a vendor certification program has been established [59–61].

Equipment Qualification.   Equipment qualification consists of determining whether the equipment is installed properly and verifying that it operates as intended. During the equipment qualification phase, all critical instruments, control devices, and components should be calibrated. Each module should be tested to verify that it performs as intended. The general outline of an installation and operational qualification protocol for packaging machinery [62–64] should include a complete description of the equipment, manufacturer's name and address, type, model, and serial numbers of the machine, a copy of the machine drawing and specifications, as well as having available an installation and operation manual.

As part of installation and identification, ensure that the pressure gauges, temperature gauges, and timers are installed and operating properly. The equipment should be identified and the required utilities (compressed air, power, and water) should be clearly described and provided. An appropriate change parts list should be prepared.

Equipment components requiring calibration should be identified and labeled. A copy of the SOPs for calibration should be provided. A calibration schedule should be established. The equipment should be checked to assure that it conforms to all the safety regulations and that the necessary safety precautions are clearly documented. A preventive maintenance schedule for the equipment should be established and documented as part of the SOP. An operational SOP and a cleaning SOP should also be written.

Operational Qualification (OQ).   As part of OQ, a placebo or a trial batch should be run to assure that the equipment operates as intended and within specifications. The exact conditions of use, timings, temperatures, pressures, materials used, and so on, should be recorded. A qualification report incorporating the description, installation, identification, safety, maintenance, SOPs, OQ, and other activities described above should be issued after review and approval by R&D, metrology, maintenance, safety and QA groups.

Packaging Process Validation.   Packaging process validation includes writing a validation protocol, data acquisition as per the written protocol, data evaluation, and documentation. The validation protocol should include the objective, operational procedure, acceptance criteria, change control procedures, revalidation procedures, and training of the personnel. The objective will depend on the specific equipment being qualified. All critical functions should be considered such as testing temperature range settings, timing devices, air volumes, speed of the machine, dwell times, and sealability. Each test should list the objectives of the test, a proposed method for conducting the test, and the acceptance criteria. Such methods should be as specific as possible, noting reasonable, quantitative ranges when possible [65,66].

A major difference between validation and simply checking the machine is that validation requires all aspects of the machine operation to be thoroughly investigated to establish safe and optimum operating parameters [67]. As part of validation, it is necessary to have a change control procedure: a formal mechanism for review and approval by qualified and authorized personnel of any changes that may affect the validation status of the system. Corrective action to restore the validation status should be implemented promptly. When any significant changes are made to the system, equipment, SOP, or software, there should be a mechanism to communicate the changes made to all the actual and potential users of the system. Such change control procedures should be incorporated in a SOP.

There should be a quality assurance system in place that requires revalidation whenever there are significant modifications to the manufacturing process, equipment, raw materials or procedures, which affect the product quality or reproducibility of the process. The quality assurance procedures should establish the circumstances under which revalidation is required. It is desirable to assign to designated persons the responsibility to review product, process, equipment, and personnel changes to determine if and when revalidation is warranted.

Finished Product Testing.   From a packaging perspective, finished product evaluation includes inspection of the container, the number of units filled in the bottle, checking the weights, checking the cap liner, seal band, and torque, labeling accountability, accuracy and placement of the label on the container, justification, and placement of containers inside the carton. In the case of unit-dose blister packaging, some of the likely defects to be examined for the packaged product are incorrect or mixed lot numbers, illegible lot numbers, absence of a lot number on the immediate container, serious contamination with oil/grease, incorrect or missed label, empty pocket in a blister, missing blister unit, chipped or broken tablet or capsule, damaged blister, scorched blister, and soiled or dirty label.

A packaging system becomes validated when at least three successive batches are packaged within specifications following a validation plan [65]. It is essential to issue a validation report that includes the test results and all SOPs pertaining to the operation of the machine/process and cleaning procedure. Regulatory agencies adopt the stance: "If it is not documented, it never happened." The report should summarize all the protocols and refer to all the documentation, such as SOPs and manuals. It should also contain the results of tests performed in accordance with the validation protocol along with the acceptance criteria.

An important part of validation effort is to provide detailed formal training on all aspects of the packaging system to the operators, managers, and support personnel. The training should include [68]:

1. *Machine training*: setup, tear-down, cleaning, product, and size changes to enable safe and effective operation of the equipment
2. *Materials training*: an understanding of the specifications and properties of both packaging materials and the products involved
3. *Troubleshooting*: application of diagnostic skills in resolving operational problems
4. *Compliance with CGMPs*: to perform all activities in accordance with CGMPs and to note and complete all documentation accurately

The appropriate SOPs together with the written manuals should be distributed to the trainees. The formal training received by the employees, in seminars, vendor-sponsored courses, and professional meetings should be documented.

V. REGULATORY CONSIDERATIONS

There are a number of regulatory considerations that influence selection of the packaging materials and operations. Among the current regulations in the United States that affect packaging are:

1. Guidelines for submitting documentation for packaging for human drugs and biologics (February 1987) [69]
2. Current good manufacturing practices (CGMPs) [70,71]
3. Guidelines for the manufacture of investigational products (February 1988)
4. Tamper-resistant packaging for OTC products

5. Poison prevention packaging act, 1970
6. Environmental regulations in the disposal of packaging waste

Of the regulations listed above, items 1 to 4 are part of CGMPs applicable to the manufacture and packaging of commercial and investigational drug products.

A. CGMPs and Related Issues

Guidelines for submitting documentation for packaging of human drugs and biologics were issued in 1987 under 21 CFR 10.90. These regulations are intended to provide guidance to pharmaceutical manufacturers on the packaging data that should be included in the NDA or ANDA submission. The regulations stipulate that:

1. The drug packaging must maintain the standards of identity, strength, quality, and purity of the drug or the drug product throughout its shelf life.
2. The packaging materials selected for use with solid oral products must be clearly identified and described in the submission and be characterized for their physicochemical properties to ensure that they meet appropriate compendial specifications.
3. The stability of the product through its labeled expiration date should be demonstrated through suitable testing using the same container/closure system at the torque pressures used in the marketed product.

In the case of investigational products, the IND should describe in detail the packaging system used for the dosage form. The selected package should assure adequate protection from the time of manufacture until the time of clinical use [72]. The NDA submission should include chemical and physical characteristics, methods used, and the test results of the container, closure, and other component parts of the drug packaging as part of the NDA requirements.

Changes in the packaging of a marketed product can be made according to the procedure outlined under 21 CFR 314.70. The applicant should provide a full description of the new packaging components and the sources and provide the specifications and test results for these components. Stability studies that were originally performed to establish the expiration date for the drug product must be expanded to include the proposed alternative container or closure to assure that the identity, strength, quality, and purity of the drug product will be maintained throughout the expiration period [69].

Packaging controls to be applied in the manufacture of finished pharmaceuticals are described under 21 CFR 211.80 entitled ''Control of Components and Drug Product Containers and Closures'' [73]. The precautions and controls used for raw materials and other components of the formulations are also applicable to packaging materials. For example, the regulations require that "there shall be written procedures describing in sufficient detail, the receipt, identification, storage, handling, sampling and testing, and approval or rejection of components and drug product containers and closures; such written procedures shall be followed. . . . Each lot of components, drug product containers, and closures shall be withheld from use until the lot has been sampled, tested, or examined, as appropriate, and released for use by the quality control unit."

Containers and closures should be tested for conformance with all appropriate written specifications by the manufacturer or, alternatively, the certificate of testing may be accepted from the supplier provided that at least a visual identification is conducted on such containers and closures by the manufacturer of the drug product coupled with establishing the reliability of the data through appropriate validation of the supplier's raw materials test results.

Retesting of containers and closures may be necessary after prolonged storage or after exposure to conditions that may adversely affect the container or closure. According to 21 CFR 211.94, (a) "Drug product containers and closures shall not be reactive, additive, or absorptive so as to alter the safety, identity, strength, quality or purity of the drug beyond the official or established requirements." (b) "Container closure systems shall provide adequate protection against foreseeable external factors in storage and use that can cause deterioration or contamination of the drug product."

B. Tamper-Resistant Packaging Requirements for OTC Human Drug Products

Tamper-resistant packaging requirements for over-the-counter (OTC) human drug products are specified in 21 CFR 211.132. According to these regulations, OTC solid oral drug products, tablets or capsules, for retail sale would be required to be packaged in a tamper-resistant package if the product is accessible to the public while held for sale. As defined in the regulations, "a tamper resistant package is one having one or more indicators or barriers to entry which, if breached or missing, can reasonably be expected to provide visible evidence to consumers that tampering has occurred." Such a package may involve an immediate-container or closure system or secondary-container carton system or any combination of systems intended to provide an indication of package integrity. The regulations further stipulate that for two-piece, hard gelatin capsule products subject to this requirement, a minimum of two tamper-resistant packaging features is required unless the capsules are sealed by a tamper-resistant technology. For all other products subject to this requirement, including two-piece, hard gelatin capsules that are sealed by a tamper-resistant technology, a minimum of one tamper-resistant feature is required.

Each retail package of a solid oral OTC drug product is required to bear a statement on the label alerting the consumer to the specific tamper-resistant feature of the package.

C. Poison Prevention Packaging Act of 1970

The Poison Prevention Packaging Act of 1970, administered by the U.S. Consumer Product Safety Commission, mandates that certain drug products, including prescription products, which are hazardous or potentially hazardous must be sold in safety packages that most children under 5 years of age cannot open. For people who cannot open such packages, a traditional easy-to-open package may be obtained upon request by a prescribing physician or the consumer. Alternatively, the manufacturer can choose to market, in addition to child-resistant packaging, one size of the product in conventional packaging with a label indicating that the package is not child-resistant. The commission of the European Communities has published a list of approved safety packages which includes both bottle and blister packages [74].

D. Environmental Regulations

In the United States, waste from plastic, paper, and glass materials is categorized as nonhazardous waste and is governed by state and local regulations and laws. The protection of human health and environment from improper management of hazardous waste is controlled by the Resource Conservation and Recovery Act (RCRA) of 1984, administered by the Environmental Protection Agency (EPA).

VI. ROLE OF PACKAGING ON THE STABILITY CHARACTERISTICS OF THE PRODUCT

A. General Considerations

Primary packaging is an important determinant of the storage stability of the product. Since many of the SR and CR products contain waxes or are coated with polymers during their preparation, packaging along with storage conditions become critical in maintaining the integrity of the product. Packaging is a primary barrier in the protection of the dosage form from the adverse effects of environmental conditions. For example, in hard gelatin capsule products, the shell normally contains 14 to 17% water. Any storage condition/container/closure combination that causes the moisture content to fall below 12% or rise above 18% can alter the physical or chemical characteristics of the product. Below 12% shell moisture, the capsules become brittle and break easily. Above 18%, they become sticky and moist [75,76].

The filled hard gelatin capsule is not gas-tight unless it is banded. Oxygen and carbon dioxide can readily pass through the overlap region of the cap and body and diffuse through the shell wall. However, the quantity of gas passing through such pathway does not pose a significant problem, except with products that are very susceptible to oxidation, and the concentration of these gases in the external environment is very high. Dry dosage forms are marketed in screw-capped, tightly closed glass or plastic bottles, with any additional free space in the container filled with cotton. Under these circumstances, the amount of gas absorbed by filled capsules or coated tablets is negligible. Therefore, during storage of these products, protection is needed primarily against ingress of moisture.

A number of published articles in the pharmaceutical literature describe the stability of commercial solid formulations as a function of packaging and storage conditions [77–81]. In one such investigation by Conine et al. the stability of commercial formulations of cephradine and cephalexin capsules was evaluated at 5°C, 25°C, and 40°C/75% RH in various packages and in open-dish experiments [80]. Cephalexin capsules showed little or no decomposition under all the conditions studied, while cephradine capsules were unstable under accelerated conditions of storage, 40°C and 75% RH, both in plastic bottles and blister packages. However, by repackaging the product in a moisture-tight amber glass container, excellent stability was attained. The data thus suggested a need for highly protective packaging for the formulation. Giacin et al. [81] describe studies wherein the chemical stability of ibuprofen tablets was evaluated in unit-dose containers, PVC, poly(chlorotrifluroethylene) (Aclar), and poly(vinylidene chloride) (Saran) blisters as well as multiple-unit containers of 30 and 50 count. The latter was studied in both closed and repetitive opening and closing conditions for up to 180 days at 22.2°C/50% RH and 25.5°C/85% RH. The findings revealed that at 22.2°C/50% RH, there was no significant gain in moisture for all packaged products; however, under stressed storage conditions, the product packaged in Aclar was most stable, and tablets packaged in plain PVC were least stable. The stability of the product in multiple-unit containers subjected to repeated opening and closing was in between unopened plain PVC and unopened Aclar. The authors assert that under certain storage conditions, blister packaging may provide a longer product shelf-life than the traditional multiple-unit bottle, depending on the permeability characteristics of the package and the moisture absorbed by the product.

In one of the most comprehensive studies, Beal et al. [82] reported on the comparative stability of pharmaceutical products packaged in plastic and glass containers. They stored a number of solid and liquid oral products in HDPE bottles and compared the stability with that of products stored in USP type III glass bottles. The stored products were tested for compatibility and stability after storage in 14 to 80% RH and a temperature range of 40 to 140°F. Some of the samples were stored at as low as −65°F for periods up to 48 weeks. The data indicated that all dry dosage forms studied were stable between 40 and 100°F at 80% RH for 48 weeks. For coated tablets, the most common failure occurred in the form of cracking and crazing of the coating at very low temperatures (−65°F). The stability of the product stored in HDPE bottles was found to be comparable to the stability of the product stored in glass containers. These studies suggest that HDPE is an excellent moisture-barrier plastic for most pharmaceutical products. As would be expected, packaging conferred no protection to the samples from the effects of heat.

An example of the importance of packaging in protecting the integrity of SR formulation is shown in Fig. 7, using an experimental formulation. While the tablets stored in HDPE bottles and foil/foil blister are well protected, samples packaged in laminated PVC/PE/PVDC (triplex) blister showed marked retardation in the dissolution rate after 3 months of storage at 37°C/75% RH (Parke-Davis files, unpublished data). There was no change in the assay values or appearance. These data reveal the need to test the stability samples for both physical and chemical characteristics in evaluating the moisture-barrier properties of the packaging materials selected.

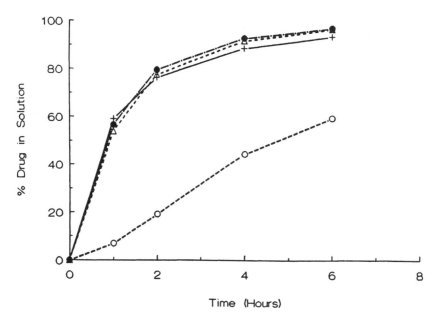

Fig. 7 Effect of packaging on the in vitro release profiles of tablets stored at 37°C/75% RH for 3 months. +, Initial; △, HDPE bottle; ○, triplex blister; ●, foil/foil/blister.

Taborsky-Urdinola et al. [83] reported the effects of packaging on the dissolution characteristics of USP prednisone calibration tablets packaged in multiunit and unit-dose containers. The tablets were stored from 3 to 6 months at 40°C/85% RH, 37°C/75% RH, and 22°C/75% RH. The packaging materials studied included a multiple-unit child-resistant polypropylene 12-dram prescription vial, a "well-closed" amber polystyrene child-resistant vial, foil-foil strip packaging, and a polyethylene bag. Samples of tablets were also stored in open containers for 7 days as negative controls. Tablets stored at 40°C/85% RH in open containers exhibited greatest reduction in dissolution; the amount of drug in solution at the 30-min time point decreased from 63% of label claim after 1 day of storage to 1% of label claim after 4 days of storage. PP vials and foil/foil strip packaging offered the greatest protection followed by PS vials. PE bags offered the least protection in retaining the initial dissolution properties. Most of the changes in the dissolution profiles occurred after a storage period of several weeks, suggesting that a minimum threshold level of moisture accumulation within the product is probably necessary before a noticeable change in the dissolution pattern begins to emerge. Such a lag time is dependent on the product, packaging materials used, and the storage conditions.

The shelf life of a given drug product is a function of the packaging or barrier properties of the container, storage conditions, and the intrinsic stability or ruggedness of the product. Packaging addresses the barrier properties by reducing moisture permeation and leakage and protecting the product against atmospheric oxygen, carbon dioxide, and light.

B. Mechanism of Permeation

There are two distinct mechanisms by which gases and water vapor can enter or leave the package: permeation and leakage. Permeation involves the passage of gases or water vapor through the barrier material by dissolution onto one surface, diffusion through the material, and desorption through the other surface. Leakage involves both convection and diffusion through discontinuities in the packaging material, such as cracks and gaps between the closure and the bottleneck finish [84].

The rate of permeation of water vapor is given by the expression [85,86]

$$\frac{dw}{dt} = \frac{K}{L} A(P_e - P_i)$$
$$= \frac{K}{L} A(H_e - H_i)$$

where

K = permeability constant of the packaging material
L = wall thickness

A = surface area of the package
P_e = partial pressure of water vapor outside the package
P_i = partial pressure of water vapor inside the package
H_e = relative humidity outside the package
H_i = relative humidity inside the package

The values P_i and H_i are obtained from a knowledge of the equilibrium moisture content of the product at various relative humidities. Knowing the permeability constant, surface area of the container, and its thickness, one can calculate the rate and amount of water transferred over a given time period.

Permeability of a given material can be reduced by the use of highly crystalline plastic (e.g., PP), addition of titanium dioxide, lacquering, or the use of combination materials such as laminates of polymeric materials with aluminum foil. In general, permeability increases with increase in temperature so that products stored at elevated temperatures are apt to experience the effects of higher permeation rates than the corresponding samples stored under ambient conditions. As described in the foregoing equation, the size of the package is a significant variable in the evaluation of permeability. The surface available for permeation in a package increases as a square function of the size, while the volume of the product that absorbs the permeant increases as a cube function. Therefore, all other factors being constant, the amount of permeant transferred to the product is less with a larger package size.

Permeation may be viewed as the flow of water vapor from an environment of higher chemical potential to a region of lower chemical potential through a matrix. Permeant molecules pass through temporary voids in the amorphous polymer regions created by the temperature-dependent segmental motion of the polymer chains. The magnitude of the diffusion rate is dependent on both the permeant and the polymer.

The barrier properties of a polymer are dependent on the tightness of molecular packing and forces that restrict the segmental movement of a polymer. The higher the degree of crystallinity in a polymer, the lower the permeability. An equation that describes the relationship between the reduction in permeability caused by crystallinity is given by Murray [87]:

$$\frac{P_c}{P_a} = a^b$$

where P_c is the permeability of the crystalline polymer, P_a the permeability of the same polymer completely amorphous, a the amorphous fraction in the polymer, and b is a constant whose value is about 2.1 for gases and 2.25 for water vapor. A quantitative knowledge of the permeability of plastic films for a specific combination of product and exposure conditions enables optimization of multilayer laminations and film thicknesses with respect to cost and performance.

Leakage can occur through poor sealing or through pinholes in the barrier film. Leakage caused by a discontinuity in the package (e.g., defective seams or gaps between the closure and the neck finish) can be reduced by modifying the design of the mechanical seal. Since container, closure, seal, torque, and so on, affect the permeability of the package, during packaging development it is important to test the product using the same packaging, container/closure/seal, and torque conditions as intended for use in the marketed package [88].

C. Mathematical Treatment of Packaging Permeability

An important objective during packaging development is to project the shelf life of the product. A number of mathematical expressions are presented in the literature relating the equilibrium moisture content of the dosage form to the storage conditions. In general, these models assume (1) that decomposition of the product results solely from the moisture transferred through the package, (2) all the moisture transferred from the environment to the inside of the container is absorbed by the product, and (3) there is a linear relationship between moisture content and one or more product quality attributes (e.g., chemical assay or dissolution). The object of mathematical modeling is to minimize experimentation and to construct a physical model to predict the stability under ambient conditions. Such modeling is usually based on (1) establishing a linear relationship between the chemical decomposition rate constant and humidity at any given temperature, and (2) determination of the amount of water that permeated through the package at a given time point. Following this line of reasoning, Nakabayashi applied Kitazawa's equation to estimate the decomposition rate constant over a humidity range of 0 to 80% RH for prednisolone tablets [89]:

$$\ln \frac{C_s}{C_s - C_t} = Kt$$

where C_s represents the total drug content of the tablets, C_t represents the amount of intact drug at time t, and K represents the degradation constant.

Using a multiple regression technique, the following empirical relationship between the moisture content and chemical assay value at different temperatures was constructed [90]:

$$\ln K = a + b \ln m - \frac{c}{T}$$

where m denotes the moisture content of the product and T the absolute temperature and a, b, and c are constants. This equation enabled the authors to predict the shelf life of the prednisolone tablets. The actual experimental data were in reasonable agreement with the predicted values.

A similar concept was applied by Amidon and Middleton, to construct a physical model that relates the amount of moisture absorbed by blister-packaged tablets to the relative humidity of the environment. The moisture content of the dosage form is given by the equation [91]

$$M_t = KC_o - (KC_o - M_0)e^{-APT/k}$$

where M_0 and M_t represent the moisture contents of the dosage form initially and at time t, respectively; C_o the external water concentration outside the packaging; A the area of the package; P the permeability of the package; T the absolute temperature; and K a measure of the capacity of the dosage form to absorb water. The next step in the model is to construct a relationship between the moisture content and a physical property such as dissolution. Such a relationship should enable an estimation of the rate constant for that physical property as a function of storage conditions and packaging.

In actual practice, temperature and humidity fluctuations in the environment and the presence of a desiccant in the container greatly modify the projected shelf life. For many preparations there is a critical moisture content below which the drug is stable. This level of moisture may be bound moisture or the water of crystallization. Decomposition of the product is initiated when the amount of moisture in the system reaches this level [92]. The presence of a desiccant will prolong the time at which the product reaches this critical moisture level, because the desiccant will take up the moisture and becomes saturated before moisture uptake by the drug will begin. Therefore, accurate prediction of shelf-life based on mathematical models is an exceedingly difficult task.

D. Practical Aspects of Stability Evaluation of Drug Products

In pharmaceutical packaging development for oral solid products, it is a common practice to test the product in HDPE bottles under accelerated conditions of storage (e.g., 37°C/75% RH or 40°C/75% RH for 3 months). The headspace in the container is filled with cotton. If the formulation is moisture sensitive, a desiccant is included in the bottle as an added protection. Similarly, if the formulation is photosensitive, a light-resistant container is selected. A tentative expiration date is projected based on evaluation of the data generated under stressed storage conditions, while parallel studies are carried out under ambient or proposed labeled storage conditions. Typically, such expiration dates range from 2 to 3 years from the time of product manufacture.

Because the products are repackaged, medications dispensed in vials, PP or PS, in retail pharmacies do not carry an expiration dating. Instead, patients may be instructed not to use the product after a certain period, often referred to as the beyond-use date. This is the time period after which the product may not retain the initial physicochemical characteristics. While setting the beyond-use date for products dispensed to the consumer, it is important to recognize that the medication is usually consumed by the patient over a relatively short time (e.g., 1 to 4 weeks), in contrast to the longer period during which the product is stored with the wholesaler, pharmacies, warehouses, or other distribution channels. In the hands of the consumer, the product may be subject to high temperatures or humidities, as for example, when a drug product is stored in medicine cabinets located in or near bathrooms. Since stability and expiration dating for a given product are dependent on the storage conditions as well as the packaging containers used, the practice of labeling the repackaged product with the expiration date on the original bulk containers is not scientifically valid and should be discouraged. In other words, the beyond-use date for the dispensed package should not be extrapolated from the expiration date shown on the bulk container. It is the responsibility of the dispensing pharmacist to set the beyond-use date, taking into account the nature of the drug product, the characteristics of the containers, and the storage conditions to which the product may be subjected. A general guideline, offered in the compendia, states that "in the absence of stability data to the contrary, the beyond-use date should not exceed, (1) 25% of the remaining time between the date of repackaging and the expiration date on the original manufacturers bulk container, or (2) a six month period from the date the drug is repackaged, whichever is earlier" [93].

In dealing with solid oral pharmaceutical products, agents that can cause physical or chemical instability are primarily moisture and less commonly oxygen and ultraviolet light [94]. Thus protection against water vapor transmission is critical to the integrity of a number of products. The data presented in Table 4 reveal that many of the packaging materials commonly used in pharmaceutical practice are effective in protecting against moisture.

VII. DISPOSAL OF WASTE PACKAGING MATERIALS

The disposal of packaging waste has been gaining a lot of attention during the last decade in the pharmaceutical industry. Based on the adage "what goes around, comes around," all of the packaging materials used in the design, fabrication and manufacture, storage, and distribution of pharmaceutical products must ultimately be disposed of as waste. Packaging waste consists primarily of plastic, glass, paper, and metals, notably aluminum. There are four well-established methods of waste disposal: incineration, use of land fills, recycling, and biodegradation. Among these, the use of landfills is rapidly declining in popularity as a means of industrial waste disposal due to lack of availability of adequate number of sites, widespread opposition from local environmental groups, not-in-my backyard (NIMBY) syndrome, and high liability for future

Superfund cases. Disposal of packaging materials is therefore accomplished mostly through incineration and recycling [95].

Incineration as a means of waste disposal is widely practiced in Japan and also in some European countries, notably Sweden and Germany. Incineration usually reduces the volume and the amount of waste to 1 to 2% of its original value. The primary problem with such a mode of disposal is the emissions discharged from incineration. For example, incineration of PVC releases hydrochloric acid, which in addition to being a pollutant, causes damage to chimneys and furnaces. Other plastics emit dioxane, posing significant risks to the environment, although current evidence suggests that dioxane emissions are not related to PVC incineration [96]. Environmental pollution can be minimized by incinerating under controlled temperatures and by fitting the incinerators with exhaust gas scrubbers and high-efficiency particle collectors. Commonly used postcombustion control devices are wet scrubbers, dry scrubbers, electrostatic precipitators, and fabric filters. Scrubbers primarily remove acid gases, including sulfur oxides and hydrogen chloride from the flue stream. Other measures to reduce pollutants include the use of lead-free PVC films, removal of plasticizers, and the elimination or minimization of vinyl chloride monomers (VCMs).

One problem with incineration is the need to separate metals and noncombustible materials from the waste prior to incineration, which increases the cost of the disposal process. A second problem is the transportation of hazardous waste when a combustion facility is located a long distance from the waste-generating plant.

EPA encourages recycling efforts (e.g., reclaiming of solvents, metals, purification, etc.). PVC, for example, can be recycled in several ways. The waste resulting from PVC processing can be immediately recycled to produce new PVC. If PVC waste is collected separately, pure PVC products can again be made from them for use in pipes. Alternatively, used PVC can be collected with other plastics, recovered, and made into new plastic products (e.g., insulation walls, wine posts, etc.) [97]. However, recycling multilayer or laminated packaging materials presents problems since different layers are not easily separable.

There is a movement in Europe to ban the use of chlorine-containing plastics such as PVC, due to the emission of hydrochloride. The use of PVC is discouraged in Germany, Switzerland, and Austria. In Germany, efforts are under way to replace PVC forming films by PP films. Most blister-packaging machines built today in Europe are designed so as to be able to run on PP. Future trends in packaging waste disposal point toward recycling, which might increase the use of aluminum or PP blister pack, where the same materials serve as forming and lidding films.

Biodegradation as a means of reducing solid waste is a topic of great current interest. For this process to be effective, the material must be moistened, turned over, and allowed to absorb heat. Aerobic organisms need plenty of oxygen to work quickly and efficiently to degrade the product. In actual practice, most of the landfills are capped and usually exclude oxygen. Consequently, the land fill is permanently anaerobic, which causes slow decomposition and produces methane with potential for explosion. In laboratory studies, the degradation process is expedited by injecting microbial organisms. However, such a technology is still in the exploratory stage and is not ready for widespread use.

Efforts to date have focused primarily on dealing with waste rather than minimizing waste generation. The environmental regulations of the future will probably emphasize waste minimization and would require a fundamental shift in the current design of packaging and will require the use of minimum amount of materials needed to assure the integrity of packaging [98–100]. A few examples of waste reduction are:

1. Elimination of excessive packaging that does not add value or contribute to the quality of the product
2. Use of printing inks that minimize volatile organic compounds
3. Avoidance of packaging and packaging components, including printing inks, dyes, pigments, adhesives, or any other additive to which lead, cadmium, mercury, or hexavalent chromium has been added

VIII. CURRENT TRENDS IN THE PACKAGING OF MULTIPARTICULATE DOSAGE FORMS

Several new trends have emerged recently in the packaging of pharmaceutical solid oral dosage forms. New technologies aimed at packaging the product at higher speeds, with improved quality at reduced cost while complying with CGMPs and appropriate environmental regulations are chartering pharmaceutical package development into new and exciting frontiers. Some of the recent advances are:

1. Automated inspection and the use of electronic sensors in monitoring packaging lines (e.g., automated vision systems, computer-integrated manufacturing operations, etc.)
2. Validation of packaging systems, including equipment installation qualification, computer validation, cleaning validation, calibration of machine tooling, and raw materials control
3. Introduction of strip packaging in the marketing of some select products (e.g., OTC analgesics)
4. Substitution of PVC with PP in some blister-packaging applications
5. Efforts to reduce the amount of packaging material entering the waste stream by eliminating unnecessary packaging

REFERENCES

1. I. Ghebre-Sellassie, in *Pharmaceutical Pellet Technology* (I. Ghebre-Sellassie, ed.), Marcel Dekker, New York, 1989, p. 7.
2. R. L. Robinson and R. G. Hollenbeck, Manufacture of spherical acetaminophen pellets: comparison of rotary processing with multiple step extrusion and spheronization, *Pharm. Technol. 15*(5):48 (1991).
3. *Physicians' Desk Reference*, 46th ed., Medical Economics Data, Montvale, N.J., 1992, pp. 403–437.
4. W. Guise, Sachet packaging, *Manuf. Chem. 57*(8):57 (1986).
5. W. Guise, Form-fill-seal pillow packs and sachets, *Manuf. Chem. 58*(12):31 (1987).
6. D. L. Smith, Compliance packaging: a patient education tool, *Am. Pharm. NS29*(2):42 (1989).
7. E. D. Martin, R. J. L. Frazer, and I. Camens, Storage of phenytoin capsules, *Med. J. Aust. 143*:634 (1985).
8. *Draft Guidelines on the Preparation of Investigational New Drug Products (Human and Animal)*, FDA, Department of Health and Human Services, Rockville, Md., February 1988.
9. G. S. Banker and C. T. Rhodes, in *Modern Pharmaceutics*, 2nd ed. (G. S. Banker and C. T. Rhodes, eds.), Marcel Dekker, New York, 1990, pp. 862–865.
10. J. P. Boehlert, Pharmaceutical quality: a 15-year retrospective, *Pharm. Technol. 16*:38 (1992).
11. W. A. Vadino, Pre-approval inspections: a research perspective, *Pharmaceutical Technology Conference*, New Brunswick, N.J., September 1991.
12. D. C. Liebe, Packaging of pharmaceutical dosage forms, in *Modern Pharmaceutics*, 2d ed. (G. S. Banker and C. T. Rhodes, eds.), Marcel Dekker, New York, 1990, pp. 695–740.
13. G. E. Reir, Excipient standardization: user's viewpoint, *Drug Dev. Ind. Pharm. 13*:2389 (1987).
14. K. S. Murthy, C. Onyiuke, M. Skrilec, and J. R. Draper, Organization and operation of a centralized raw materials management unit in pharmaceutical product development, *Pharm. Technol. 15*(3):142 (1991).
15. Tamper-resistant packaging requirements for over-the-counter human drug products, *Current Good Manufacturing Practice for Finished Pharmaceuticals, Including Revisions through 1986*, CFR 211.132, FDA, U.S. Department of Health and Human Services, Rockville, Md.
16. Environmental liability, *Manuf. Chem. 61*(5):49 (1990).
17. I. C. Canadine, Impact of safety and environmental concern, *Manuf. Chem. 60*(1):33 (1989).
18. *Incineration and Energy Recovery: An Environmentally Sound Approach to the Solid Waste Problem*, Council on Plastics and Packaging in the Environment, Washington, D.C., April 1989.
19. D. A. Dean, Stability aspects of packaging, *Drug Dev. Ind. Pharm. 10*:1463 (1984).

20. R. Gerstman, When to redesign a pharmaceutical package, *Pharm. Technol.* 6(12):40 (1982).
21. F. Reiterer, Blister packaging for the pharmaceutical industry, *Pharm. Technol.* 15:74 (1991).
22. C. P. Croce, A. Fischer, and R. H. Thomas, in *Theory and the Practice of Industrial Pharmacy* (L. Lachman, H. A. Lieberman, and J. L. Kanig, eds.), Lea & Febiger, Philadelphia, 1986, pp. 711–732.
23. W. Miller, in *Encyclopedia of Polymer Science and Engineering*, Vol. 3, 2nd ed. (H. F. Mark, N. Bikales, C. G. Overberger, G. Menges, and J. I. Kroschwitz, eds.), Wiley, New York, 1985, pp. 463–491.
24. P. E. Campbell, Plastics in drug packaging, *19th National Meeting of the Academy of Pharmaceutical Sciences*, Atlanta, November 1975.
25. B. Guise, Variations on blister packaging, *Manuf. Chem.* 62:(5):35 (1991).
26. C. D. Marotta, in *Encyclopedia of Polymer Science and Engineering*, Vol. 10, 2nd ed. (H. F. Mark, N. Bikales, C. G. Overberger, G. Menges, and J. I. Kroschwitz, eds.), Wiley, New York, 1985, pp. 684–721.
27. *United States Pharmacopoeia XXII, National Formulary XVII*, United States Pharmacopeial Convention, Rockville, Md., 1990.
28. *Poison Prevention Packaging Act of 1970*, Regulations CRF § 1700.15, 1981.
29. FDA, *Fed. Regist.*, 53(87): 16192 (1988).
30. M. Bakker, ed., *Wiley Encyclopedia of Packaging Technology*, Wiley, New York, 1986.
31. Multiform Desiccants, Inc., Buffalo, New York, Moisture in packaging: selecting the right desiccant, *Packag. Eng.*, June 1985.
32. R. L. Dobson, Protection of pharmaceutical and diagnostic products through desiccant technology, *J. Packag. Technol. 1*, August 1987.
33. J. T. Carstensen, *Drug Stability: Principles and Practices*, Marcel Dekker, New York, 1990, pp. 301–336.
34. J. P. Ausikaitis, in *Kirk-Othmer's Encyclopedia of Chemical Technology*, Vol. 8, 3rd ed. (H. F. Mark, D. F. Othmer, C. G. Overberger, and G. T. Seaborg, eds.), Wiley, New York, 1979, pp. 114–130.
35. Drug packaging, *Manuf. Chem.* 60(2):37 (1989).
36. E. D. Hancock, Unit dose blister packaging of pharmaceuticals for healthcare compliance, Interphex USA, *Proceedings of the 1991 Technical Program*, New York, April 1991.
37. M. Larson, Drug packaging hospitals want, *Packaging 36*, (1) (1991).
38. B. Guise, Blister packaging today, *Manuf. Chem.* 55(11) (1984).
39. K. A. Christiansen, The most commonly used blister films for pharmaceutical packaging, *Caribe Pack '90 Sponsored by Klockner-Pentaplast of Latin America*, San Juan, Puerto Rico, September 1990.
40. J. T. Reamer and L. T. Grady, Moisture permeation of newer unit dose repackaging materials, *Am. J. Hosp. Pharm.* 35:787 (1978).
41. R. L. Giles and R. W. Pecino, In *Remington's Pharmaceutical Sciences*, 18th ed. (A. R. Gennaro, ed.), Mack Publishing, Easton, Pa. 1990, pp. 1499–1503.
42. B. Guise, Flexible packaging, *Manuf. Chem.* 53(4):63 (1982).

43. R. B. McClosky, Strip packaging of tablets and capsules, *Proceedings of a Symposium on Packaging Technology*, Medical Manufacturing Tech Source, Key Biscayne, Fla., March 1990.

44. Strip packaging at Watford General, *Pharm. J. 224*:634 1980.

45. S. Sachrow, The effect of pinholes on the barrier properties of aluminum foil, *Pharm. Manuf. 2*(9):14 (1985).

46. C. G. Davis, *Introduction to Packaging Machinery*. Packaging Machinery Manufacturing Institute, Washington, D.C., 1990.

47. Uhlmanns Pac-Systems GmbH & Co., Laupheim, Germany.

48. *United States Pharmacopoeia XXII, National Formulary XVII*, United States Pharmacopeial Convention, Rockville, Md., 1990, p. 1573.

49. S. J. Borchert, G. A. Kelley, and E. A. Hardwidge, A program for identification testing of package materials, *Pharm. Technol. 7*:72 (1983).

50. R. C. Griffin, Jr., S. Sacharow, and A. L. Brody, eds., *Principles of Package Development*, 2nd ed., AVI, Westport, Conn., 1985.

51. *Guidelines on General Principles of Process Validation*, Center for Drugs and Biologics and Center for Devices and Radiological Health, FDA, Rockville, Md., May 1987.

52. E. A. Leonard, in *Packaging, Specifications, Purchasing and Quality Control* (E. A. Leonard, ed.), Marcel Dekker, New York, 1987.

53. PMA's Validation Advisory Committee, Process validation concepts for drug products, *Pharm. Technol. 9*(9):78 (1985).

54. G. K. Estes and G. H. Luttrell, An approach to process validation in a multiproduct pharmaceutical plant, *Pharm. Technol. 7*(4):74 (1983).

55. R. C. Branning and L. D. Torbeck, Validation and experimentation, *Pharm Tech Conference '91 Proceedings*, Aster Publishing, Eugene, Oreg., 1991, pp. 549–561.

56. R. Kieffer and J. Nally, Implementing total quality in the pharmaceutical industry, *Pharm. Technol. 15*(9):130 (1991).

57. I. R. Berry, Process validation: practical applications to pharmaceutical products, *Drug Dev. Ind. Pharm. 14*:377 (1988).

58. I. R. Berry, in *Pharmaceutical Process Validation* (B. T. Loftus and R. A. Nash, eds.), Marcel Dekker, New York, 1984, pp. 203–250.

59. W. Chambliss, The characterization of raw materials, *Pharm. Technol. 8*(6):83 (1984).

60. J. P. Boehlert and R. Gomez, Establishing a vendor certification program, *Pharm. Technol. 12*(11):54 (1988).

61. D. A. VonBehren, Certified vendor programs: a vendor's perspective, *Pharm Tech Conference Proceedings '90*, New Brunswick, N.J., September 1990.

62. P. Carr, The role of packaging materials play in the design and validation of packaging equipment, *Proceedings Validating Pharmaceutical Packaging/Diagnostics Processes and Related Equipment Applications*, seminar sponsored by Avalon Communications, East Brunswick, N.J., May 1992.

63. J. F. Nykanen, Protocol development for validation of pharmaceutical packaging

equipment, *Proceedings Validating Pharmaceutical Packaging/Diagnostics Processes and Related Equipment Applications*, seminar sponsored by Avalon Communications, East Brunswick, N.J., May 1992.

64. J. C. Ball, Validation of torque/retorque test instrumentation, *Proceedings Validating Pharmaceutical Packaging/Diagnostics Processes and Related Equipment Applications*, seminar sponsored by Avalon Communications, East Brunswick, N.J., May 1992.

65. M. Liszkay, Qualification and validation in pharmaceutical packaging, *Pharm. Ind. 47*(3):303 (1985).

66. K. F. Popp, Organizing technology transfer from research to production, *Drug Dev. Ind. Pharm. 13*(13):2339 (1987).

67. A. Isaacs, Validating machinery with electronic control systems, *Manuf. Chem. 60*(2):19 (1992).

68. A. R. Kroll, New working practices for blister packaging lines, *Manuf. Chem. 60*(11):33 (1989).

69. *Guidelines for submitting documentation for packaging for human drugs and biologics*, FDA, U.S. Department of Health and Human Services, Washington, D.C., February 1987.

70. J. L. Turner, GMP for Europe: how to stay ahead of the pack, *Manuf. Chem. 60*(12):24 (1989).

71. M. H. Anisfeld, *International Drug GMP's*, 4th ed., Interpharm Press, Buffalo Grove, Ill., 1990.

72. J. Y. Lee, GMP Compliance for clinical packaging, *Pharm. Manuf. 2*:35 (1985).

73. S. H. Willig, M. M. Tuckerman, and W. S. Hitchings, eds., *Good Manufacturing Practices for Pharmaceuticals: A Plan for Total Quality Control*, 2nd ed., Marcel Dekker, New York, 1982, pp. 99–132.

74. *Pharmaceuticals, Veterinary Medicines*, Commission of the European Communities, Brussels, 1987.

75. L. L. Augsburger, in *Modern Pharmaceutics*, 2nd ed. (G. S. Banker and C. T. Rhodes, eds.), Marcel Dekker, New York, 1990, pp. 441–490.

76. B. E. Jones, in *Hard Capsules: Development and Technology* (K. Ridgway, ed.), Pharmaceutical Press, London, 1987, pp. 61–67.

77. M. W. Gouda, M. A. Moustafa, and A. M. Molokhia, Effect of storage conditions on erythromycin tablets marketed in Saudi Arabia, *Int. J. Pharm. 5*:345 (1980).

78. M. D. Kentala, H. E. Lockhart, J. R. Giacin, and R. Adams, Computer-aided simulation of quality degradation of oral solid drugs following repackaging, *Pharm. Technol. 6*(12):46 (1982).

79. M. Veillard, R. Bentejac, D. Duchene, and J. T. Carstensen, Moisture transfer tests in blister packaging, *Drug Dev. Ind. Pharm. 5*:227 (1979).

80. J. W. Conine, D. W. Johnson, and D. L. Coleman, A comparison of the stability of commercial cephradine and cephalexin capsules, *Curr. Ther. Res. 24*:967 (1978).

81. J. R. Guicin, C. L. Pires, and H. E. Lockhart, Predicting packaged product shelf life: experimental and mathematical models, *Pharm. Technol. 15*(8):98 (1991).

82. H. M. Beal, R. J. Dicenzo, P. J. Jannke, H. A. Palmer, J. Pinsky, M. Salame, and T. J. Spealer, Stability of pharmaceuticals stored in plastic containers, *J. Pharm. Sci.* *56*:1310 (1967).

83. C. J. Taborsky-Urdinola, V. A. Gray, and L. T. Grady, Effect of packaging and storage on the dissolution of model prednisone tablets, *Am. J. Hosp. Pharm.* *38*(9):1322 (1981).

84. M. A. Amini, Testing permeation and leakage rates of pharmaceutical containers, *Pharm. Technol.* *5*(12):39 (1982).

85. M. Bakker, ed., *The Wiley Encyclopedia of Packaging Technology*, Wiley, New York, 1986, pp. 581.

86. R. Lefaux, Permeability of polymeric materials for gases and organic vapours. Paper presented at the International Symposium on Packaging of Pharmaceutical Products, Brussels, Belgium, June 1971.

87. L. J. Murray in *Plastic Film Technology: High Barrier Plastic Films for Packaging*, Vol. 1 (K. M. Finlayson, ed.), Technomic, Lancaster, Pa., 1989, pp. 21–31.

88. R. D. Cilento, Methods for evaluating hermetic closures for screw-capped bottles, *J. Pharm. Sci.* *66*:333 (1977).

89. K. Nakabayashi, T. Shimamoto, and H. Mima, Stability of packaged solid dosage forms. I. Shelf-life prediction for packaged tablets liable to moisture damage, *Chem. Pharm. Bull.* *28*(4):1090 (1980).

90. K. Nakabayashi, S. Hanatani, and T. Shimamota, Stability of packaged solid dosage forms. VI. Shelf-life prediction of packaged prednisolone tablets in relation to dissolution properties, *Chem. Pharm. Bull.* *29*(7):2057 (1981).

91. G. Amidon and K. R. Middleton, Accelerated physical stability testing and long-term predictions of changes in the crushing strength of tablets stored in blister packages, *Int. J. Pharm.* *45*:79 (1988).

92. J. T. Carstensen, M. Slotsky, and D. Dolfini, Aspects of packaging from NDA to IND, *Drug Dev. Ind. Pharm.* *15*(13):2131 (1989).

93. *United States Pharmacopoeia XXII, National Formulary XVII*, United States Pharmacopeial Convention, Rockville, Md. 1990 p. 10.

94. A. N. Narukar, P. C. Sheen, D. F. Bernstein, and M. A. Augustine, Studies on the light stability of flordipine tablets in amber blister packaging material, *Drug Dev. Ind. Pharm.* *12*(8–9):1241 (1986).

95. *The Nation's Solid Waste Crisis: An Overview*, Council on Plastics and Packaging in the Environment, Washington, D.C., April 1989.

96. M. Haney and J. Casler, eds., *RCRA Handbook: A Guide to Permitting, Compliance, Closure and Corrective Action Under the Resource Conservation and Recovery Act*, 3rd ed., ENSR Consulting and Engineering, Acton, Mass., October 1990.

97. E. Chenoweth, PVC recycling, *Chem. Week.* *20*:14 (1991).

98. Solving the waste management puzzle, *Manuf. Chem.* *61*(11):55 (1990).

99. J. C. Mullen, Rigid vinyl packaging and the environment, *Interphex: USA Proceedings of the 1991 Technical Program*, New York, April 1991.

100. A. K. S. Poderman, Waste disposal in the pharmaceutical industry, *Pharm. Weekbl.* *119*:951 (1984).

16

Marketing Considerations for Multiparticulate Drug Delivery Systems

Roland Daumesnil

Capsugel AG
Arlesheim, Switzerland

I. INTRODUCTION

The goal of pharmaceutical research is to find innovative drugs with desirable therapeutic properties and low risks of undesirable side effects: the new chemical entities (NCEs). A NCE is the best means to differentiate a product and to protect its market. But developing a NCE is a costly process [1] (Table 1). Very few companies can afford the growing costs of the development of a NCE (Fig. 1). These figures include the cost of failures and the preclinical research expenditures. The current attrition rate of 40 to 50 NCEs per year cannot provide enough profit to sustain the growth of more than 20 major companies [2] (Fig. 2). Of the 512 NCEs launched internationally between 1979 and 1988, only 25% managed to achieve sales of $25 million a year. The remaining 75% are unlikely to have recouped their R&D investment [3]. Other strategies not based on total innovation must be developed.

The companies that will survive by the end of the century will be, of course, the fittest companies: in other words, the companies that will develop internationally competitive new drugs and improvements of existing products [2]. It must also be considered that (1) 60 to 70% of sales from research-based companies are from sales of non-patent-protected products, and (2) the cost of launching and establishing a new product is usually far greater than the cost of maintaining one already in existence. Although innovation is, undoubtedly, a prerequisite for survival, it is not enough and more is needed.

Table 1 Annual R&D Spendings to Generate One
Average NCE Approval Each Year

Stage	Expenditure (millions of dollars)	Percent of total expenditure
Discovery	45	23
Preclinical	30	15
Phase I	40	21
Phase II	45	23
Phase III	30	15
NDA	5	3
Total	195	

Developing controlled-release or modified-release pharmaceutical dosage forms is a clear strategy to improve existing products. International competitiveness must now include rational drug design, drug delivery systems, and quality of life. A concept that is now well understood by researchers of major pharmaceutical companies: "I believe that pharmaceutical companies can fulfill their social mission of contributing to people's happiness and welfare through development of new drugs. For this purpose, it is essential for them to compete and

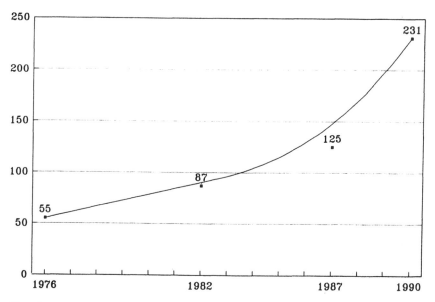

Fig. 1 Cost, in millions of dollars, to develop a new chemical entity from 1976 to 1990. (From the U.S. Pharmaceutical Manufacturers' Association.)

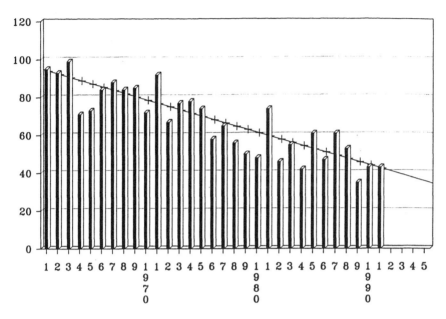

Fig. 2 Number of new chemical entities launched per year from 1961 to 1991. (From *Scrip.*)

cooperate with each other at the same time and maintain a good balance between development of innovative new drugs and improvement of existing products with well-defined advantages'' [4].

II. RATIONALE FOR MARKETING CONTROLLED-RELEASE DOSAGE FORMS

The development of controlled delivery systems is prompted in part by inadequacies in conventional delivery systems. Other factors driving the development of controlled delivery systems are the medical, health care, and business benefits.

The medical and health care benefits are well known:

1. Replacing infusion system
2. Better drug utilization by enabling a smaller drug dose to produce the same clinical effect as the conventional drug form
3. Decreasing of toxicity and occurrence of adverse reactions by providing a more uniform blood concentration
4. Controlling the rate and site of release
5. Once- or twice-a-day formulations, increasing patient compliance

According to a study of medication use by hypertensive patients conducted by Lee Strandberg, associate professor of pharmacy at Oregon State University, the cost of patient noncompliance in the use of medication is artificially increasing U.S. health care spending by $7 billion to $10 billion in terms of increased hospitalization and physician costs [5]. The business benefits are equally important:

1. Means of extending the financial return of the original investment
2. Establishment of product differentiation
3. Improvement in product effectiveness and competitiveness
4. Increase in gross profit margin

In addition, experience shows that many "well-known" active substances (i.e., those used for decades) are not so well known. Developing a controlled-release medication using new techniques has enabled the industry to "rediscover" the pharmacokinetics of some of these "old products."

III. MARKET PROJECTIONS

For all the reasons noted above, nearly all major companies worldwide are now developing controlled-release products. Taking into consideration the oral solid dosage forms (hard gelatin capsules, soft gelatin capsules, and tablets), oral controlled delivery systems could by the century's end account for 8% of the $260 billion conventional drug delivery market in the United States, Japan, and the main European countries (Fig. 3). By the late 1990s in the United States, drug treatments that are administered through alternative delivery systems could displace 10 to 15% of the $40 billion conventional drug delivery prescription market. According to the FIND/SVP report (data from FIND/SVP, New York), sales are expected to rise at a compound annual average growth rate of 21.9% between 1989 and 1997, resulting in a $2.2 billion market in the United States. In 1990, oral controlled-release systems (including osmotic and erodible) accounted for 35% of the alternative drug market in the United States.

The total sales of oral controlled delivery systems reached $4 billion worldwide in 1990. Sales are expected to rise at a compound annual average growth rate (CAGR) of 17% between 1990 and 2000, resulting in a $20 billion market by century's end, roughly three times faster than the conventional dosage forms (Fig. 3). In the 1990s the main segments in which companies developed controlled release were:

1. Cardiovascular
2. Gastrointestinal
3. Respiratory

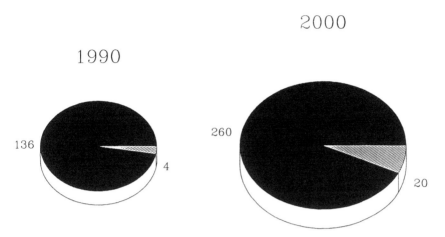

Fig. 3 Market potential, in billions of dollars, of controlled-release medication from 1990 to 2000.

4. Anti-infective
5. Cancer drugs
6. CNS

The 1990s can be described as the years of delivery systems, as evidenced by the impact of patches, nasal and inhalant formulations, osmotic systems, and multiple units.

IV. MULTIPARTICULATES

Among the newer drug delivery systems, by 1997 more than 50% will be controlled-release system. A great many dosage forms have been developed to achieve controlled-release medications. The number of contract research companies or patients pertaining to controlled-release medication is steadily increasing. Every company is trying to develop its patented system (Table 2). Today it is possible to classify these drug delivery systems into four groups:

1. Single units
2. Multiparticulates
3. Multiparticulates in tablets
4. Minitablets in hard capsules

Single-unit formulations are popular in the controlled-release market. The main advantages come from the market share of the conventional tablet and the

Table 2 Solid Oral Controlled-Release Available Systems

Company	Trademark	Tablets	Multi-particulates in capsules	Multi-particulates in tablets	Minitablets in capsules
Alza, U.S.	Oros	×			
Benzon, Denmark	Repro-Dose		×	×	
Biovail, Switzerland	Timelets		×		
Elan, Ireland	Modas	×			
	Sodas		×		
	Sodas HC			×	
	Indas	×			
	Zonap	×			
	Pharmazome		×	×	
Ethypharm, France	Sparklets		×		
Euderma, Italy	Eucaps		×		
Eurand, U.S.	Diffutabs	×			
	Diffucaps		×		
	Minicaps				×
	Microcaps		×		
Forest, U.S.	Synchron	×			
Gacel, Sweden	Multipor	×			
Ivax, U.S.	Snaplets		×		
Jago, Switzerland	Geomatrix	×			×
KV, U.S.	KV 24	×			
	Microrelease		×	×	
	Meter Release		×	×	
Lejus, Sweden	MUDF		×		
Mepha, Switzerland	Cardotabs	×	×	×	
	Depotabs	×			
	Depocaps				
Nova, U.S.	Symetry	×			
Prographarm, France	Flashtab			×	
3M, U.K.	3M Beadlets		×		
RPR/APS, U.K.	Osat		×		
Verex, U.S.	Verex	×			

Table 3 Characteristics of Multiple- and Single-Unit Dosage Forms[a]

Multiple-unit	Single-unit
More predictable gastic emptying	Highly variable gastric emptying
Gastric emptying is less dependent on nutritional state	Gastric emptying is highly dependent on nutritional state
More reproducible absorption	Inter- and intrasubject variability in rate and extent of absorption
Limited risk of dose dumping or local irritation	Risk of dosage dumping or local irritation
Complex manufacturing processes	More simple technologies involved

[a]An approximate diameter of 5 mm is generally an accepted limit between the two types.

expertise of almost all companies in the production of such a form. Multiparticulate formulations present several advantages and are therefore used increasingly by pharmaceutical development and marketing managers to:

1. Reduce inter- and intrasubject variability
2. Reduce local irritation
3. Improve bioavailability
4. Improve stability
5. Improve patient comfort and compliance
6. Achieve a unique release pattern
7. Create line extension
8. Extend patent protection
9. Globalize product
10. Overcome competition

Compared to single-unit formulations, multiparticulates present additional advantages [6] (Table 3). The benefits listed above explain the growing popularity of multiparticulates. The results of the survey carried out during the 15th International Symposium on Controlled Release (Basel, 1988) corroborate the recognition of multiparticulates by the pharmaceutical industry. A vast majority of participants expressed their confidence in the future of this form.

V. SPECIFIC EXAMPLES OF MULTIPARTICULATE MARKETED PRODUCTS

Cardiovascular and respiratory products present a major drawback: medication noncompliance. It affects the patient's well-being and prevents manufacturers from taking full advantage of their pharmaceutical innovations. For example, nearly one-half of the U.S. adult population will develop high blood pressure.

Unfortunately, half of the patients who receive prescriptions for hypertension are noncompliant within 1 year. Most of these patients stop therapy completely. About two-thirds of the patients who are compliant skip doses and take less than the recommended regimen. In other words, only 17% of patients are compliant.

For such asymptomatic disease, any pharmaceutical company should seek to improve patient compliance by developing sustained-release medication. Calcium inhibitors are the leading products in the cardiovascular market. But patient compliance with calcium antagonists is a serious problem because they often have a short half-life and require up to four-times-a-day dosing. The best competitors to calcium inhibitors are the angiotensin converting enzyme (ACE) inhibitors. They have a longer half-life and sometimes require only once-a-day dosing. The solution to both these issues was to develop oral sustained-release products. By the end of the 1980s, calcium inhibitors became the largest sustained-release market in the United States. The products that led the growth are mainly nifedipine, verapamil, and diltiazem. From 1987 to 1995, the total sales of calcium channel blockers will have a CAGR of 13%. During the same period, sustained-release drugs will increase by 31% and conventionally released drugs by only 3% (Fig. 4). Indeed, enhancing patient compliance with calcium inhibitors is a win–win strategy that helps the manufacturer to maintain a firm foot in the hypotensive market.

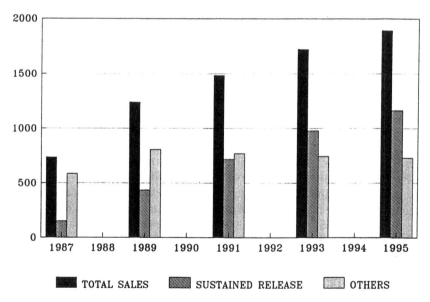

Fig. 4   Sales in the United States, from 1987 to 1995, in millions of dollars per year, of oral conventional and sustained-release calcium antagonists. (From *Scrip* and Frost and Sullivan.)

The calcium inhibitor market is a battlefield in which generic companies expect to achieve substantial growth. Defending key products by extending the effective length of patent protection is a strategy that is now well established in the pharmaceutical industry. This strategy includes novel dosage forms and changes in schedules of administration. Planning for patent expiry is not something that should be done at the last minute. It is one of the few issues for which there is a 17-year warning [7]. A typical program of life extension strategy development should begin no later than 3 years after first launch and certainly 5 to 7 years before patent expiry in major countries [7]. This approach is evidenced in the strategy developed by Sandoz and Yamanouchi to protect their major calcium antagonist: nicardipine. The original patent for a three-times-a-day tablet expired in 1992. In 1988, Sandoz and Yamanouchi launched a sustained-release multiparticulate form in France and Japan and consequently extended the effective length of patent protection. Thanks to a twice-a-day formulation, both were able to maintain their share in their respective market. Furthermore, by the end of 1991, the multiparticulate formulation in hard gelatin capsules represented 60% of the total sales of nicardipine (Fig. 5). By the end of 1992, 70 to 80% of the nicardipine market was in sustained-release form. In addition to providing better protection against ACE inhibitors, this strategy protected the product against

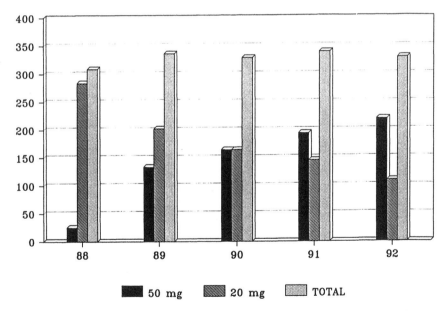

Fig. 5 Multiparticulate sales of nicardipine, in millions of doses per year, from 1988 to 1991 in France.

generics. When the tablet loses its patent protection, the replacement product will be a different one.

One of the greatest potential applications of controlled-release products is to be found in drugs with a poor therapeutic index. Theophylline is the most prescribed xanthine derivative in the bronchodilator segment. In 1990, controlled-release dosage forms accounted for over 90% of the entire theophylline market in France, Germany, Italy, and the United Kingdom. In these countries, as in the United States and now in Japan, the theophylline market became a sustained-release market led by multiple-unit formulations (Fig. 6). In 1976, Key Pharmaceuticals began selling Theo-Dur, a sustained-release theophylline. By 1981 the product accounted for $34 million of the company's 41 million sales. This new formulation of theophylline is a great improvement of an existing product with well-defined benefits. it is also an example on how a new delivery system can offer major opportunities for smaller companies.

Eryc, a coated-pellet formulation of erythromycin, has completely changed the complection of Faulding. The erythromycin base is the active form of the macrolide, but it is neutralized by acidic gastric juices. Consequently, most of the existing erythromycin products on the market were coated tablets or esters with a low bioavailability. To improve bioavailability and reduce patient-to-

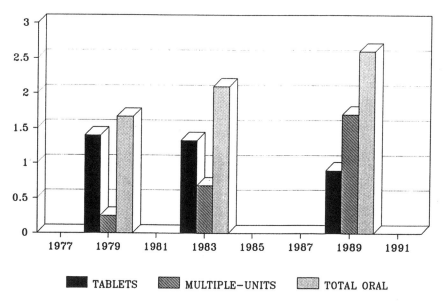

Fig. 6   Sales, in billions of doses per year, from 1979 to 1989 in France, Germany, Italy, and the United Kingdom, of theophylline in tablet and multiparticulate form. (From IMS.)

patient variation, Faulding developed enteric-coated pellets. Furthermore, plasma concentrations of erythromycin arising from the enteric-coated multiparticulate formulations showed that in each subject studied, the erythromycin plasma levels were considerably higher than the minimum inhibitory concentration (MIC) for the antibiotic under fasted and nonfasted conditions. In contrast, however, when an equivalent dose of erythromycin was administered in an enteric-coated tablet form, in 50% of the subjects the antibiotic failed to achieve the MIC under nonfasted conditions. Gamma scintigraphy confirmed that the multiparticulate formulation emptied from the stomach gradually and predictably within 2 h. The tablet, however, exhibited large intrapatient variation [8]. Such an increased therapeutic value led to the success of erythromycin pellets. The product is now marketed in more than 40 countries. Its share of the U.S. market is around 25%, including the generic versions. Licensed to Warner-Lambert for several countries, the product was launched on the U.S. market in 1981. By 1988 the product achieved sales of $45 million—a bright success that paled on introduction of the generics.

Doxycycline can create gastric irritation with a nausea-inducing potential up to 25% of patients. To minimize this potential, Faulding designed an enteric-coated multiparticulate (Doryx) to retard the release of doxycycline in the acidic environment of the stomach while ensuring rapid and complete release in the more alkaline duodenum, where most doxycycline absorption occurs. A study by Pharmakinetics Laboratories (Baltimore, 1984) showed that there was no statistically significant difference between Doryx and placebo ($p < 0.05$). In contrast, the incidence of nausea associated with the conventional U.S. doxycycline capsules was significantly greater in this study than that of both Doryx and placebo ($p < 0.01$). This minimal risk of nausea led to increased patient comfort, hence improved patient compliance. This notably improved product achieved a significant market share in the doxycycline market and has maintained a steady growth since its launch in 1986.

Pancreatic enzyme supplementation is fundamental to the management of pancreatic exocrine insufficiency. The deficiency in pancreatic lipase results in a failure to absorb essential fatty acids, fat-soluble vitamins, and a valuable source of calories. The clinical manifestation, steatorrhoea, can be a considerable social embarrassment to the patient. To be efficient a pancreatic lipase formulation must be protected against the acidity of the gastric juice and must be adequately dispersed in the chyme to produce sufficient available enzyme in the duodenum. Consideration of these factors implies that to be an effective therapy the pancreatin supplement given should reach the duodenum and the jejunum while retaining high degree of enzyme activity [9].

A comparison of simple pancreatin preparations, enteric-coated tablets, or multiparticulates in hard gelatin capsules showed that the enteric-coated multiparticulate is most efficient [9]. The very nature of the formulation provides the basis for an effective and convincing presentation for physicians and led to the European market dominance by Kreon. This multiparticulate form of pancreatic lipase, developed by Kalichemie, is now the leading brand in Europe (Fig. 7).

The conventional way to prevent acid-labile substance from contact with the gastric juice is to coat the dosage form. A group of compounds exerting these stability properties are substituted benzimidazoles. A common feature of these compounds are that they are transformed into biologically active compounds via rapid degradation/transformation in acid media. It is an inherent property of these compounds to be activated to the active moiety in the acid environment within the parietal cells. The activated compound interacts with the enzyme in the

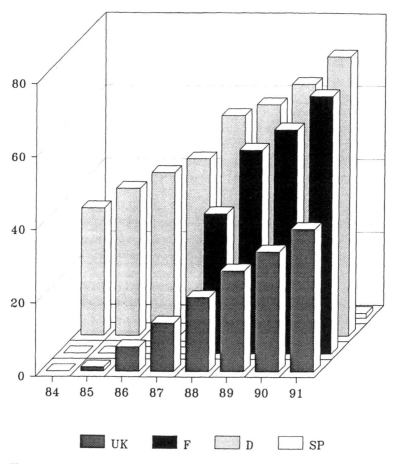

Fig. 7 Sales from 1984 to 1991, in millions of doses per year in France, Germany, Spain, and the United Kingdom, of Kreon, modified-release pancreatic lipase.

parietal cells, which mediates the production of hydrochloric acid in the gastric mucosa. If covered with conventional enteric coating, made of acidic substances, the acid-labile substance rapidly decomposes by direct or indirect contact with it. In addition, to be well protected from contact with the gastric juice, the formulation must dissolve rapidly in the small intestine. This short description is included in the background of the invention described in a European patent application covering omeprazole, the first of a group of chemicals based on benzimidazole which inhibits hydrogen/potassium adenosine triphosphatase, otherwise known as the proton pump [10]. The "special multiple units" described in the patent permit a rapid release of the drug in the proximal part of the small intestine and a once-a-day formulation.

The formulation enhanced the tremendous potential of this new chemical entity. Launched in March 1988 in Sweden, the product records outstanding acceptance and sales. Approved in more than 40 countries, sales of omeprazole, including license, exceeded $1 billion in 1992 (Fig. 8). It must also be emphasized that omeprazole is the first new chemical entity launched as a controlled-release formulation. So far the technology has been restricted to improving the therapeutic values of well-known drugs. Including such a concept in the development of a new chemical entity will change the types of drugs that become available as well as their use.

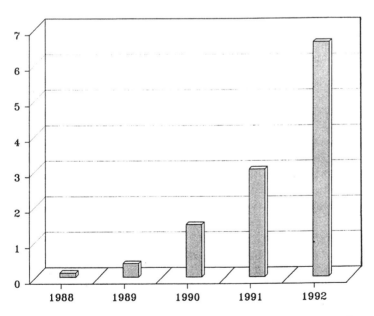

Fig. 8 Sales per year, worldwide, in billions of Swedish krona, of omeprazole.

Improving patient comfort is a well-known strategy to improve patient compliance. Whatever the oral dosage form, children have a lot of difficulty in swallowing. This is why nearly all products for children are formulated in syrups or suppositories. For essential drugs such as theophylline and valproic acid, the exact dose should be administered while achieving a high degree of patient compliance. Sprinkles has been developed for this purpose. Mixed with a semisolid food, they are easy to swallow. Furthermore, filled into hard gelatin capsules or foil-to-foil sachets, the exact dose can easily be administered. This is a convenient dosage form that will improve children comfort and compliance.

VI. OPPORTUNITIES AND FUTURE TRENDS

Health authorities are more and more aware of the difficulty of controlling individual patient's gastrointestinal variability. Therefore, the European and Japanese health authorities have published guidelines for the design and evaluation of oral prolonged-release dosage forms. They recommend the study and evaluation of the effect of food, which is well known to affect transit rate, disintegration, and release of the drug in the gastrointestinal tract.

During the 17th International Symposium on Controlled Release of Bioactive Materials that took place in July 1990 in Reno, Nevada, Luth Research, a market research firm from San Diego, California, surveyed the attendees on multiparticulates in hard gelatine capsules and tablets. The results were analyzed by McCarthy & Associates of Syracuse, New York. The purpose of the survey was to investigate and rate the importance of the factors considered when a company chooses the drug form design of a multiparticulate formulation.

Three factors are taken into consideration: (1) technical or pharmaceutical factors, (2) marketing, and (3) production. Indeed, each factor has a different degree of importance. As expected, the most important factor is pharmaceutical followed by marketing. The least important factor is production.

The breakdown of each factor into its elements, ranked by importance, highlights some interesting results. The pharmaceutical elements give a clear advantage to hard gelatin capsules. The principal reasons given for the preference for hard gelatin capsules relates to the pharmaceutical response, the ability to combine different release patterns, and indeed, less sensitivity to compression.

The Luth survey corroborated the well-known advantages of hard gelatin capsules in the area of patient compliance, patient preference, and in product identification. On the other hand, all elements of the production factor are rated more positively for tablets. For R&D this result hampers the decision to choose hard gelatin capsules freely in the design of multiparticulate dosage forms. With the worldwide simplification and acceptance of pellet technology, this limitation will become increasingly less important.

Table 4 Factors Driving the Drug Form Design of Multiparticulates

| | More positive rating | | No rating |
Element	Capsules	Tablets	difference
Pharmaceutical			
Pharmaceutical response	×		
Combination of release patterns	×		
Sensitivity of compression	×		
Stability			×
Uniformity of response			×
Ease of formulation			×
Marketing			
Patient compliance	×		
Patient preference	×		
Product Identification	×		
Competitive pressure			×
Masking taste			×
Production			
Manufacturing limitations		×	
Machinery constraints		×	
Lack of technical expertise		×	
Cost of goods		×	
Company precedent		×	

Nevertheless, if the most important factors are considered, a clear advantage is provided to formulators for hard gelatin capsules. Such a preference explains why a growing majority of multiparticulates are filled into capsules (Table 4). Thanks to the new technologies used for the manufacturing of multiparticulates, whose development entered the growth phase in the beginning of the 1980s, more and more products are launched or are under development worldwide. From 1990 to 2000, sales of multiparticulates in hard gelatin capsules are expected to rise at a compound annual average growth of 13%, resulting in a $8.5 billion market. Figure 9 shows that multiparticulates in hard gelatin capsules is the leading delivery system in oral controlled-release medication (Fig. 9). The primary segments in which companies developed multiparticulates are equivalent to those of controlled-release medication, with a leading segment for cardiovascular. The proton group pump inhibitors (omeprazole, lanzoprazole) are expected to capture a substantial portion of the multiparticulates market (Fig. 10).

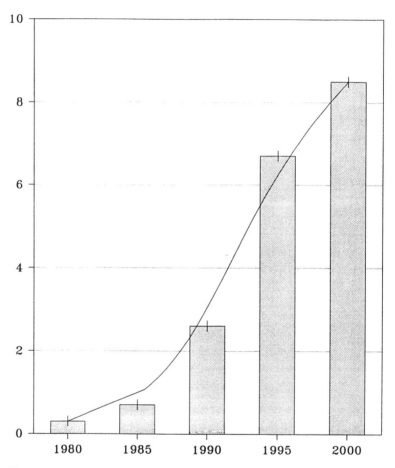

Fig. 9 Multiparticulate sales worldwide, in billions of dollars, from 1980 to 2000.

VII. CONCLUSIONS

For the pharmaceutical industry, innovation in new products is not the only way to survive. Anything that enhances the effectiveness of therapy increases the utility of the company's product to patients. Furthermore, product life extension should be an integral part of the product strategy from the beginning. Oral controlled-release medication is a means to achieve this objective. In addition, clinically important difference in drug effects may arise from changes in dosage program and can help old products to enter in additional therapeutic classes.

So far the technology has been restricted to improving the therapeutic properties of well-known drugs. If the concept of rate-controlled delivery be-

MARKET VALUE $6.5 BILLION

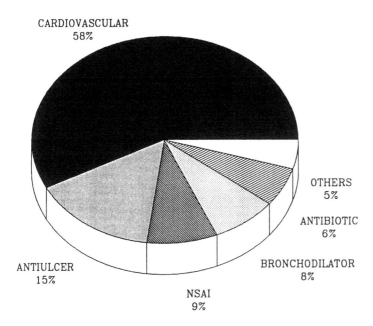

Fig. 10 Value in millions of dollars and leading market segments in percentage of the multiparticulate prescription market projected to the year 1995.

comes more common in the development of new drugs, it may change the types of drugs that become available as well as their use. Among all possible oral controlled-delivery systems currently on the market, multiparticulates are becoming more and more popular. They offer a convenient form to pharmaceutical development managers and an attractive form to the patient. Altogether, it represents one of the most desirable options from a profit perspective.

REFERENCES

1. Booz-Allen & Hamilton, *Pharm. View Point*, October (1990).
2. J. Drews, *Pharm. Bus. News*, November 15, p. 9 (1991).
3. K. Mansford, *Scrip*, March p. 28 (1992).
4. Seiichi Saito, *Pharma Japan 1289*, February 3, p. 9 (1992).
5. *Scrip 1523*, June 15, p. 23 (1990).

6. A. C. Vial-Vernasconi, E. Doelker, and P. Buri, *STP Pharma 4*(5): 397 (1988).
7. M. Marber, *Scrip*, March, p. 24 (1992).
8. G. Digenis, The in vivo behavior of multiparticulate versus single unit dose formulations, *Capsugel Symposium*, Tokyo, 1990.
9. A. M. Whitehead, *Pharm. Weekbl. Sci. Ed. 10*(1):12 (1988).
10. Astra Hässle, European Patent 0,244,380.

Index